Excel

SUCCESS ONE® HSC*

MATHEMATICS EXTENSION 1

Past HSC papers and worked answers 2012–2024

PLUS

Topic index of past HSC questions

PASCAL PRESS

ISBN 978 1 74125 767 0

Pascal Press
PO Box 250
Glebe NSW 2037
(02) 9198 1748
www.pascalpress.com.au

Publisher: Vivienne Joannou
Commissioning and series editor: Mark Dixon
Project editor: Rosemary Peers
Edited by Rosemary Peers
Answers checked by Peter Little
Cover design by Michael Sherman
Typesetting by Julianne Billington
Cover photo by The *Sydney Morning Herald*
Printed by Vivar Printing/Green Giant Press

Contents

2012 HSC Mathematics Extension 1
Examination paper **1**
Replacement questions 14
Worked answers 18

2013 HSC Mathematics Extension 1
Examination paper **28**
Replacement questions 43
Worked answers 47

2014 HSC Mathematics Extension 1
Examination paper **59**
Replacement questions 72
Worked answers 76

2015 HSC Mathematics Extension 1
Examination paper **86**
Replacement questions 99
Worked answers 103

2016 HSC Mathematics Extension 1
Examination paper **114**
Replacement questions 127
Worked answers 130

2017 HSC Mathematics Extension 1
Examination paper **142**
Replacement questions 156
Worked answers 159

2018 HSC Mathematics Extension 1
Examination paper **172**
Replacement questions 185
Worked answers 190

2019 HSC Mathematics Extension 1
Examination paper **204**
Replacement questions 218
Worked answers 220

2020 HSC Mathematics Extension 1
Examination paper **230**
Worked answers 244

2021 HSC Mathematics Extension 1
Examination paper **252**
Worked answers 264

2022 HSC Mathematics Extension 1
Examination paper **272**
Worked answers 290

2023 HSC Mathematics Extension 1
Examination paper **297**
Worked answers 312

2024 HSC Mathematics Extension 1
Examination paper **321**
Worked answers 334

Reference sheet **340**

PAST HSC EXAMINATION PAPERS

WORKED ANSWERS

The worked answers contained in this publication are examples of answers which the authors believe would score full marks. They are not necessarily ideal or model answers, nor are they the only answers which would score full marks. They are not endorsed by NESA.

Worked answers for the 2012–2020 HSC Examination papers were written by Lyn Baker.

Questions and answers to replace the questions not in the latest syllabus were written by Allyn Jones.

Worked answers for the 2021 and 2022 HSC Examination papers were written by Mitchell Jones.

Worked answers for the 2023 and 2024 HSC Examination paper were written by Melinda Amaral.

THE HSC MATHEMATICS EXTENSION 1 EXAMINATION

Since 2020 the HSC Mathematics Extension 1 Examination paper has consisted of two sections.

Section I has 10 objective-response questions worth 10 marks in total. Section II has questions that may contain parts. There are 23 to 28 items with a total of 60 marks. At least one of these items is worth 4 or 5 marks.

There is a reading time of 10 minutes and working time of 2 hours.

Students should note that marks are shown for each item and should allocate their time accordingly.

IMPORTANT: PLEASE READ before you start

For the first time, in **2020**, the new syllabus for the **Mathematics Extension 1** course was examined.

- Most of the HSC Examination papers in this book reflect the previous HSC Mathematics Extension 1 course.
- In order to help you revise for the HSC Mathematics Extension 1 Examination we have gone through every single question in those papers and put a cross (✗) next to the questions which are no longer examinable.
- We have also provided you with **Replacement questions** after each HSC Examination paper in this book for questions that are no longer examinable.
- This will mean that for each paper you can complete all the questions that do not have a cross next to them and then turn to the Replacement question section (it is earmarked with a grey margin for easy location) and complete a replacement question for each question that is marked with a cross.
- This will ensure that you have a complete paper to revise for each year.
- The Replacement questions for each paper:
  - consist of questions from topics in the Mathematics Extension 1 course that were not in the previous syllabus.
  - will replace all the questions with a cross next to them (questions that are not applicable to the new syllabus).
- This book includes the 2020–2024 HSC Examination papers of the new Mathematics Extension 1 course.

2012

HIGHER SCHOOL CERTIFICATE EXAMINATION

Mathematics Extension 1

General Instructions

- Reading time – 5 minutes
- Working time – 2 hours
- Write using black or blue pen Black pen is preferred
- Board-approved calculators may be used
- A table of standard integrals is provided at the back of this paper
- In Questions 11–14, show relevant mathematical reasoning and/or calculations

Total marks – 70

Section I

10 marks

- Attempt Questions 1–10
- Allow about 15 minutes for this section

Section II

60 marks

- Attempt Questions 11–14
- Allow about 1 hour and 45 minutes for this section

Section I

10 marks
Attempt Questions 1–10
Allow about 15 minutes for this section

Use the multiple-choice answer sheet for Questions 1–10.

---

1 Which expression is a correct factorisation of $x^3 - 27$?

(A) $(x-3)(x^2-3x+9)$

(B) $(x-3)(x^2-6x+9)$

(C) $(x-3)(x^2+3x+9)$

(D) $(x-3)(x^2+6x+9)$

2 The point P divides the interval from $A(-2, 2)$ to $B(8, -3)$ internally in the ratio $3:2$.

What is the x-coordinate of P?

(A) 4

(B) 2

(C) 0

(D) –1

3 A polynomial equation has roots α, β and γ where

$$\alpha + \beta + \gamma = -2,\ \alpha\beta + \alpha\gamma + \beta\gamma = 3 \text{ and } \alpha\beta\gamma = 1.$$

Which polynomial equation has the roots α, β and γ?

(A) $x^3 + 2x^2 + 3x + 1 = 0$

(B) $x^3 + 2x^2 + 3x - 1 = 0$

(C) $x^3 - 2x^2 + 3x + 1 = 0$

(D) $x^3 - 2x^2 + 3x - 1 = 0$

4 Which function best describes the following graph?

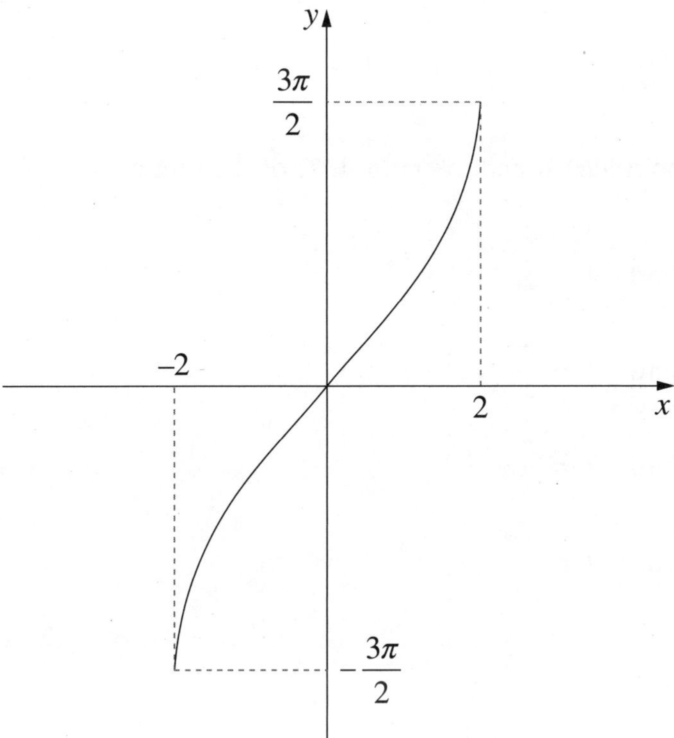

(A) $y = 3\sin^{-1} 2x$

(B) $y = \dfrac{3}{2}\sin^{-1} 2x$

(C) $y = 3\sin^{-1} \dfrac{x}{2}$

(D) $y = \dfrac{3}{2}\sin^{-1} \dfrac{x}{2}$

5 How many arrangements of the letters of the word OLYMPIC are possible if the C and the L are to be together in any order?

(A) $5!$

(B) $6!$

(C) $2 \times 5!$

(D) $2 \times 6!$

 **6** A particle is moving in simple harmonic motion with displacement x. Its velocity v is given by

$$v^2 = 16\left(9 - x^2\right).$$

What is the amplitude, A, and the period, T, of the motion?

(A) $A = 3$ and $T = \frac{\pi}{2}$

(B) $A = 3$ and $T = \frac{\pi}{4}$

(C) $A = 4$ and $T = \frac{\pi}{3}$

(D) $A = 4$ and $T = \frac{2\pi}{3}$

7 Which expression is equal to $\int \sin^2 3x\,dx$?

(A) $\frac{1}{2}\left(x - \frac{1}{3}\sin 3x\right) + C$

(B) $\frac{1}{2}\left(x + \frac{1}{3}\sin 3x\right) + C$

(C) $\frac{1}{2}\left(x - \frac{1}{6}\sin 6x\right) + C$

(D) $\frac{1}{2}\left(x + \frac{1}{6}\sin 6x\right) + C$

8 When the polynomial $P(x)$ is divided by $(x+1)(x-3)$, the remainder is $2x + 7$.

What is the remainder when $P(x)$ is divided by $x - 3$?

(A) 1

(B) 7

(C) 9

(D) 13

9 What is the derivative of $\cos^{-1}(3x)$?

(A) $\dfrac{1}{3\sqrt{1-9x^2}}$

(B) $\dfrac{-1}{3\sqrt{1-9x^2}}$

(C) $\dfrac{3}{\sqrt{1-9x^2}}$

(D) $\dfrac{-3}{\sqrt{1-9x^2}}$

✗ **10** The points A, B and P lie on a circle centred at O. The tangents to the circle at A and B meet at the point T, and $\angle ATB = \theta$.

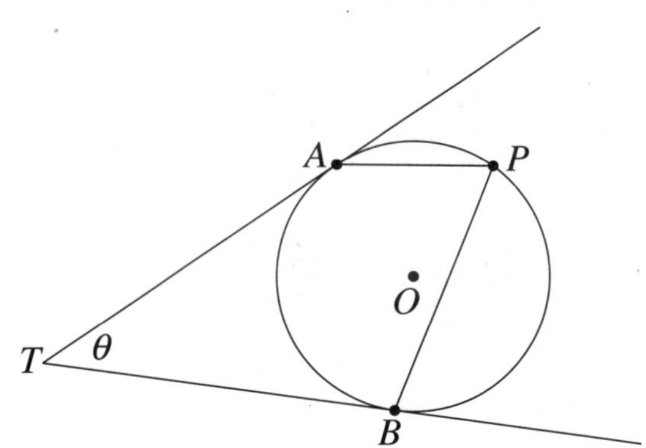

What is $\angle APB$ in terms of θ?

(A) $\dfrac{\theta}{2}$

(B) $90° - \dfrac{\theta}{2}$

(C) θ

(D) $180° - \theta$

Section II

60 marks
Attempt Questions 11–14
Allow about 1 hour and 45 minutes for this section

Answer each question in a SEPARATE writing booklet. Extra writing booklets are available.

In Questions 11–14, your responses should include relevant mathematical reasoning and/or calculations.

Question 11 (15 marks) Use a SEPARATE writing booklet.

(a) Evaluate $\displaystyle\int_0^3 \frac{1}{9+x^2}\,dx$. **3**

(b) Differentiate $x^2 \tan x$ with respect to x. **2**

(c) Solve $\dfrac{x}{x-3} < 2$. **3**

(d) Use the substitution $u = 2 - x$ to evaluate $\displaystyle\int_1^2 x(2-x)^5\,dx$. **3**

(e) In how many ways can a committee of 3 men and 4 women be selected from a group of 8 men and 10 women? **1**

(f) (i) Use the binomial theorem to find an expression for the constant term in the expansion of $\left(2x^3 - \dfrac{1}{x}\right)^{12}$. **2**

(ii) For what values of n does $\left(2x^3 - \dfrac{1}{x}\right)^n$ have a non-zero constant term? **1**

Question 12 (15 marks) Use a SEPARATE writing booklet.

(a) Use mathematical induction to prove that $2^{3n} - 3^n$ is divisible by 5 for $n \geq 1$. **3**

(b) Let $f(x) = \sqrt{4x - 3}$.

(i) Find the domain of $f(x)$. **1**

(ii) Find an expression for the inverse function $f^{-1}(x)$. **2**

(iii) Find the points where the graphs $y = f(x)$ and $y = x$ intersect. **1**

(iv) On the same set of axes, sketch the graphs $y = f(x)$ and $y = f^{-1}(x)$ showing the information found in part (iii). **2**

(c) Kim and Mel play a simple game using a spinner marked with the numbers 1, 2, 3, 4 and 5.

The game consists of each player spinning the spinner once. Each of the five numbers is equally likely to occur.

The player who obtains the higher number wins the game.

If both players obtain the same number, the result is a draw.

(i) Kim and Mel play one game. What is the probability that Kim wins the game? **1**

(ii) Kim and Mel play six games. What is the probability that Kim wins exactly three games? **2**

Question 12 continues on the following page

Question 12 (continued)

(d) Let $A(0, -k)$ be a fixed point on the y-axis with $k > 0$. The point $C(t, 0)$ is on the x-axis. The point $B(0, y)$ is on the y-axis so that $\triangle ABC$ is right-angled with the right angle at C. The point P is chosen so that $OBPC$ is a rectangle as shown in the diagram.

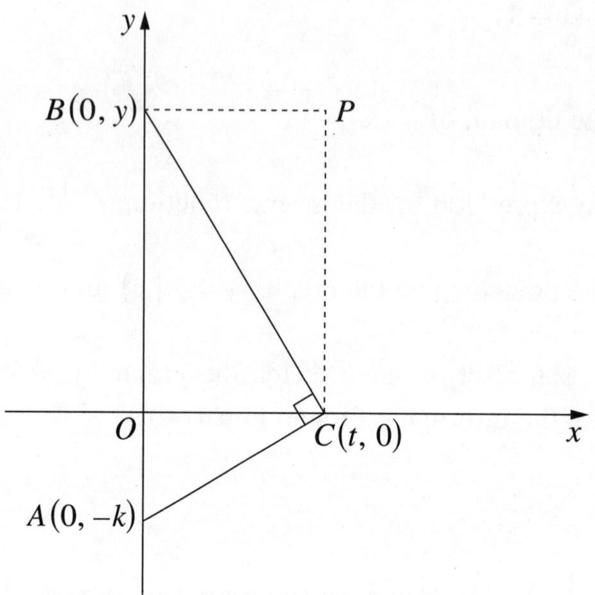

(i) Show that P lies on the parabola given parametrically by **2**

$$x = t \quad \text{and} \quad y = \frac{t^2}{k}.$$

(ii) Write down the coordinates of the focus of the parabola in terms of k. **1**

End of Question 12

Question 13 (15 marks) Use a SEPARATE writing booklet.

(a) Write $\sin\left(2\cos^{-1}\left(\frac{2}{3}\right)\right)$ in the form $a\sqrt{b}$, where a and b are rational. **2**

(b) (i) Find the horizontal asymptote of the graph $y = \dfrac{2x^2}{x^2+9}$. **1**

(ii) Without the use of calculus, sketch the graph $y = \dfrac{2x^2}{x^2+9}$, showing the asymptote found in part (i). **2**

(c) A particle is moving in a straight line according to the equation

$$x = 5 + 6\cos 2t + 8\sin 2t,$$

where x is the displacement in metres and t is the time in seconds.

(i) Prove that the particle is moving in simple harmonic motion by showing that x satisfies an equation of the form $\ddot{x} = -n^2(x-c)$. **2**

(ii) When is the displacement of the particle zero for the first time? **3**

Question 13 continues on the following page

Question 13 (continued)

(d) The concentration of a drug in the blood of a patient t hours after it was administered is given by

$$C(t) = 1.4te^{-0.2t},$$

where $C(t)$ is measured in mg/L.

(i) Initially the concentration of the drug in the blood of the patient increases until it reaches a maximum, and then it decreases. **3**

Find the time when this maximum occurs.

(ii) Taking $t = 20$ as a first approximation, use one application of Newton's method to find approximately when the concentration of the drug in the blood of the patient reaches 0.3 mg/L. **2**

End of Question 13

Question 14 (15 marks) Use a SEPARATE writing booklet.

(a) The diagram shows a large semicircle with diameter AB and two smaller semicircles with diameters AC and BC, respectively, where C is a point on the diameter AB. The point M is the centre of the semicircle with diameter AC.

The line perpendicular to AB through C meets the largest semicircle at the point D. The points S and T are the intersections of the lines AD and BD with the smaller semicircles. The point X is the intersection of the lines CD and ST.

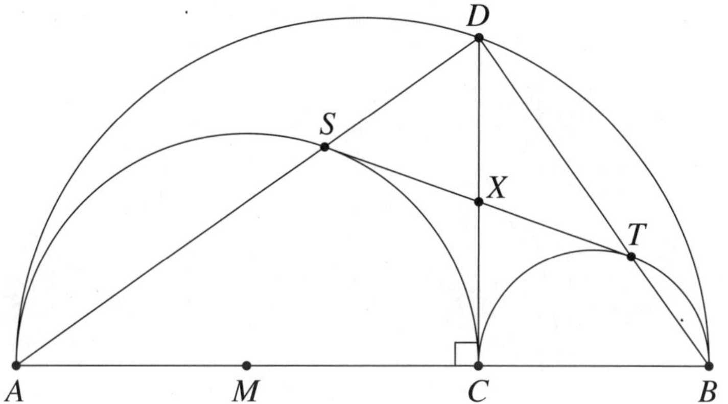

Copy or trace the diagram into your writing booklet.

(i) Explain why $CTDS$ is a rectangle. **1**

(ii) Show that $\triangle MXS$ and $\triangle MXC$ are congruent. **2**

(iii) Show that the line ST is a tangent to the semicircle with diameter AC. **1**

Question 14 continues on the following page

Question 14 (continued)

* (b) A firework is fired from O, on level ground, with velocity 70 metres per second at an angle of inclination θ. The equations of motion of the firework are

$$x = 70t\cos\theta \text{ and } y = 70t\sin\theta - 4.9t^2. \text{ (Do NOT prove this.)}$$

The firework explodes when it reaches its maximum height.

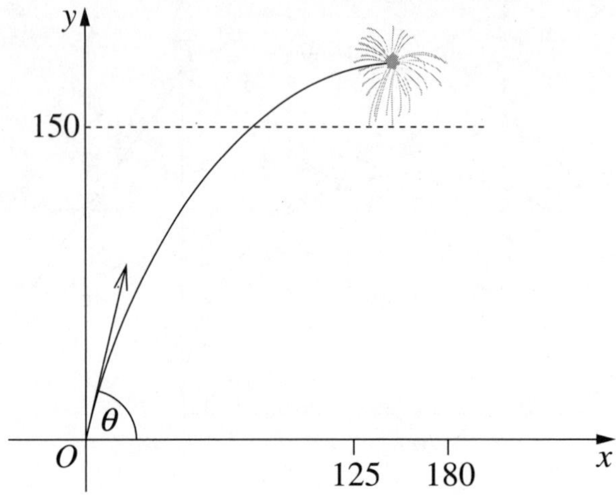

(i) Show that the firework explodes at a height of $250\sin^2\theta$ metres. **2**

(ii) Show that the firework explodes at a horizontal distance of $250\sin 2\theta$ metres from O. **1**

(iii) For best viewing, the firework must explode at a horizontal distance between 125 m and 180 m from O, and at least 150 m above the ground. **3**

For what values of θ will this occur?

* In the new Mathematics Extension 1 course, Projectile Motion questions will be expressed in vector form:

$x = Vt\cos\theta$ and $y = Vt\sin\theta - \frac{1}{2}gt^2$ is expressed as a position vector:

$$\underset{\sim}{r}(t) = (Vt\cos\theta)\underset{\sim}{i} + \left(Vt\sin\theta - \frac{1}{2}gt^2\right)\underset{\sim}{j}$$

Question 14 continues on the following page

Question 14 (continued)

(c) A plane P takes off from a point B. It flies due north at a constant angle α to the horizontal. An observer is located at A, 1 km from B, at a bearing 060° from B. Let u km be the distance from B to the plane and let r km be the distance from the observer to the plane. The point G is on the ground directly below the plane.

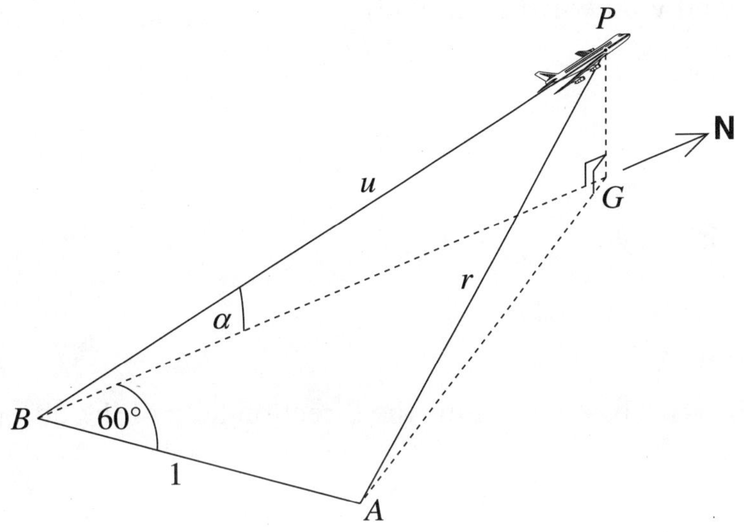

(i) Show that $r = \sqrt{1 + u^2 - u\cos\alpha}$. **3**

(ii) The plane is travelling at a constant speed of 360 km/h. **2**

At what rate, in terms of α, is the distance of the plane from the observer changing 5 minutes after take-off?

End of paper

Replacement questions

with content from the most up-to-date syllabus

Marks

Question 1 (1 mark)

Given that $\dfrac{dy}{dx} = y$ and $y = 1$ when $x = 1$, then

(A) $y = e^x$

(B) $y = e^{x-1}$

(C) $y = e^{1-x}$

(D) $y = e^x - 1$ **1**

Question 2 (1 mark)

Which of these diagrams best represents the direction field of the differential equation

$\dfrac{dy}{dx} = \dfrac{x}{y}$? **1**

(A)

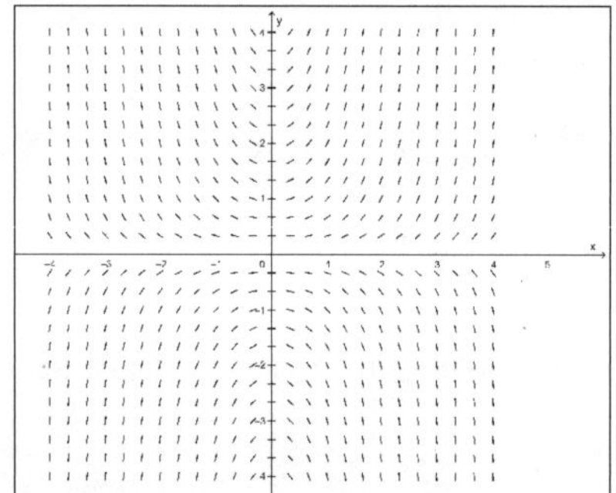

(B)

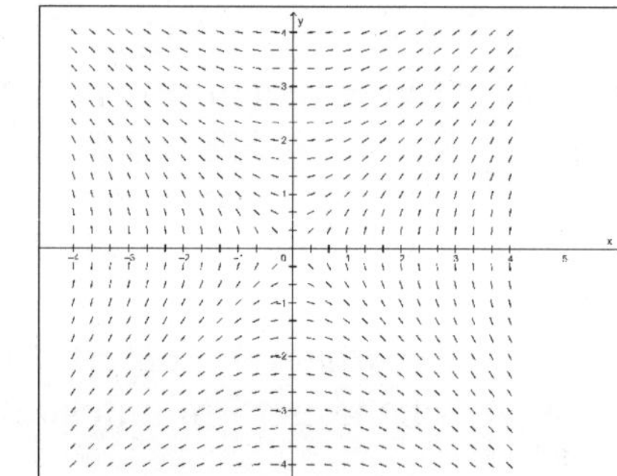

(C)

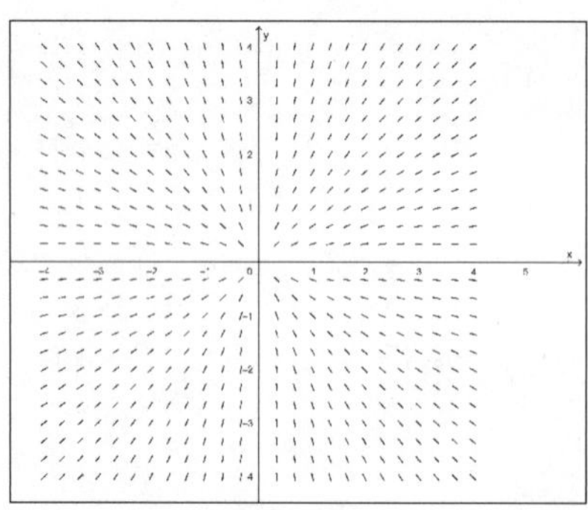

(D)

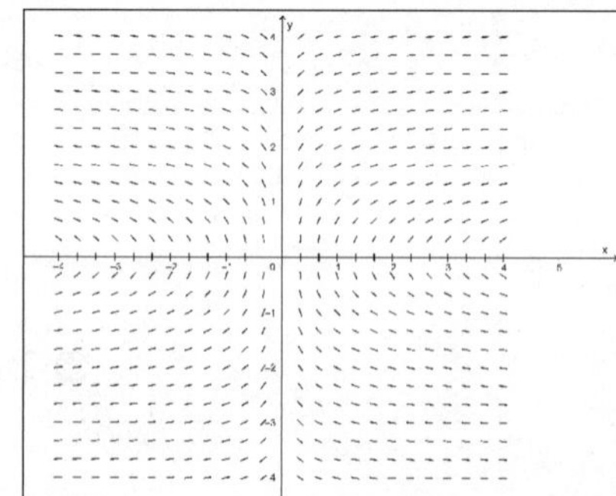

REPLACEMENT QUESTIONS

Marks

Question 6 (1 mark)

If $\underset{\sim}{a} = 3\underset{\sim}{i} - 2\underset{\sim}{j}$ and $\underset{\sim}{b} = -\underset{\sim}{i} + 5\underset{\sim}{j}$, find $2\underset{\sim}{a} - 3\underset{\sim}{b}$. **1**

(A) $6\underset{\sim}{i} - 19\underset{\sim}{j}$

(B) $9\underset{\sim}{i} - 19\underset{\sim}{j}$

(C) $9\underset{\sim}{i} - 11\underset{\sim}{j}$

(D) $6\underset{\sim}{i} - 11\underset{\sim}{j}$

Question 10 (1 mark)

An examination consists of 20 multiple-choice questions, each question having four possible answers.

Marie guesses the answer to each question.

Let X be the number of correct answers.

Which of these is the value of $Var(X)$? **1**

(A) $\frac{5}{4}$

(B) $\frac{15}{4}$

(C) 5

(D) $\frac{25}{4}$

Question 12 (3 marks)

(d) If $x = \sin t + \cos t$ and $y = \frac{1}{2}\sin 2t$, show that $\frac{dy}{dx} = x$. **3**

Marks

Question 13 (7 marks)

(c)* In the diagram $\overrightarrow{OA} = \underset{\sim}{a}$ and $\overrightarrow{OB} = \underset{\sim}{b}$. 

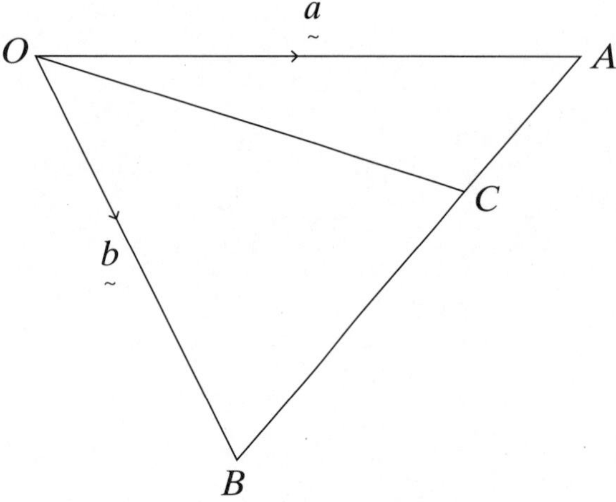

C lies on AB where $AC:CB = p:q$.

Express OC in terms of $\underset{\sim}{a}$, $\underset{\sim}{b}$, p and q.

(c)** Use the substitution $u = \cos 2x$ to evaluate $\int_0^{\frac{\pi}{8}} 2\cos^3 2x \sin 2x \, dx$. **3**

(d) (ii) Solve $\dfrac{|1-4x|}{2} \leq x$ **2**

Marks

Question 14 (4 marks)

(a) The population of a gold mining town is initially 8000 people. This population would increase at a rate of 1.5% per year, except that there is a steady flow of people leaving the town.

The population P after t years may be modelled by the differential equation $\frac{dP}{dt} = \frac{3P}{200} - k$, where k is the number of people leaving per year minus the number of people arriving per year.

(i) Show that, when $k = 600$, $P = 8000(5 - 4e^{0.015t})$ satisfies $\frac{dP}{dt} = \frac{3P}{200} - 600$. **2**

(ii) For $k = 600$, determine the number of years until the population is zero. **2**

REPLACEMENT QUESTIONS

2012 Higher School Certificate
Worked answers

Section I

(*Total 10 marks*)

1. $x^3 - 27 = (x - 3)(x^2 + 3x + 9)$

Answer C

[Check by expanding.]

2. A(−2, 2), B(8, −3), 3:2

$$x = \frac{kx_2 + lx_1}{k + l}$$

$$= \frac{3 \times 8 + 2 \times -2}{3 + 2}$$

$$= 4$$

Answer A

3. $ax^3 + bx^2 + cx + d = 0$

[Note: all the choices have $a = 1$.]

Now $\alpha + \beta + \gamma = \dfrac{-b}{a}$

$\therefore \alpha + \beta + \gamma = -b$ when $a = 1$

But $\alpha + \beta + \gamma = -2$

$\therefore -b = -2$

$b = 2$

$\alpha\beta + \alpha\gamma + \beta\gamma = \dfrac{c}{a}$

$\therefore \alpha\beta + \alpha\gamma + \beta\gamma = c$ when $a = 1$

But $\alpha\beta + \alpha\gamma + \beta\gamma = 3$

$\therefore c = 3$

$\alpha\beta\gamma = \dfrac{-d}{a}$

$\therefore \alpha\beta\gamma = -d$ when $a = 1$

But $\alpha\beta\gamma = 1$

$\therefore -d = 1$

$d = -1$

So $b = 2, c = 3$ and $d = -1$

$\therefore x^3 + 2x^2 + 3x - 1 = 0$ has roots α, β and γ.

Answer B

4.

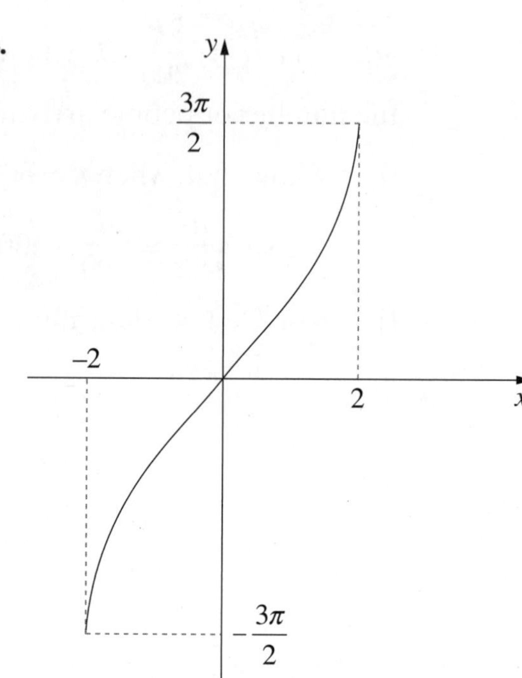

$y = \sin^{-1}u$ has domain $-1 \leq u \leq 1$

Here, $-2 \leq x \leq 2$

$$\therefore -1 \leq \frac{x}{2} \leq 1$$

So the equation is of the form $y = a\sin^{-1}\dfrac{x}{2}$.

When $x = 2, y = \dfrac{3\pi}{2}$

$$\frac{3\pi}{2} = a\sin^{-1}1$$

$$= a \times \frac{\pi}{2}$$

$\therefore a = 3$

The equation is $y = 3\sin^{-1}\dfrac{x}{2}$.

Answer C

5. If the C and L are kept together they count as a single element. But the C and L can be arranged in 2 ways.

Total arrangements = 2 × 6!

Answer D

6. $v^2 = 16(9 - x^2)$

The equation is of the form $v^2 = n^2(A^2 - x^2)$.

$A^2 = 9$

$\therefore A = 3 \quad (A > 0)$

$n^2 = 16$

$\therefore n = 4 \quad (n > 0)$

$$T = \frac{2\pi}{n} = \frac{2\pi}{4} = \frac{\pi}{2}$$

$\therefore A = 3$ and $T = \dfrac{\pi}{2}$

Answer A

7.
$$\cos 2x = \cos^2 x - \sin^2 x = 1 - 2\sin^2 x$$
$$2\sin^2 x = 1 - \cos 2x$$
$$\sin^2 x = \frac{1}{2}(1 - \cos 2x)$$
$$\int \sin^2 3x\, dx = \int \frac{1}{2}(1 - \cos 6x)\, dx = \frac{1}{2}\left(x - \frac{1}{6}\sin 6x\right) + C$$

Answer C

8. $P(x) = (x + 1)(x - 3)Q(x) + 2x + 7$

$P(3) = (3 + 1)(3 - 3)Q(3) + 2(3) + 7$
$= 13$

$\therefore$ the remainder is 13.

Answer D

9. $y = \cos^{-1}(3x)$

$$\frac{dy}{dx} = \frac{-1}{\sqrt{1 - (3x)^2}} \times 3 = \frac{-3}{\sqrt{1 - 9x^2}}$$

Answer D

10.

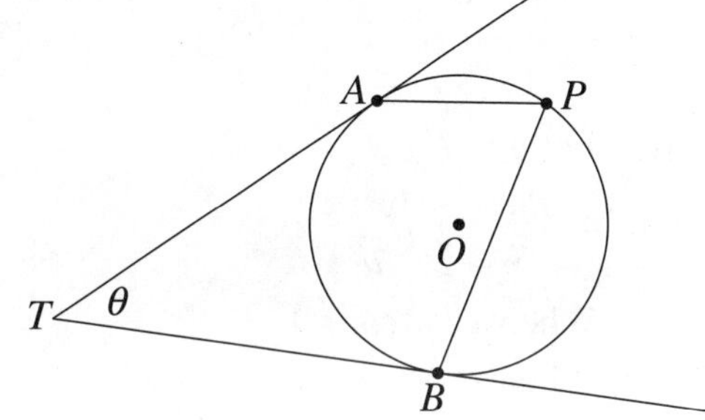

$\angle TAO = \angle TBO = 90°$ (tangents perpendicular to the radius)

$\angle AOB = 180° - \theta$ (angle sum of quadrilateral is 360°)

$\angle APB = \frac{1}{2}(180° - \theta)$ (angle at centre is twice angle at circumference standing on same arc)

$\therefore \angle APB = 90° - \dfrac{\theta}{2}$

Answer B

Section II

QUESTION 11

(a)
$$\int_0^3 \frac{1}{9 + x^2}\, dx = \left[\frac{1}{3}\tan^{-1}\frac{x}{3}\right]_0^3 = \frac{1}{3}\tan^{-1}1 - \frac{1}{3}\tan^{-1}0 = \frac{1}{3} \times \frac{\pi}{4} - \frac{1}{3} \times 0 = \frac{\pi}{12}$$

(3 marks)

(b) $y = x^2 \tan x$

$$\frac{dy}{dx} = x^2 \times \sec^2 x + (\tan x) \times 2x = x(x\sec^2 x + 2\tan x)$$

(2 marks)

(c)
$$\frac{x}{x - 3} < 2$$
$$\frac{x}{x - 3}(x - 3)^2 < 2(x - 3)^2$$
$$x(x - 3) < 2(x^2 - 6x + 9)$$
$$x^2 - 3x < 2x^2 - 12x + 18$$
$$x^2 - 9x + 18 > 0$$
$$(x - 3)(x - 6) > 0$$

Let $y = (x - 3)(x - 6)$

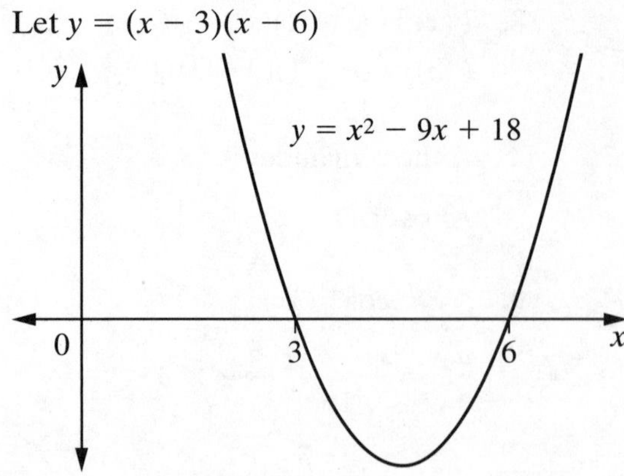

From the diagram, $y > 0$ when $x < 3$ or $x > 6$.

$\therefore \dfrac{x}{x-3} < 2$ when $x < 3$ or $x > 6$

(3 marks)

(d) Let $u = 2 - x$

$du = -dx$

$x = 2 - u$

When $x = 1, u = 1$

When $x = 2, u = 0$

$$\int_1^2 x(2-x)^5 dx = \int_1^0 -(2-u)u^5 du$$
$$= \int_0^1 (2u^5 - u^6) du$$
$$= \left[\frac{u^6}{3} - \frac{u^7}{7}\right]_0^1$$
$$= \frac{1}{3} - \frac{1}{7} - 0$$
$$= \frac{4}{21}$$

(3 marks)

(e) Number of ways $= {}^8C_3 \times {}^{10}C_4$

$= 11\,760$ *(1 mark)*

(f) (i) $\left(2x^3 - \dfrac{1}{x}\right)^{12}$

General term $= {}^{12}C_k(2x^3)^k\left(-\dfrac{1}{x}\right)^{12-k}$

$= {}^{12}C_k\, 2^k\, x^{3k}(-1)^{12-k}\, x^{k-12}$

$= {}^{12}C_k\,(-1)^k\, 2^k\, x^{4k-12}$

Constant term occurs when $4k - 12 = 0$

$4k = 12$

$k = 3$

Constant term $= {}^{12}C_3(-2)^3$

$[= -1760]$ *(2 marks)*

(ii) $\left(2x^3 - \dfrac{1}{x}\right)^n$

General term $= {}^nC_k(2x^3)^k\left(-\dfrac{1}{x}\right)^{n-k}$

$= {}^nC_k\,(-1)^{n-k}\,2^k\,x^{4k-n}$

Constant term occurs when $4k - n = 0$

$n = 4k$

$\therefore \left(2x^3 - \dfrac{1}{x}\right)^n$ will have a non-zero constant term when n is a multiple of 4.

(1 mark)

QUESTION 12

(a) Aim to prove that $2^{3n} - 3^n$ is divisible by 5 for $n \geq 1$.

When $n = 1$,

$2^3 - 3^1 = 8 - 3 = 5$

So it is true for $n = 1$.

Assume true for $n = k$.

i.e. assume $2^{3k} - 3^k = 5m$

If $n = k + 1$,

$$\begin{aligned}
2^{3(k+1)} - 3^{k+1} &= 2^{3k+3} - 3^{k+1} \\
&= 2^{3k}.2^3 - 3^k.3^1 \\
&= 8(2^{3k}) - 3(3^k) \\
&= (5+3)(2^{3k}) - 3(3^k) \\
&= 5(2^{3k}) + 3(2^{3k}) - 3(3^k) \\
&= 5(2^{3k}) + 3(2^{3k} - 3^k) \\
&= 5(2^{3k}) + 3(5m) \\
&= 5(2^{3k} + 3m)
\end{aligned}$$

So, if true for $n = k$ it is also true for $n = k + 1$.

It is true for $n = 1$, so it is true for $n = 1 + 1 = 2$, so it is true for $n = 2 + 1 = 3$ and so on.

By the process of induction, $2^{3n} - 3^n$ is divisible by 5 for all integers n greater than or equal to 1.

(3 marks)

(b) $f(x) = \sqrt{4x - 3}$

(i) $4x - 3 \geq 0$

$4x \geq 3$

$x \geq \dfrac{3}{4}$

The domain is all real numbers $x \geq \dfrac{3}{4}$.

(1 mark)

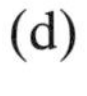

(ii) $y = \sqrt{4x - 3}$

Range is $y \geq 0$

Inverse function: $x = \sqrt{4y - 3}$

$$x^2 = 4y - 3$$
$$4y = x^2 + 3$$
$$y = \frac{x^2 + 3}{4}$$

So $f^{-1}(x) = \dfrac{x^2 + 3}{4}, x \geq 0$ *(2 marks)*

(iii) Graphs intersect when $f(x) = x$

$$\sqrt{4x - 3} = x$$
$$4x - 3 = x^2$$
$$x^2 - 4x + 3 = 0$$
$$(x - 1)(x - 3) = 0$$
$$x = 1 \text{ or } x = 3$$

So the points of intersection are $(1, 1)$ and $(3, 3)$. *(1 mark)*

(iv)

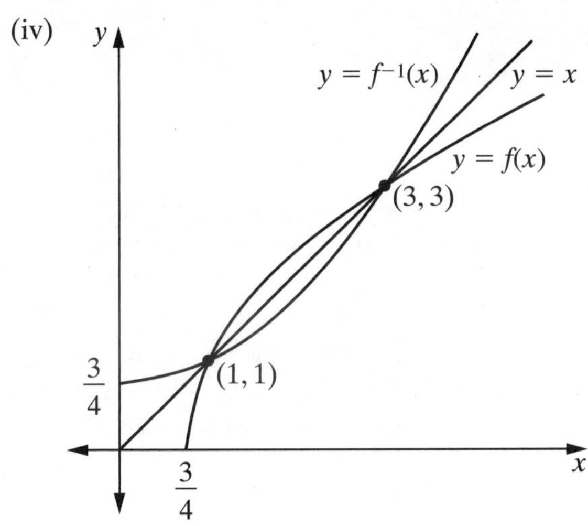

(2 marks)

(c) (i) $P(\text{draw}) = \dfrac{1}{5}$

[P(Second player spins the same number as the first player did)]

$$P(\text{not a draw}) = \frac{4}{5}$$

$$P(\text{Kim wins}) = \frac{1}{2} \times \frac{4}{5} = \frac{2}{5}$$ *(1 mark)*

(ii) $P(\text{Kim does not win}) = \dfrac{3}{5}$

P(Kim wins exactly 3 games out of 6)

$$= {}^6C_3\left(\frac{2}{5}\right)^3\left(\frac{3}{5}\right)^3$$
$$= \frac{864}{3125}$$ *(2 marks)*

(d)

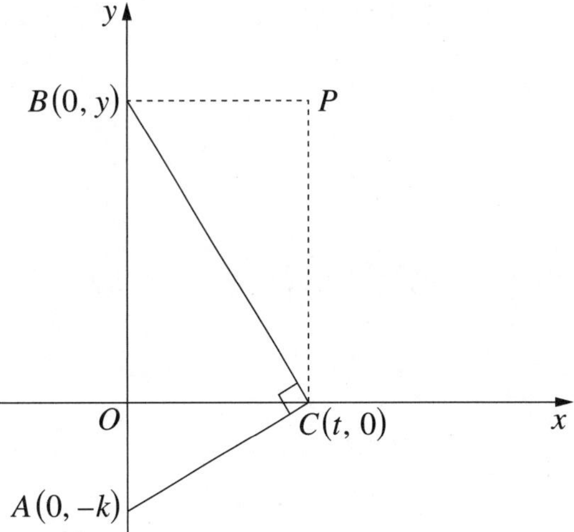

(i) $OBPC$ is a rectangle.

So PC is parallel to the y-axis.

So at $P, x = t$

$$m = \frac{y_2 - y_1}{x_2 - x_1}$$

$$m_{AC} = \frac{0 - (-k)}{t - 0} = \frac{k}{t}$$

$$m_{BC} = \frac{y - 0}{0 - t} = -\frac{y}{t}$$

Now AC is perpendicular to BC.

So $\dfrac{k}{t} \times -\dfrac{y}{t} = -1$

$$ky = t^2$$
$$y = \frac{t^2}{k}$$

BP is parallel to the x-axis so, at P,

$y = \dfrac{t^2}{k}$.

$\therefore$ P lies on the parabola given parametrically by

$x = t$ and $y = \dfrac{t^2}{k}$. *(2 marks)*

(ii) $x = t, \ y = \dfrac{t^2}{k}$

$$y = \frac{x^2}{k}$$
$$ky = x^2$$

So the parabola is of the form $x^2 = 4ay$.

Now $4a = k$

$$a = \frac{k}{4}$$

The focus of the parabola is $(0, \dfrac{k}{4})$.

(1 mark)

QUESTION 13

(a) $\sin\left(2\cos^{-1}\left(\frac{2}{3}\right)\right)$

Let $x = \cos^{-1}\left(\frac{2}{3}\right)$

$\therefore \cos x = \frac{2}{3}$

$\cos^2 x = \frac{4}{9}$

$\sin^2 x = 1 - \cos^2 x$

$= 1 - \frac{4}{9}$

$= \frac{5}{9}$

$\sin x = \pm\frac{\sqrt{5}}{3}$

But $0 \le \cos^{-1} u \le \pi$ for all u

$\therefore 0 \le x \le \pi$

$\therefore \sin x \ge 0$

So $\sin x = \frac{\sqrt{5}}{3}$

$$\sin\left(2\cos^{-1}\left(\frac{2}{3}\right)\right) = \sin 2x$$
$$= 2\sin x\cos x$$
$$= 2 \times \frac{\sqrt{5}}{3} \times \frac{2}{3}$$
$$= \frac{4\sqrt{5}}{9}$$

(2 marks)

(b) (i) $y = \frac{2x^2}{x^2+9}$

$y(x^2 + 9) = 2x^2$

$x^2y + 9y = 2x^2$

$2x^2 - x^2y = 9y$

$x^2(2 - y) = 9y$

$x^2 = \frac{9y}{2-y}$

Now $2 - y \ne 0$

$y \ne 2$

The horizontal asymptote is $y = 2$.

(1 mark)

(ii) $x^2 \ge 0$ for all values of x

$\therefore \frac{2x^2}{x^2+9} \ge 0$ for all values of x

$f(x) = \frac{2x^2}{x^2+9}$

$f(0) = 0$

$$f(-x) = \frac{2(-x)^2}{(-x)^2+9}$$
$$= \frac{2x^2}{x^2+9}$$
$$= f(x)$$

$\therefore$ the function is even

$$f(x) = \frac{2}{1+\frac{9}{x^2}}$$

$\therefore$ as $x \to \infty$, $f(x) \to 2^-$

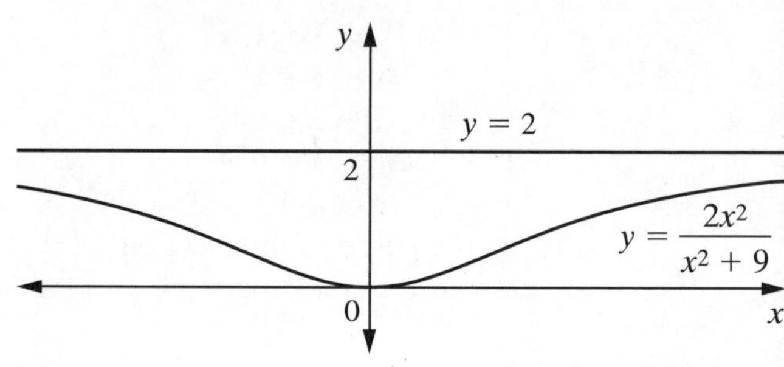

(2 marks)

(c) (i) $x = 5 + 6\cos 2t + 8\sin 2t$

$\dot{x} = -12\sin 2t + 16\cos 2t$

$\ddot{x} = -24\cos 2t - 32\sin 2t$

$= -4(6\cos 2t + 8\sin 2t)$

$= -4(x - 5)$

$\therefore$ the particle is undergoing simple harmonic motion as the acceleration is directly proportional and of opposite sign to the displacement.

(2 marks)

(ii) If $x = 0$,

$5 + 6\cos 2t + 8\sin 2t = 0$

$6\cos 2t + 8\sin 2t = -5$

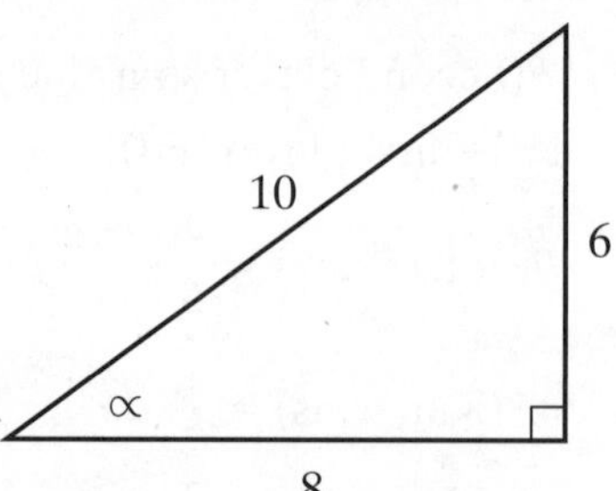

$$\frac{6}{10}\cos 2t + \frac{8}{10}\sin 2t = -\frac{1}{2}$$

$$\sin\alpha\cos 2t + \cos\alpha\sin 2t = -\frac{1}{2}$$

$$\therefore \sin(\alpha + 2t) = -\frac{1}{2}$$

where $\sin\alpha = \dfrac{6}{10}$

$\alpha = 0.643\,5011\ldots$

Now $t \geq 0$, so $\alpha + 2t > 0$

$\alpha + 2t = \dfrac{7\pi}{6}, \dfrac{11\pi}{6}, \ldots$

The displacement will first be negative when

$$\alpha + 2t = \frac{7\pi}{6}$$

$$\begin{aligned} 2t &= \frac{7\pi}{6} - 0.643\,5011\ldots \\ &= 3.021\,6903\ldots \end{aligned}$$

$$\begin{aligned} t &= 1.510\,845\ldots. \\ &= 1.5 \qquad \text{[1 d.p.]} \end{aligned}$$

$\therefore$ the displacement of the particle will first be zero after 1.5 seconds.

(3 marks)

(d) $C(t) = 1.4te^{-0.2t}$

(i) $$\begin{aligned} \frac{dC}{dt} &= 1.4t \times -0.2e^{-0.2t} + e^{-0.2t} \times 1.4 \\ &= -0.28te^{-0.2t} + 1.4e^{-0.2t} \\ &= e^{-0.2t}(1.4 - 0.28t) \end{aligned}$$

Stationary point occurs when $\dfrac{dC}{dt} = 0$

i.e. $e^{-0.2t}(1.4 - 0.28t) = 0$

$e^{-0.2t} = 0$ or $1.4 - 0.28t = 0$

No solution $\quad 0.28t = 1.4$

$t = 5$

If $t < 5$, $\dfrac{dC}{dt} > 0$

If $t > 5$, $\dfrac{dC}{dt} < 0$

$\therefore$ the maximum occurs when $t = 5$

$\therefore$ the concentration of the drug in the blood is a maximum after 5 hours.

(3 marks)

(ii) $C(t) = 0.3$

$\therefore 1.4te^{-0.2t} - 0.3 = 0$

Let $f(t) = 1.4te^{-0.2t} - 0.3$

$$f'(t) = e^{-0.2t}(1.4 - 0.28t)$$

$$t_2 = t_1 - \frac{f(t_1)}{f'(t_1)} \text{ where } t_1 = 20$$

$$t_2 = 20 - \frac{1.4 \times 20e^{-0.2\times 20} - 0.3}{e^{-0.2\times 20}(1.4 - 0.28 \times 20)}$$

$$= 22.766\,798\ldots$$

The concentration of the drug in the blood will be approximately 0.3 mg/L after 22 hours and 46 minutes.

(2 marks)

QUESTION 14

(a)

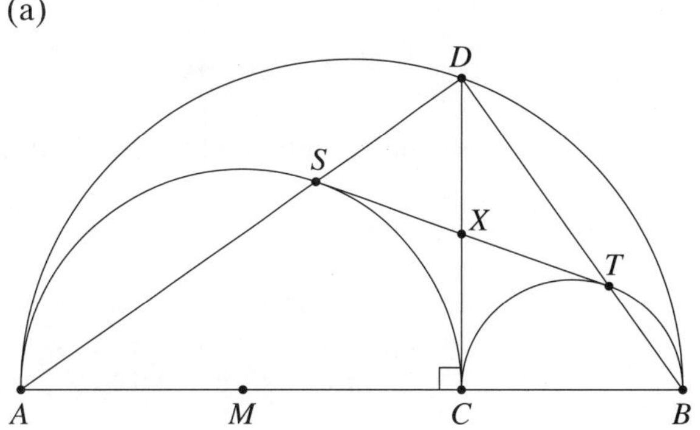

(i) $\angle SDT = 90°$ (angle in semi-circle, diameter AB)

$\angle ASC = 90°$ (angle in semi-circle, diameter AC)

$\therefore \angle DSC = 90°$ (AD is a straight line)

$\angle CTB = 90°$ (angle in a semi-circle, diameter CB)

$\therefore \angle CTD = 90°$ (DB is a straight line)

$\therefore \angle SCT = 90°$ (angle sum of quadrilateral)

$\therefore$ $CTDS$ is a rectangle (quadrilateral with all angles right angles)

(1 mark)

(ii) In $\triangle MXS$, $\triangle MXC$

$MX = MX$ (common side)

$MS = MC$ (radii of circle)

$XS = XC$ (diagonals of rectangle are equal, and bisect each other)

$\therefore \triangle MXS \equiv \triangle MXC$ SSS

(2 marks)

(iii) $\angle MCX = 90°$ (given)

$\angle MSX = \angle MCX$ (corresponding angles of congruent triangles)

$\therefore \angle MSX = 90°$

$\therefore$ ST is a tangent to the circle (meets radius on the circumference at right angles)

(1 mark)

(b) (i) $y = 70t\sin\theta - 4.9t^2$

$\frac{dy}{dt} = 70\sin\theta - 9.8t$

Stationary point occurs when $\frac{dy}{dt} = 0$

i.e. $9.8t = 70\sin\theta$

$t = \frac{50}{7}\sin\theta$

$\frac{d^2y}{dt^2} = -9.8 \quad (< 0)$

$\therefore$ the maximum height is reached when

$t = \frac{50}{7}\sin\theta$

When $t = \frac{50}{7}\sin\theta$,

$y = 70 \times \frac{50}{7}\sin\theta \times \sin\theta - 4.9 \times (\frac{50}{7}\sin\theta)^2$

$= 500\sin^2\theta - 250\sin^2\theta$

$= 250\sin^2\theta$

$\therefore$ the firework explodes at a height of $250\sin^2\theta$ metres. *(2 marks)*

(ii) $x = 70t\cos\theta$

When $t = \frac{50}{7}\sin\theta$,

$x = 70 \times \frac{50}{7}\sin\theta \times \cos\theta$

$= 500\sin\theta\cos\theta$

$= 250 \times 2\sin\theta\cos\theta$

$= 250\sin 2\theta$

$\therefore$ the firework explodes at a horizontal distance of $250\sin 2\theta$ metres from O.

(1 mark)

(iii) $0° < \theta < 90°$

Height ≥ 150 m

$250\sin^2\theta \geq 150$

$\sin^2\theta \geq 0.6$

$\sin\theta \geq 0.7745966\ldots \quad (\sin\theta > 0)$

$\theta \geq 50.768479\ldots°$

$0° < 2\theta < 180°$

125 m < horizontal distance < 180 m

So $250\sin 2\theta > 125$

$\sin 2\theta > 0.5$

$30° < 2\theta < 150°$

$15° < \theta < 75°$

And $250\sin 2\theta < 180$

$\sin 2\theta < 0.72$

$2\theta < 46.054\ldots°$ or $2\theta > 133.945\ldots°$

$\theta < 23.027\ldots°$ or $\theta > 66.972\ldots°$

$\therefore$ the firework will be at the best position for viewing when $67° \leq \theta < 75°$.

(3 marks)

(c)

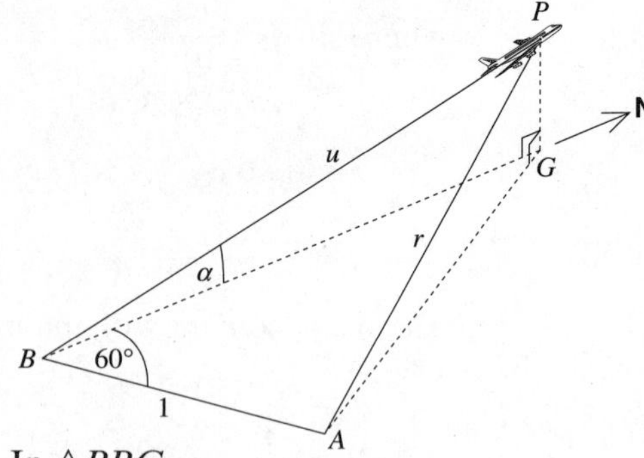

(i) In $\triangle PBG$,

$\sin\alpha = \frac{PG}{u}$

$\therefore PG = u\sin\alpha$

$\cos\alpha = \frac{BG}{u}$

$\therefore BG = u\cos\alpha$

In $\triangle ABG$, by the cosine rule,

$AG^2 = (u\cos\alpha)^2 + 1^2 - 2 \times u\cos\alpha \times 1 \times \cos 60°$

$= u^2\cos^2\alpha + 1 - u\cos\alpha$

In $\triangle PGA$, by Pythagoras' theorem,

$PA^2 = PG^2 + AG^2$

$r^2 = u^2\sin^2\alpha + u^2\cos^2\alpha + 1 - u\cos\alpha$

$= u^2(\sin^2\alpha + \cos^2\alpha) + 1 - u\cos\alpha$

$= u^2 + 1 - u\cos\alpha$

$= 1 + u^2 - u\cos\alpha$

$\therefore r = \sqrt{1 + u^2 - u\cos\alpha} \quad (r > 0)$

(3 marks)

(ii) $r = \sqrt{1 + u^2 - u\cos\alpha}$

$= (1 + u^2 - u\cos\alpha)^{\frac{1}{2}}$

$\frac{dr}{du} = \frac{1}{2}(1 + u^2 - u\cos\alpha)^{-\frac{1}{2}}(2u - \cos\alpha)$

$= \frac{2u - \cos\alpha}{2\sqrt{1 + u^2 - u\cos\alpha}}$

Now 360 km/h = 6 km/min

So $\frac{du}{dt} = 6$

$\frac{dr}{dt} = \frac{dr}{du} \cdot \frac{du}{dt}$

$= \frac{2u - \cos\alpha}{2\sqrt{1 + u^2 - u\cos\alpha}} \times 6$

$= \frac{3(2u - \cos\alpha)}{\sqrt{1 + u^2 - u\cos\alpha}}$

5 minutes after take off $u = 5 \times 6$

$= 30$

$\therefore \frac{dr}{dt} = \frac{3(2 \times 30 - \cos\alpha)}{\sqrt{1 + 30^2 - 30\cos\alpha}}$

$= \frac{3(60 - \cos\alpha)}{\sqrt{901 - 30\cos\alpha}}$ km/min

[or $\frac{60 - \cos\alpha}{20\sqrt{901 - 30\cos\alpha}}$ km/s] *(2 marks)*

Solutions to replacement questions

QUESTION 1

$$\frac{dy}{dx} = y$$

$$\frac{dy}{dx} = \frac{1}{y}$$

$$x = \int \frac{1}{y}\, dy$$

$$= \ln|y| + c$$

Substitute $y = 1$ when $x = 1$:

$$1 = \ln|1| + c$$

$$c = 1$$

$$x = \ln|y| + 1$$

$$\ln|y| = x - 1$$

$$y = e^{x-1}$$

Alternatively, consider each option:

For $y = e^x$ and $y = e^{x-1}$, $\frac{dy}{dx} = y$. So possible answers are A and B.

Also, the point $(1, 1)$ lies on $y = e^{x-1}$. Hence, the answer is B.

Answer B

(1 mark)

QUESTION 2

At the points $(1, 1)$, $(2, 2)$ etc gradient = 1. So possible solutions are B and C.

At $(2, 0)$, the gradient is undefined. So the solution is B.

Hence,

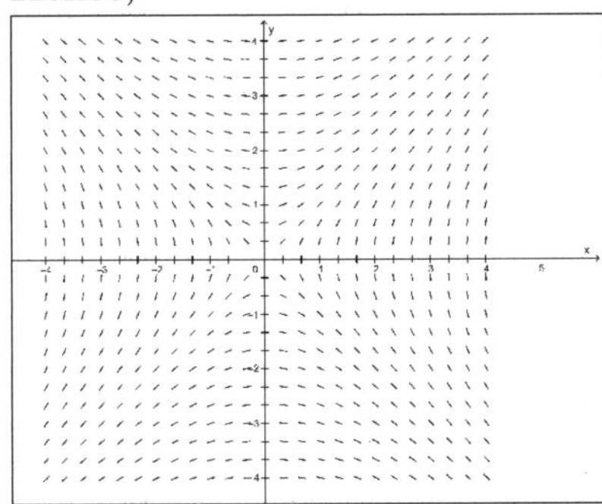

Answer B

(1 mark)

QUESTION 6

$$2\underset{\sim}{a} - 3\underset{\sim}{b} = 2(3\underset{\sim}{i} - 2\underset{\sim}{j}) - 3(-\underset{\sim}{i} + 5\underset{\sim}{j})$$

$$= 6\underset{\sim}{i} - 4\underset{\sim}{j} + 3\underset{\sim}{i} - 15\underset{\sim}{j}$$

$$= 9\underset{\sim}{i} - 19\underset{\sim}{j}$$

Answer B

(1 mark)

QUESTION 10

$$E(X) = \mu = np$$

$$= 20 \times \frac{1}{4}$$

$$= 5$$

$$Var(X) = np(1 - p)$$

$$= 5\left(1 - \frac{1}{4}\right)$$

$$= 5\left(\frac{3}{4}\right)$$

$$= \frac{15}{4}$$

Answer B

(1 marks)

QUESTION 12

(d)

$$x = \sin t + \cos t$$

$$x^2 = (\sin t + \cos t)^2$$

$$= \sin^2 t + 2 \sin t \cos t + \cos^2 t$$

$$= 1 + 2 \sin t \cos t$$

$$= 1 + \sin 2t$$

$$\sin 2t = x^2 - 1$$

But $y = \frac{1}{2} \sin 2t$

$$\therefore y = \frac{1}{2}(x^2 - 1)$$

$$= \frac{x^2}{2} - \frac{1}{2}$$

$$\therefore \frac{dy}{dx} = x$$

Alternatively,

$$\frac{dy}{dx} = \frac{dy}{dt} \times \frac{dt}{dx}$$

$$\frac{dy}{dt} = \cos 2t$$

$$\frac{dt}{dx} = \cos t - \sin t$$

$$\frac{dt}{dx} = \frac{1}{\cos t - \sin t}$$

$$\frac{dy}{dx} = \cos 2t \times \frac{1}{\cos t - \sin t}$$

$$= \frac{\cos^2 t - \sin^2 t}{\cos t - \sin t}$$

$$= \frac{(\cos t - \sin t)(\cos t + \sin t)}{\cos t - \sin t}$$

$$= \cos t + \sin t$$

$$= x$$

(3 marks)

QUESTION 13

(c)* $\overrightarrow{OA} + \overrightarrow{AB} = \overrightarrow{OB}$

$$\underset{\sim}{a} + \overrightarrow{AB} = \underset{\sim}{b}$$

$$\overrightarrow{AB} = \underset{\sim}{b} - \underset{\sim}{a}$$

Now, $AB = p + q$

$$\therefore \overrightarrow{AC} = \frac{p}{p+q}(\underset{\sim}{b} - \underset{\sim}{a})$$

Now, as $\overrightarrow{OC} = \underset{\sim}{a} + \overrightarrow{AC}$

$$\therefore \overrightarrow{OC} = \underset{\sim}{a} + \frac{p}{p+q}(\underset{\sim}{b} - \underset{\sim}{a})$$

[Alternatively, $\overrightarrow{OC} = \underset{\sim}{a} + \frac{q}{p+q}(\underset{\sim}{a} - \underset{\sim}{b})$]

(2 marks)

(c)** $\int_0^{\frac{\pi}{8}} 2\cos^3 2x \sin 2x \, dx.$

$u = \cos 2x \qquad \frac{du}{dx} = -2\sin 2x \qquad dx = \frac{du}{-2\sin 2x}$

$x = \frac{\pi}{8}$, then $u = \frac{1}{\sqrt{2}}$; $x = 0$, then $u = 1$

$$= \int_1^{\frac{1}{\sqrt{2}}} 2u^3 \sin 2x . \frac{du}{-2\sin 2x}$$

$$= \int_{\frac{1}{\sqrt{2}}}^1 u^3 du$$

$$= \left[\frac{u^4}{4}\right]_{\frac{1}{\sqrt{2}}}^1$$

$$= \frac{1}{4}\left[1^4 - \left(\frac{1}{\sqrt{2}}\right)^4\right]$$

$$= \frac{1}{4}\left[1 - \frac{1}{4}\right]$$

$$= \frac{3}{16}$$

(3 marks)

(d) (ii) $\frac{|1-4x|}{2} \le x$

$$-x \le \frac{1-4x}{2} \le x$$

Two cases:

$-2x \le 1 - 4x$	$1 - 4x \le 2x$
$2x \le 1$	$6x \ge 1$
$x \le \frac{1}{2}$	$x \ge \frac{1}{6}$

$$\therefore \frac{1}{6} \le x \le \frac{1}{2}$$

(2 marks)

QUESTION 14

(a) (i)

$$P = 8000(5 - 4e^{0.015t})$$
$$= 40000 - 32000e^{0.015t} \ldots ①$$
$$32000e^{0.015t} = 40\,000 - P$$
$$e^{0.015t} = \frac{1}{32\,000}(40000 - P)$$

Now, using ①:

$$\frac{dP}{dt} = -480e^{0.015t}$$
$$= -480 \times \frac{1}{32\,000}(40000 - P)$$
$$= -0.015(40000 - P)$$
$$= \frac{3P}{200} - 600$$

(2 marks)

(ii) Let $P = 0$:

$$0 = 8000(5 - 4e^{0.015t})$$
$$5 - 4e^{0.015t} = 0$$
$$4e^{0.015t} = 5$$
$$e^{0.015t} = \frac{5}{4}$$
$$0.015t = \log_e 1.25$$
$$t = \frac{\log_e 1.25}{0.015}$$
$$= 14.87623675\ldots$$
$$= 14.88 \text{ (2 dec. pl.)}$$

∴ It will take 14.88 years.

(2 marks)

BOARD OF STUDIES
NEW SOUTH WALES

2013
HIGHER SCHOOL CERTIFICATE EXAMINATION

Mathematics Extension 1

General Instructions

- Reading time – 5 minutes
- Working time – 2 hours
- Write using black or blue pen
  Black pen is preferred
- Board-approved calculators may be used
- A table of standard integrals is provided at the back of this paper
- In Questions 11–14, show relevant mathematical reasoning and/or calculations

Total marks – 70

Section I

10 marks

- Attempt Questions 1–10
- Allow about 15 minutes for this section

Section II

60 marks

- Attempt Questions 11–14
- Allow about 1 hour and 45 minutes for this section

Section I

10 marks
Attempt Questions 1–10
Allow about 15 minutes for this section

Use the multiple-choice answer sheet for Questions 1–10.

---

1 The polynomial $P(x) = x^3 - 4x^2 - 6x + k$ has a factor $x - 2$.

What is the value of k?

(A) 2

(B) 12

(C) 20

(D) 36

2 The diagram shows the graph $y = f(x)$.

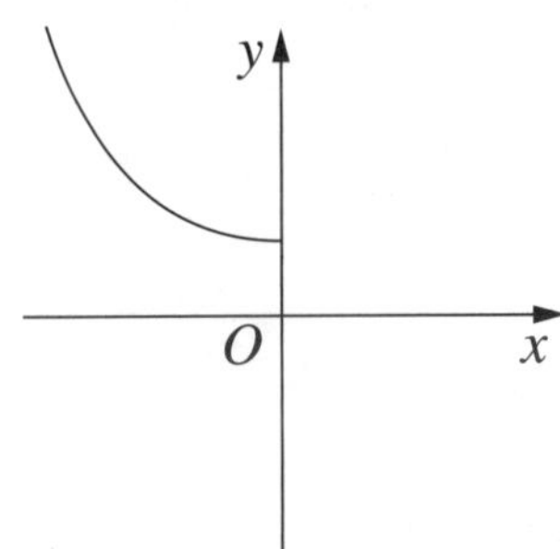

Which diagram shows the graph $y = f^{-1}(x)$?

(A)

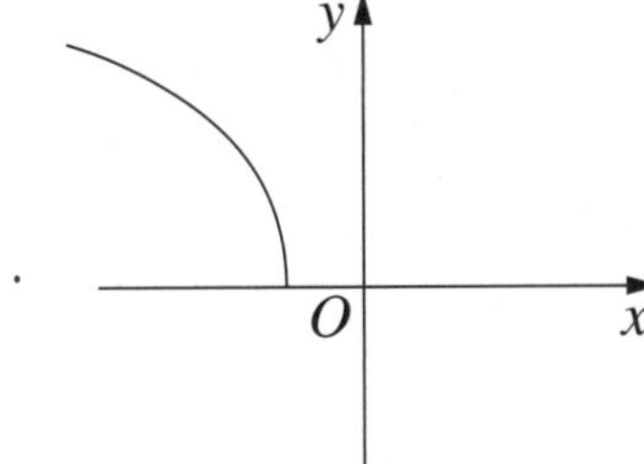

(B)

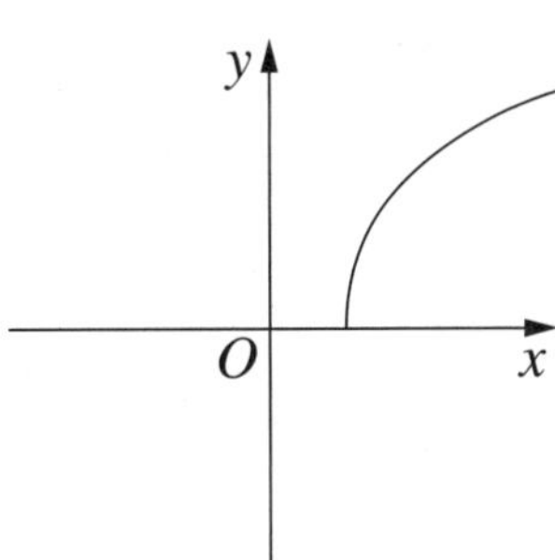

(C)

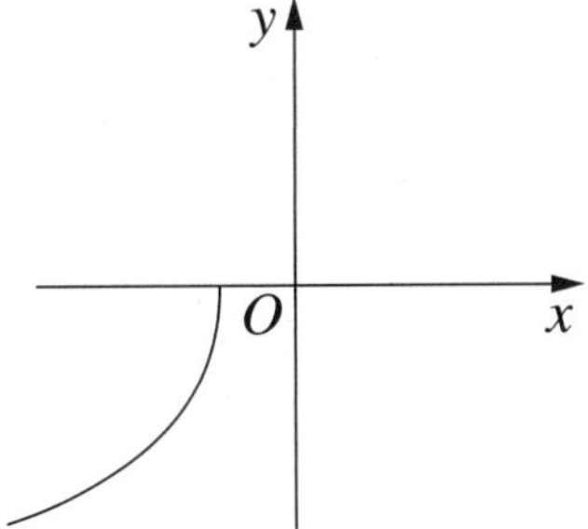

(D)

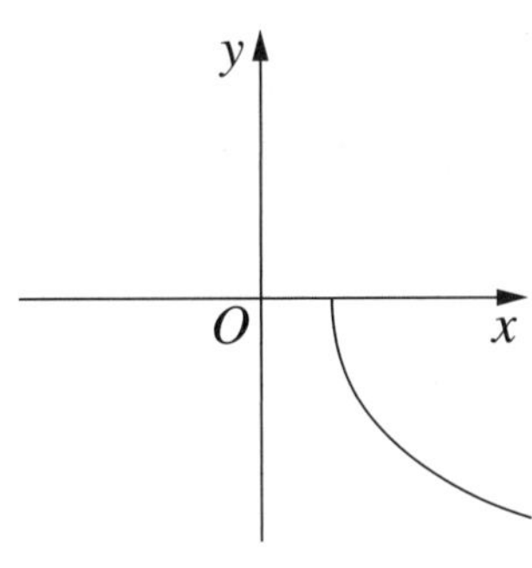

✗ **3** The points A, B and C lie on a circle with centre O, as shown in the diagram.

The size of $\angle AOC$ is $\frac{3\pi}{5}$ radians.

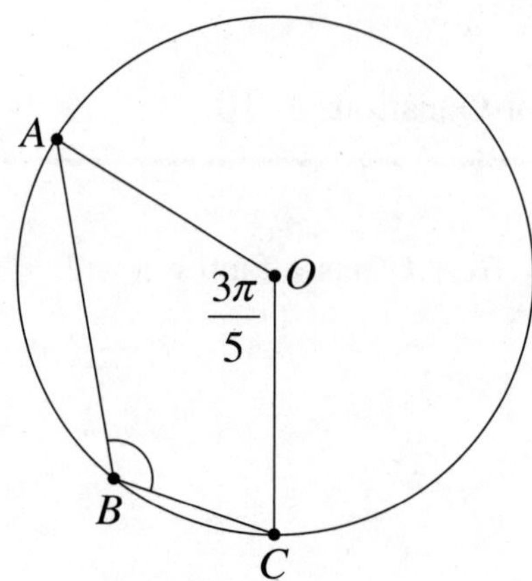

NOT TO SCALE

What is the size of $\angle ABC$ in radians?

(A) $\frac{3\pi}{10}$

(B) $\frac{2\pi}{5}$

(C) $\frac{7\pi}{10}$

(D) $\frac{4\pi}{5}$

4 Which diagram best represents the graph $y = x(1-x)^3(3-x)^2$?

(A)

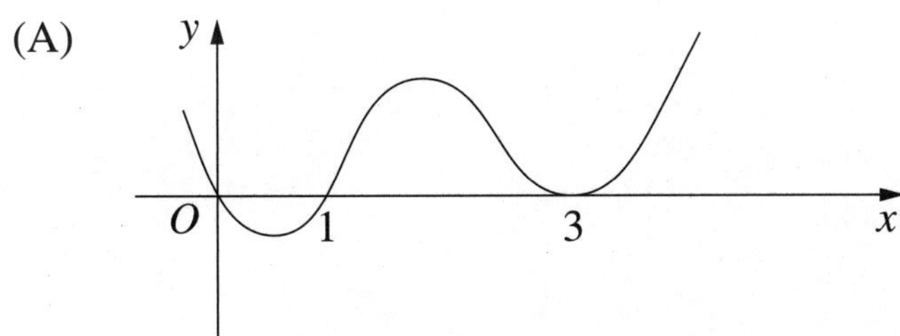

(B)

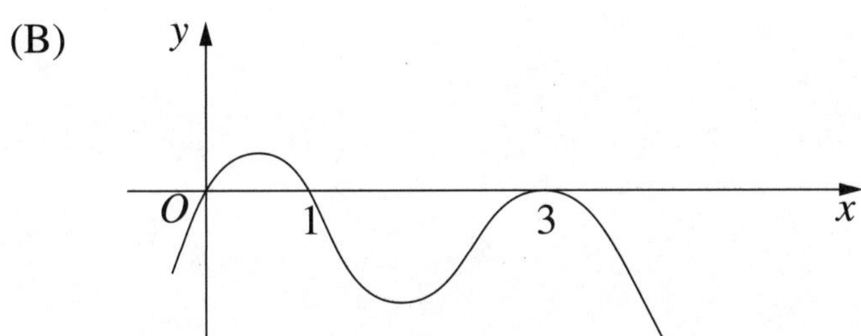

(C)

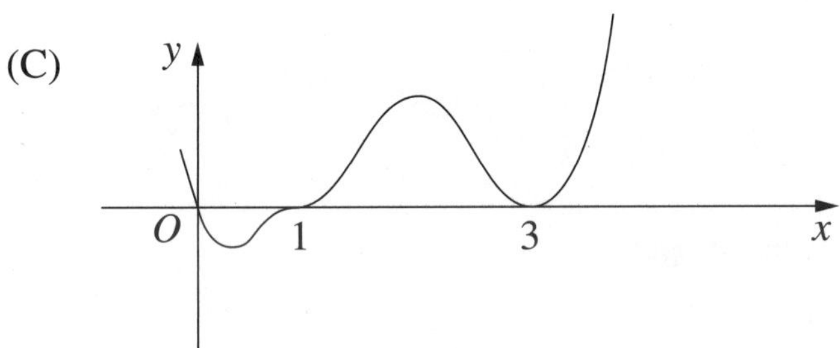

(D)

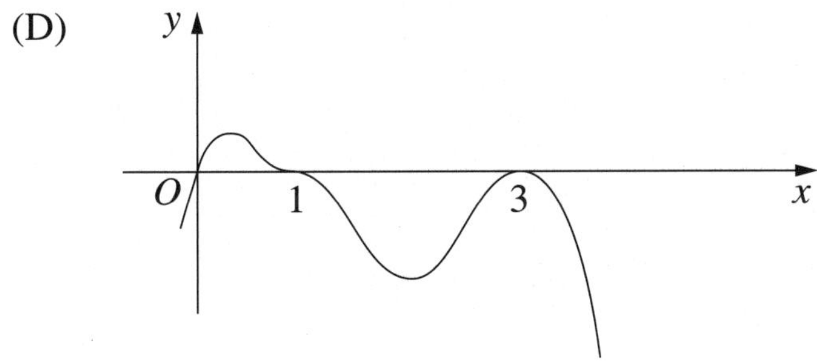

5 Which integral is obtained when the substitution $u = 1 + 2x$ is applied to $\int x\sqrt{1+2x}\,dx$?

(A) $\frac{1}{4}\int (u-1)\sqrt{u}\,du$

(B) $\frac{1}{2}\int (u-1)\sqrt{u}\,du$

(C) $\int (u-1)\sqrt{u}\,du$

(D) $2\int (u-1)\sqrt{u}\,du$

✗ **6** Let $|a| \le 1$. What is the general solution of $\sin 2x = a$?

(A) $x = n\pi + (-1)^n \frac{\sin^{-1} a}{2}$, n is an integer

(B) $x = \frac{n\pi + (-1)^n \sin^{-1} a}{2}$, n is an integer

(C) $x = 2n\pi \pm \frac{\sin^{-1} a}{2}$, n is an integer

(D) $x = \frac{2n\pi \pm \sin^{-1} a}{2}$, n is an integer

7 A family of eight is seated randomly around a circular table.

What is the probability that the two youngest members of the family sit together?

(A) $\dfrac{6!2!}{7!}$

(B) $\dfrac{6!}{7!2!}$

(C) $\dfrac{6!2!}{8!}$

(D) $\dfrac{6!}{8!2!}$

8 The angle θ satisfies $\sin\theta = \dfrac{5}{13}$ and $\dfrac{\pi}{2} < \theta < \pi$.

What is the value of $\sin 2\theta$?

(A) $\dfrac{10}{13}$

(B) $-\dfrac{10}{13}$

(C) $\dfrac{120}{169}$

(D) $-\dfrac{120}{169}$

9 The diagram shows the graph of a function.

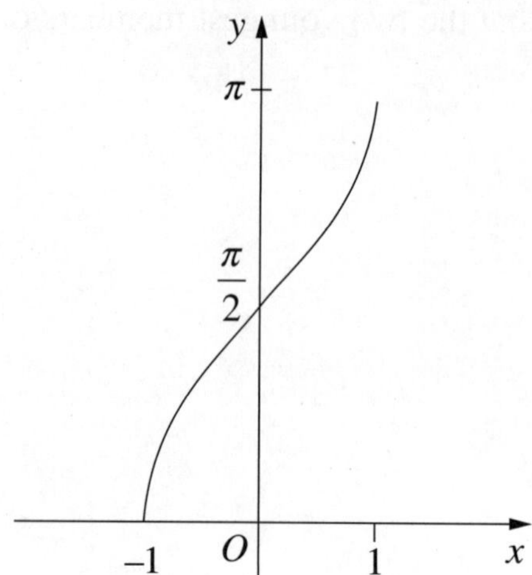

Which function does the graph represent?

(A) $y = \cos^{-1} x$

(B) $y = \frac{\pi}{2} + \sin^{-1} x$

(C) $y = -\cos^{-1} x$

(D) $y = -\frac{\pi}{2} - \sin^{-1} x$

10 Which inequality has the same solution as $|x+2| + |x-3| = 5$?

(A) $\frac{5}{3-x} \geq 1$

(B) $\frac{1}{x-3} - \frac{1}{x+2} \leq 0$

(C) $x^2 - x - 6 \leq 0$

(D) $|2x - 1| \geq 5$

Section II

60 marks
Attempt Questions 11–14
Allow about 1 hour and 45 minutes for this section

Answer each question in a SEPARATE writing booklet. Extra writing booklets are available.

In Questions 11–14, your responses should include relevant mathematical reasoning and/or calculations.

Question 11 (15 marks) Use a SEPARATE writing booklet.

(a) The polynomial equation $2x^3 - 3x^2 - 11x + 7 = 0$ has roots α, β and γ. **1**

Find $\alpha\beta\gamma$.

(b) Find $\int \frac{1}{\sqrt{49 - 4x^2}}\,dx$. **2**

(c) An examination has 10 multiple-choice questions, each with 4 options. In each question, only one option is correct. For each question a student chooses one option at random. **2**

Write an expression for the probability that the student chooses the correct option for exactly 7 questions.

(d) Consider the function $f(x) = \frac{x}{4 - x^2}$.

(i) Show that $f'(x) > 0$ for all x in the domain of $f(x)$. **2**

(ii) Sketch the graph $y = f(x)$, showing all asymptotes. **2**

Question 11 continues on page 9

Question 11 (continued)

(e) Find $\lim_{x \to 0} \dfrac{\sin \frac{x}{2}}{3x}$. **1**

(f) Use the substitution $u = e^{3x}$ to evaluate $\int_0^{\frac{1}{3}} \dfrac{e^{3x}}{e^{6x}+1}\, dx$. **3**

(g) Differentiate $x^2 \sin^{-1} 5x$. **2**

End of Question 11

Question 12 (15 marks) Use a SEPARATE writing booklet.

(a) (i) Write $\sqrt{3}\cos x - \sin x$ in the form $2\cos(x+\alpha)$, where $0 < \alpha < \frac{\pi}{2}$. **1**

(ii) Hence, or otherwise, solve $\sqrt{3}\cos x = 1 + \sin x$, where $0 < x < 2\pi$. **2**

(b) The region bounded by the graph $y = 3\sin\frac{x}{2}$ and the x-axis between $x = 0$ and $x = \frac{3\pi}{2}$ is rotated about the x-axis to form a solid. **3**

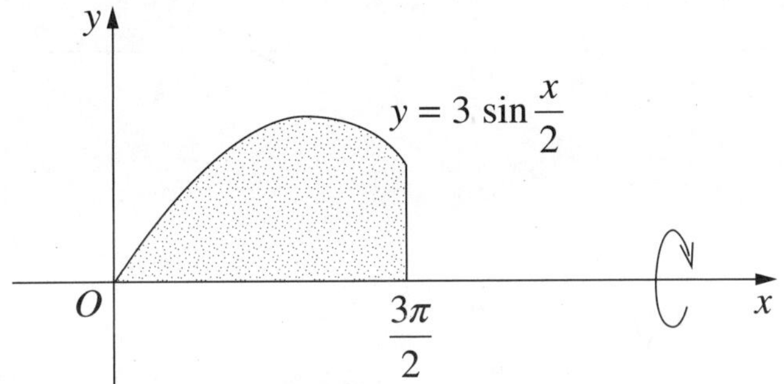

Find the exact volume of the solid.

(c) A cup of coffee with an initial temperature of 80°C is placed in a room with a constant temperature of 22°C. **3**

The temperature, T°C, of the coffee after t minutes is given by

$$T = A + Be^{-kt},$$

where A, B and k are positive constants. The temperature of the coffee drops to 60°C after 10 minutes.

How long does it take for the temperature of the coffee to drop to 40°C? Give your answer to the nearest minute.

Question 12 continues on page 11

Question 12 (continued)

(d) The point $P(t, t^2 + 3)$ lies on the curve $y = x^2 + 3$. The line ℓ has equation $y = 2x - 1$. The perpendicular distance from P to the line ℓ is $D(t)$.

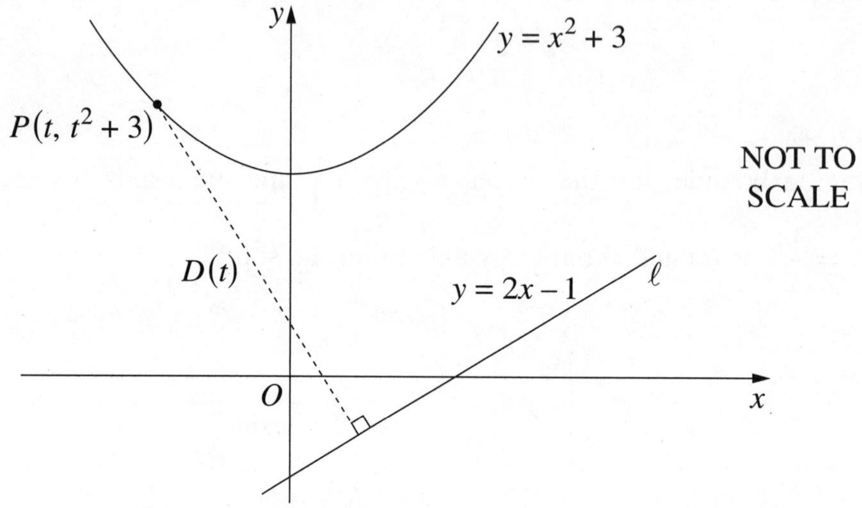

(i) Show that $D(t) = \dfrac{t^2 - 2t + 4}{\sqrt{5}}$. **2**

(ii) Find the value of t when P is closest to ℓ. **1**

(iii) Show that, when P is closest to ℓ, the tangent to the curve at P is parallel to ℓ. **1**

(e) A particle moves along a straight line. The displacement of the particle from the origin is x, and its velocity is v. The particle is moving so that $v^2 + 9x^2 = k$, where k is a constant. **2**

Show that the particle moves in simple harmonic motion with period $\dfrac{2\pi}{3}$.

End of Question 12

Question 13 (15 marks) Use a SEPARATE writing booklet.

(a) A spherical raindrop of radius r metres loses water through evaporation at a rate that depends on its surface area. The rate of change of the volume V of the raindrop is given by

$$\frac{dV}{dt} = -10^{-4} A,$$

where t is time in seconds and A is the surface area of the raindrop. The surface area and the volume of the raindrop are given by $A = 4\pi r^2$ and $V = \frac{4}{3}\pi r^3$ respectively.

(i) Show that $\frac{dr}{dt}$ is constant. **1**

(ii) How long does it take for a raindrop of volume $10^{-6}\ \text{m}^3$ to completely evaporate? **2**

(b) The point $P(2ap, ap^2)$ lies on the parabola $x^2 = 4ay$. The tangent to the parabola at P meets the x-axis at $T(ap, 0)$. The normal to the tangent at P meets the y-axis at $N(0, 2a + ap^2)$.

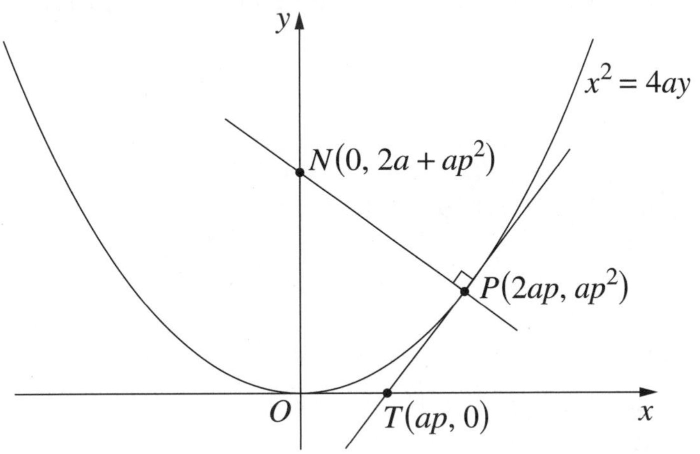

The point G divides NT externally in the ratio $2:1$.

(i) Show that the coordinates of G are $(2ap, -2a - ap^2)$. **2**

(ii) Show that G lies on a parabola with the same directrix and focal length as the original parabola. **2**

Question 13 continues on page 13

Question 13 (continued)

* (c) Points A and B are located d metres apart on a horizontal plane. A projectile is fired from A towards B with initial velocity u m s^{-1} at angle α to the horizontal.

At the same time, another projectile is fired from B towards A with initial velocity w m s^{-1} at angle β to the horizontal, as shown on the diagram.

The projectiles collide when they both reach their maximum height.

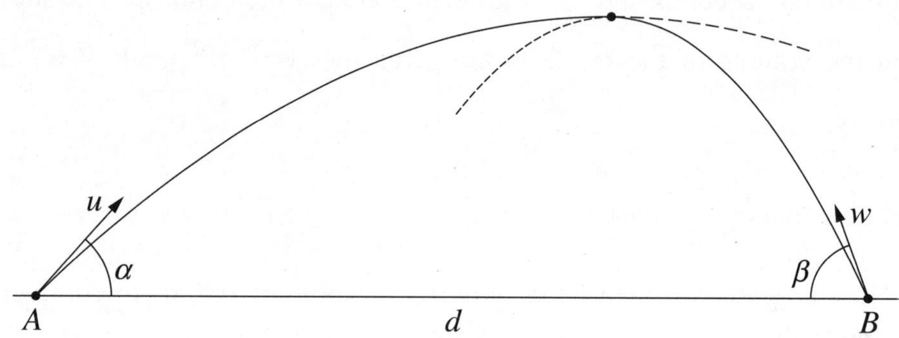

The equations of motion of a projectile fired from the origin with initial velocity V m s^{-1} at angle θ to the horizontal are

$$x = Vt\cos\theta \quad \text{and} \quad y = Vt\sin\theta - \frac{g}{2}t^2. \qquad \text{(Do NOT prove this.)}$$

(i) How long does the projectile fired from A take to reach its maximum height? **2**

(ii) Show that $u\sin\alpha = w\sin\beta$. **1**

(iii) Show that $d = \dfrac{uw}{g}\sin(\alpha + \beta)$. **2**

* In the new Mathematics Extension 1 course, Projectile Motion questions will be expressed in vector form:

$x = Vt\cos\theta$ and $y = Vt\sin\theta - \frac{1}{2}gt^2$ is expressed as a position vector:

$$\underset{\sim}{r}(t) = (Vt\cos\theta)\underset{\sim}{i} + \left(Vt\sin\theta - \frac{1}{2}gt^2\right)\underset{\sim}{j}$$

Question 13 (continued)

(d) The circles C_1 and C_2 touch at the point T. The points A and P are on C_1. The line AT intersects C_2 at B. The point Q on C_2 is chosen so that BQ is parallel to PA. **3**

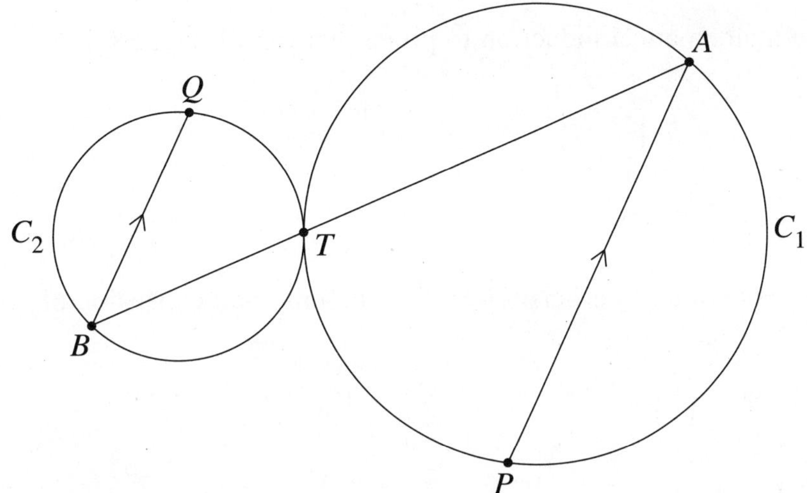

Copy or trace the diagram into your writing booklet.

Prove that the points Q, T and P are collinear.

End of Question 13

Question 14 (15 marks) Use a SEPARATE writing booklet.

(a) (i) Show that for $k > 0$, $\dfrac{1}{(k+1)^2} - \dfrac{1}{k} + \dfrac{1}{k+1} < 0$. **1**

(ii) Use mathematical induction to prove that for all integers $n \geq 2$, **3**

$$\frac{1}{1^2} + \frac{1}{2^2} + \frac{1}{3^2} + \cdots + \frac{1}{n^2} < 2 - \frac{1}{n}.$$

(b) (i) Write down the coefficient of x^{2n} in the binomial expansion of $(1+x)^{4n}$. **1**

(ii) Show that $\left(1 + x^2 + 2x\right)^{2n} = \displaystyle\sum_{k=0}^{2n} \binom{2n}{k} x^{2n-k} (x+2)^{2n-k}$. **2**

(iii) It is known that **3**

$$x^{2n-k}(x+2)^{2n-k} = \binom{2n-k}{0} 2^{2n-k} x^{2n-k} + \binom{2n-k}{1} 2^{2n-k-1} x^{2n-k+1}$$

$$+ \cdots + \binom{2n-k}{2n-k} 2^0 x^{4n-2k}.$$

(Do NOT prove this.)

Show that

$$\binom{4n}{2n} = \sum_{k=0}^{n} 2^{2n-2k} \binom{2n}{k} \binom{2n-k}{k}.$$

(c) The equation $e^t = \dfrac{1}{t}$ has an approximate solution $t_0 = 0.5$.

(i) Use one application of Newton's method to show that $t_1 = 0.56$ is another approximate solution of $e^t = \dfrac{1}{t}$. **2**

(ii) Hence, or otherwise, find an approximation to the value of r for which the graphs $y = e^{rx}$ and $y = \log_e x$ have a common tangent at their point of intersection. **3**

End of paper

Replacement questions

with content from the most up-to-date syllabus

Marks

Question 3 (1 mark)

Let $O = (0, 0)$, $A = (4, -3)$, $B = (-2, 7)$ and $C = (5, 6)$.

Which of these is $\overrightarrow{BC}$?

(A) $7\underset{\sim}{i} - \underset{\sim}{j}$

(B) $5\underset{\sim}{i} - \underset{\sim}{j}$

(C) $7\underset{\sim}{i} - 13\underset{\sim}{j}$

(D) $-7\underset{\sim}{i} + 13\underset{\sim}{j}$ 1

Question 6 (1 mark)

A binomial random variable X has a probability of success of p after 4 trials.

Use the table for the binomial distribution to find the value of p, correct to 2 decimal places. 1

x	0	1	2	3	4
$P(X = x)$	0.03419	0.18128	0.36044	0.31853	0.10556

(A) $p = 0.51$

(B) $p = 0.53$

(C) $p = 0.57$

(D) $p = 0.61$

Question 11 (1 mark)

(e) Find the range of values of k for which the expression $(1 - 2k) + 2x - x^2$ is always negative. 1

REPLACEMENT QUESTIONS

Marks

Question 12 (6 marks)

(d) The diagram shows the graph of a function $y = f(x)$.

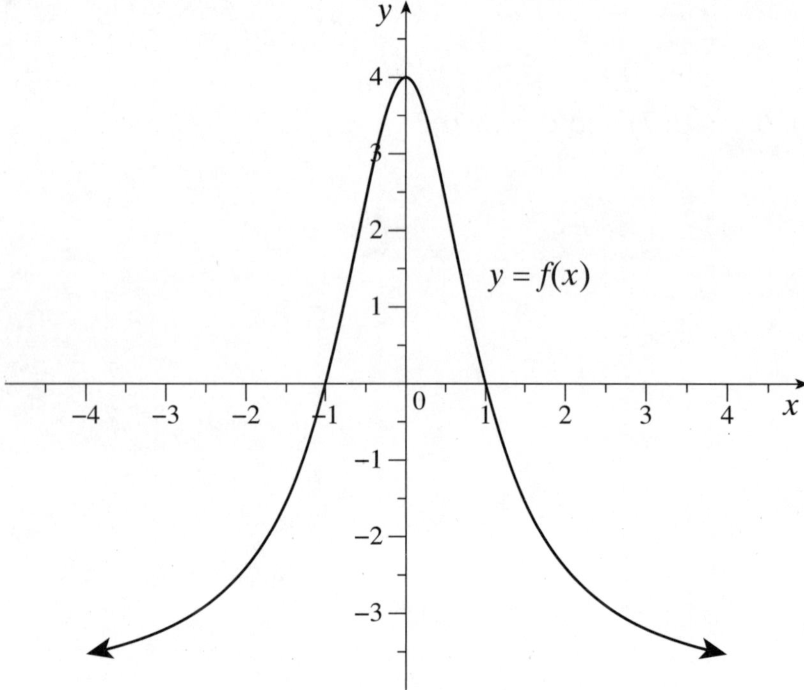

On the diagrams below draw the graphs of the functions:

(i) $y = \sqrt{f(x)}$ **2**

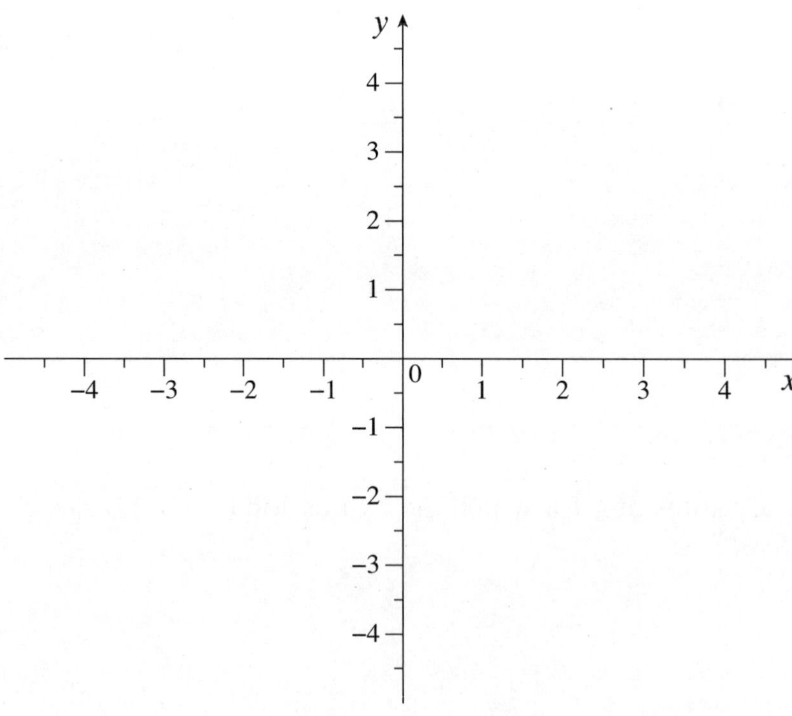

Marks

(ii) $y = \dfrac{1}{f(x)}$ **2**

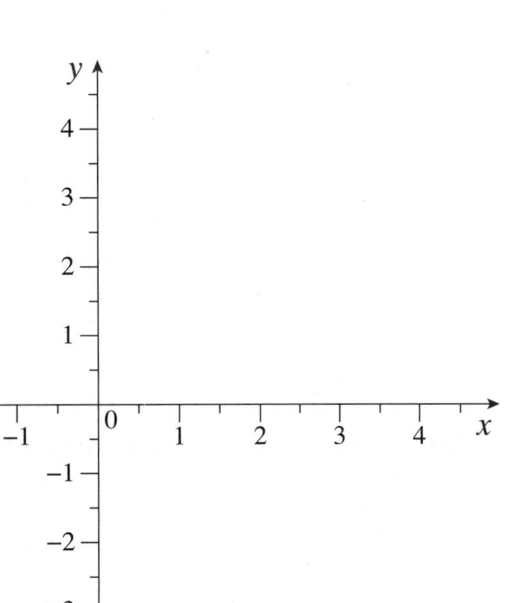

(e) Determine whether the vectors $\underset{\sim}{a} = 4\underset{\sim}{i} - 10\underset{\sim}{j}$ and $\underset{\sim}{b} = 5\underset{\sim}{i} + 2\underset{\sim}{j}$ are parallel, perpendicular or neither. **2**

Question 13 (7 marks)

(b) Evaluate $\int_0^{\frac{\pi}{6}} \cos^2 2x \sin 2x \, dx$, using the substitution $u = \cos 2x$. **4**

(d) Solve $\cos 5\theta + \cos 3\theta + \cos\theta = 0$, where $0 \leq \theta \leq \pi$. **3**

REPLACEMENT QUESTIONS

Marks

Question 14 (11 marks)

(b)* The population in a particular city is known to be 50% female.

What is the probability that a random sample of 100 people will contain less than 45% female? **2**

(b)** Show that the area bounded by the curve $y = -x^3 + 7x - 6$ and the x-axis is 32.75 units^2. **4**

(c)* A curve has parametric equations $x = \frac{t}{2}, y = 3t^2$. **2**

Find the gradient of the tangent to the curve at the point where $t = 1$.

(c)** Express y in terms of x for the differential equation

$(1 + x^2)\frac{dy}{dx} = 4xy$, where $x = 1$ when $y = 2$. **3**

REPLACEMENT QUESTIONS

2013 Higher School Certificate
Worked answers

Section I (*Total 10 marks*)

1. C	**2.** D	**3.** C	**4.** D	**5.** A
6. B	**7.** A	**8.** D	**9.** B	**10.** C

1. $P(x) = x^3 - 4x^2 - 6x + k$

$P(2) = 2^3 - 4 \times 2^2 - 6 \times 2 + k$

$= -20 + k$

But $P(2) = 0$

$\therefore k = 20$

Answer C

2. The graph of $y = f^{-1}(x)$ is the reflection of the graph of $y = f(x)$ in the line $y = x$.

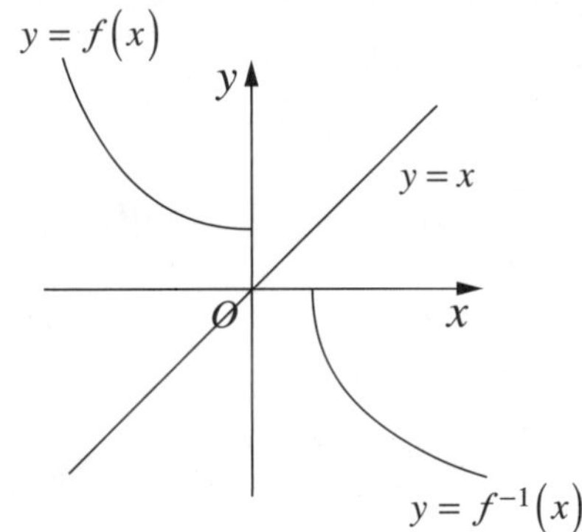

Answer D

3.

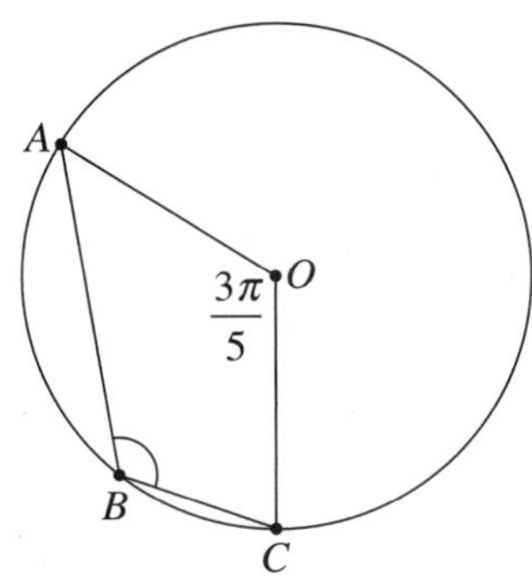

$\angle AOC = \dfrac{3\pi}{5}$

$\therefore$ Reflex $\angle AOC = 2\pi - \dfrac{3\pi}{5}$

$= \dfrac{7\pi}{5}$

But, the angle at the centre is twice the angle at the circumference.

$\therefore \angle ABC = \dfrac{7\pi}{10}$

Answer C

4. $y = x(1 - x)^3(3 - x)^2$

As $x \to \infty, y \to -\infty$

The polynomial $y = P(x)$ has a triple root at $x = 1$ and a double root at $x = 3$.

So the curve will have stationary points at $x = 1$ and $x = 3$.

Only graph D will satisfy these requirements.

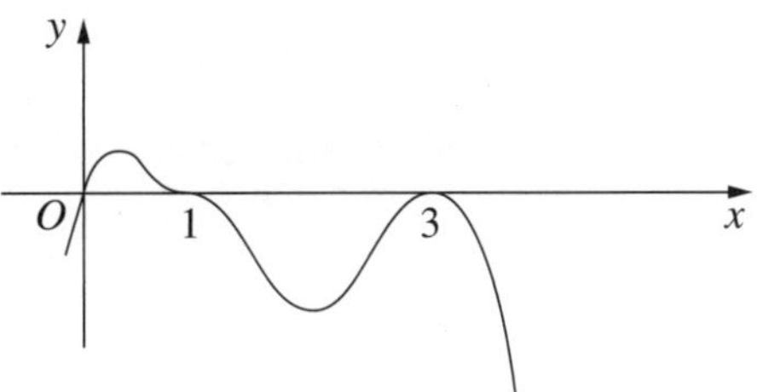

Answer D

5. Let $u = 1 + 2x$

$2x = u - 1$

$x = \dfrac{1}{2}(u - 1)$

$dx = \dfrac{1}{2}du$

$$\int x\sqrt{1 + 2x}\,dx = \int \frac{1}{2}(u - 1)\sqrt{u}\,\frac{1}{2}du$$

$$= \frac{1}{4}\int (u - 1)\sqrt{u}\,du$$

Answer A

6. $\sin 2x = a$

$2x = n\pi + (-1)^n \sin^{-1} a$

$x = \dfrac{n\pi + (-1)^n \sin^{-1} a}{2}$

Answer B

7. The youngest family member can sit anywhere. The next youngest can sit in any of 7 places, 2 of which are beside the youngest.

$\therefore P(\text{2 youngest sit together}) = \dfrac{2}{7}$

Of the options, only $\dfrac{6!2!}{7!} = \dfrac{2}{7}$

[The first person can sit anywhere. There are 7! possible arrangements of the remaining family members. Two particular members can sit together in 2! ways, while the remaining 6 can sit in 6! ways, so the number of arrangements of the family members with the 2 youngest sitting together is 6!2!]

Answer A

8. $\sin\theta = \dfrac{5}{13}$

Now $\sin^2\theta + \cos^2\theta = 1$

$$\begin{aligned} \text{So } \cos^2\theta &= 1 - \sin^2\theta \\ &= 1 - \left(\frac{5}{13}\right)^2 \\ &= \frac{144}{169} \\ \therefore \cos\theta &= -\frac{12}{13} \quad \left(\frac{\pi}{2} < \theta < \pi\right) \end{aligned}$$

$$\begin{aligned} \sin 2\theta &= 2\sin\theta\cos\theta \\ &= 2 \times \frac{5}{13} \times -\frac{12}{13} \\ &= -\frac{120}{169} \end{aligned}$$

[Or simply use your calculator to find the value of θ and hence 2θ and $\sin 2\theta$.]

Answer D

9. Consider the graphs of $y = \sin^{-1}x$ and $y = \cos^{-1}x$.

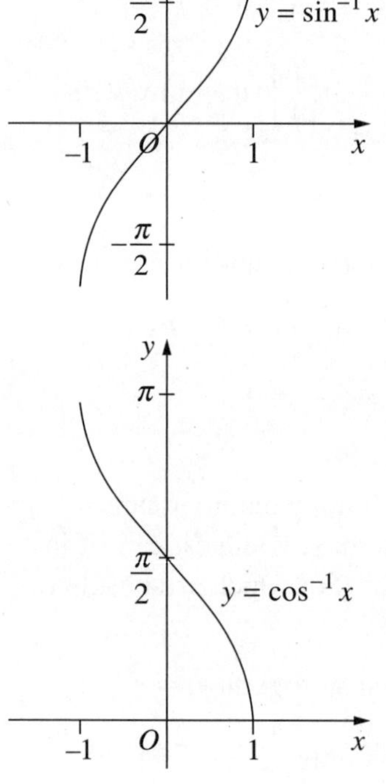

The graph in the question is that of $y = \sin^{-1}x$ moved up by $\dfrac{\pi}{2}$ units.

So it is the graph of $y = \dfrac{\pi}{2} + \sin^{-1}x$.

Answer B

10. $|x + 2| + |x - 3| = 5$

[Both options A and B are not defined when $x = 3$, but by inspection $x = 3$ is a solution of the original equation. Large values of x will be solutions to the inequality in option D, but not to the original equation. The answer could only be option C.]

$|x + 2| + |x - 3| = 5$

If $x < -2$,

$|x - 3| > 5$

$\therefore$ no solution for $x < -2$

If $-2 \leqslant x \leqslant 3$,

$$\begin{aligned} x + 2 - (x - 3) &= 5 \\ x + 2 - x + 3 &= 5 \\ 5 &= 5 \end{aligned}$$

$\therefore$ true for all $-2 \leqslant x \leqslant 3$

If $x > 3$,

$|x + 2| > 5$

$\therefore$ no solution for $x > 3$

$$\begin{aligned} x^2 - x - 6 &\leqslant 0 \\ (x + 2)(x - 3) &\leqslant 0 \end{aligned}$$

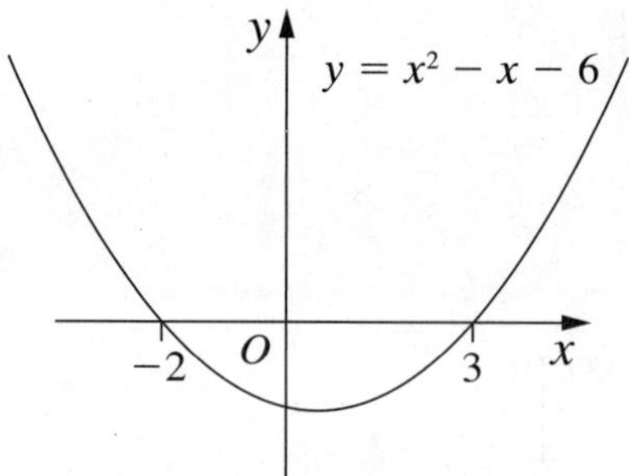

From the graph, $x^2 - x - 6 \leqslant 0$ for $-2 \leqslant x \leqslant 3$

Answer C

Section II

QUESTION 11

(a) $P(x) = 2x^3 - 3x^2 - 11x + 7$

$$\alpha\beta\gamma = -\frac{d}{a} = -\frac{7}{2}$$

(1 mark)

(b) $$\int \frac{1}{\sqrt{49 - 4x^2}}\,dx = \frac{1}{2}\int \frac{1}{\sqrt{\frac{49}{4} - x^2}}\,dx = \frac{1}{2}\sin^{-1}\frac{2x}{7} + C$$

(2 marks)

(c) $p = P(\text{correct answer}) = \dfrac{1}{4}$

$q = P(\text{incorrect answer}) = \dfrac{3}{4}$

$$P(\text{exactly 7 correct in 10}) = \binom{10}{7}\left(\frac{1}{4}\right)^7\left(\frac{3}{4}\right)^3$$

(2 marks)

(d) (i) $f(x) = \dfrac{x}{4 - x^2}$

$$f'(x) = \frac{(4 - x^2) \times 1 - x \times -2x}{(4 - x^2)^2} = \frac{4 + x^2}{(4 - x^2)^2}$$

Now $(4 - x^2)^2 > 0$ for all x in the domain

$x^2 \ge 0$

$4 + x^2 \ge 4$

$\therefore f'(x) > 0$ for all x in the domain

(2 marks)

(ii) The function is undefined when

$4 - x^2 = 0$

$x^2 = 4$

$x = \pm 2$

As $x \to \infty, y \to 0^-$

As $x \to 2^+, y \to -\infty$; as $x \to 2^-, y \to \infty$

As $x \to -2^+, y \to -\infty$; as $x \to -2^-, y \to \infty$

As $x \to -\infty, y \to 0^+$

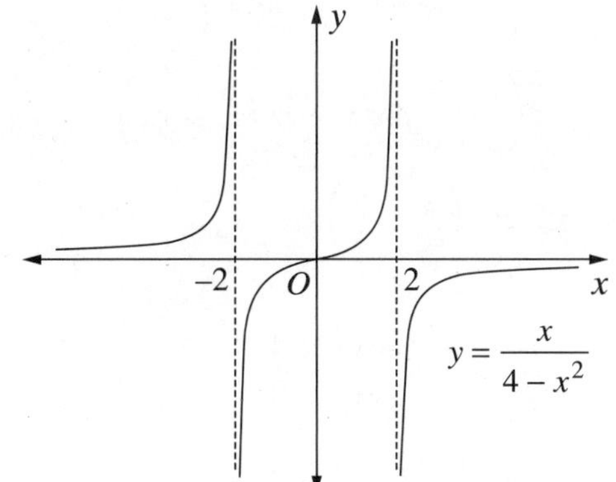

(2 marks)

(e) $$\lim_{x\to 0}\frac{\sin\frac{x}{2}}{3x} = \frac{1}{6}\lim_{x\to 0}\frac{\sin\frac{x}{2}}{\frac{x}{2}} = \frac{1}{6}$$

(1 mark)

(f) $u = e^{3x}$

$u^2 = (e^{3x})^2 = e^{6x}$

$du = 3e^{3x}\,dx$

When $x = 0, u = 1$

When $x = \dfrac{1}{3}, u = e$

$$\int_0^{\frac{1}{3}} \frac{e^{3x}}{e^{6x} + 1}\,dx = \frac{1}{3}\int_1^e \frac{du}{u^2 + 1} = \frac{1}{3}[\tan^{-1}u]_1^e = \frac{1}{3}(\tan^{-1}e - \tan^{-1}1) = 0.144\,2949\ldots = 0.144 \text{ [3 d.p.]}$$

(3 marks)

(g) $y = x^2\sin^{-1}5x$

$$\frac{dy}{dx} = x^2 \times \frac{1}{\sqrt{1 - (5x)^2}} \times 5 + (\sin^{-1}5x) \times 2x = \frac{5x^2}{\sqrt{1 - 25x^2}} + 2x\sin^{-1}5x$$

(2 marks)

QUESTION 12

(a) (i) $\sqrt{3}\cos x - \sin x$

$$= 2\left(\frac{\sqrt{3}}{2}\cos x - \frac{1}{2}\sin x\right)$$

$$= 2\left(\cos\frac{\pi}{6}\cos x - \sin\frac{\pi}{6}\sin x\right)$$

$$= 2\cos\left(x + \frac{\pi}{6}\right)$$

(1 mark)

(ii) $\sqrt{3}\cos x = 1 + \sin x$

$$\sqrt{3}\cos x - \sin x = 1$$

$$2\cos\left(x + \frac{\pi}{6}\right) = 1$$

$$\cos\left(x + \frac{\pi}{6}\right) = \frac{1}{2}$$

$$\therefore x + \frac{\pi}{6} = \frac{\pi}{3} \quad \text{or} \quad x + \frac{\pi}{6} = \frac{5\pi}{3}$$

$(0 < x < 2\pi)$

$$x = \frac{\pi}{6} \quad \text{or} \quad x = \frac{3\pi}{2}$$

(2 marks)

(b) $$V = \int_0^{\frac{3\pi}{2}} \pi\left(3\sin\frac{x}{2}\right)^2 dx$$

$$= 9\pi\int_0^{\frac{3\pi}{2}} \sin^2\left(\frac{x}{2}\right) dx$$

$$= 9\pi\int_0^{\frac{3\pi}{2}} \frac{1}{2}(1 - \cos x)\, dx$$

$$= \frac{9\pi}{2}\left[x - \sin x\right]_0^{\frac{3\pi}{2}}$$

$$= \frac{9\pi}{2}\left(\frac{3\pi}{2} - \sin\frac{3\pi}{2} - (0 - \sin 0)\right)$$

$$= \frac{9\pi}{2}\left(\frac{3\pi}{2} + 1\right)$$

$$= \frac{9\pi}{4}(3\pi + 2)\ \text{units}^3$$

(3 marks)

(c) $T = A + Be^{-kt}$

$A = 22$

When $t = 0$, $T = 80$

$80 = 22 + Be^{-k \times 0}$

$58 = B$

When $t = 10$, $T = 60$

$$60 = 22 + 58e^{-k \times 10}$$

$$38 = 58e^{-10k}$$

$$e^{-10k} = \frac{38}{58}$$

$$-10k = \ln\left(\frac{38}{58}\right)$$

$$k = -\frac{1}{10}\ln\left(\frac{38}{58}\right)$$

$[\ = 0.042\,2856\ldots]$

When $T = 40$,

$$40 = 22 + 58e^{-kt}$$

$$18 = 58e^{-kt}$$

$$e^{-kt} = \frac{18}{58}$$

$$-kt = \ln\left(\frac{18}{58}\right)$$

$$t = -\frac{1}{k}\ln\left(\frac{18}{58}\right)$$

$= 27.670\,623\ldots$

$= 28$ [nearest unit]

To the nearest minute, it will take 28 minutes for the coffee to cool to 40 °C.

(3 marks)

(d) (i) $y = 2x - 1$

$2x - y - 1 = 0 \quad (t, t^2 + 3)$

$$D(t) = \left|\frac{Ax_1 + By_1 + C}{\sqrt{A^2 + B^2}}\right|$$

$$= \left|\frac{2t - (t^2 + 3) - 1}{\sqrt{2^2 + (-1)^2}}\right|$$

$$= \left|\frac{2t - t^2 - 4}{\sqrt{5}}\right|$$

$$= \left|\frac{-(t^2 - 2t + 4)}{\sqrt{5}}\right|$$

Now $t^2 - 2t + 4 = (t - 1)^2 + 3$

≥ 3 for all values of t

i.e. $-(t^2 - 2t + 4) < 0$ for all t

$\therefore |-(t^2 - 2t + 4)| = t^2 - 2t + 4$

$$D(t) = \frac{t^2 - 2t + 4}{\sqrt{5}}$$

(2 marks)

(ii) $\dfrac{dD}{dt} = \dfrac{2t-2}{\sqrt{5}}$

Stationary value when $\dfrac{dD}{dt} = 0$

i.e. $2t - 2 = 0$

$t = 1$

$\dfrac{d^2D}{dt^2} = \dfrac{2}{\sqrt{5}}\ (>0)$

$\therefore$ the minimum distance will occur when $t = 1$.

P is closest to l when $t = 1$.

(1 mark)

(iii) l is the line $y = 2x - 1$

$\therefore l$ has gradient 2.

$P(1, 4)$ is the point on the parabola closest to l.

$y = x^2 + 3$

$\dfrac{dy}{dx} = 2x$

When $x = 1$, $\dfrac{dy}{dx} = 2 \times 1 = 2$

$\therefore$ the gradient of the tangent at P is 2.

$\therefore$ the tangent is parallel to line l.

(1 mark)

(e) $v^2 + 9x^2 = k$

$v^2 = k - 9x^2$

$\dfrac{1}{2}v^2 = \dfrac{1}{2}k - \dfrac{9}{2}x^2$

$\dfrac{d}{dx}\left(\dfrac{1}{2}v^2\right) = -9x$

$\therefore \ddot{x} = -9x$

$\therefore$ the particle is moving in simple harmonic motion.

$n^2 = 9$

$\therefore n = 3 \quad (n > 0)$

$T = \dfrac{2\pi}{n}$

$= \dfrac{2\pi}{3}$

$\therefore$ the particle is moving in simple harmonic motion with period $\dfrac{2\pi}{3}$.

(2 marks)

QUESTION 13

(a) (i) $\dfrac{dV}{dt} = -10^{-4}A$

$V = \dfrac{4}{3}\pi r^3$

$\dfrac{dV}{dr} = 4\pi r^2$

$= A$

$\dfrac{dr}{dt} = \dfrac{dr}{dV}\cdot\dfrac{dV}{dt}$

$= \dfrac{1}{A}(-10^{-4}A)$

$= -10^{-4}$

$\therefore \dfrac{dr}{dt}$ is a constant.

(1 mark)

(ii) $\dfrac{dr}{dt} = -10^{-4}$

$r = -10^{-4}t + C$

When $t = 0$, $V = 10^{-6}$

$\dfrac{4}{3}\pi r^3 = 10^{-6}$

$r^3 = \dfrac{3 \times 10^{-6}}{4\pi}$

$r = \sqrt[3]{\dfrac{3 \times 10^{-6}}{4\pi}}$

$\therefore C = \sqrt[3]{\dfrac{3 \times 10^{-6}}{4\pi}}$

$\therefore r = -10^{-4}t + \sqrt[3]{\dfrac{3 \times 10^{-6}}{4\pi}}$

When the raindrop has completely evaporated $r = 0$.

i.e $-10^{-4}t + \sqrt[3]{\dfrac{3 \times 10^{-6}}{4\pi}} = 0$

$10^{-4}t = \sqrt[3]{\dfrac{3 \times 10^{-6}}{4\pi}}$

$t = 10^4 \times \sqrt[3]{\dfrac{3 \times 10^{-6}}{4\pi}}$

$= 62.035\,049\ldots$

$\therefore$ it will take 62 seconds, to the nearest second, for the raindrop to completely evaporate.

(2 marks)

(b) (i) $N(0, 2a + ap^2)$, $T(ap, 0)$

divided externally in the ratio 2:1

$k = 2$ and $l = -1$

$$x = \frac{kx_2 + lx_1}{k + l} = \frac{2ap - 0}{2 - 1} = 2ap$$

$$y = \frac{ky_2 + ly_1}{k + l} = \frac{0 - (2a + ap^2)}{2 - 1} = -2a - ap^2$$

$\therefore G$ is the point $(2ap, -2a - ap^2)$

[Or use the fact that T is the midpoint of NG.]

(2 marks)

(ii) The parabola $x^2 = 4ay$ has focal length a units and the directrix is $y = -a$.

$G(2ap, -2a - ap^2)$

$$\begin{aligned} x^2 &= (2ap)^2 \\ &= 4a^2p^2 \\ &= -4a(-ap^2) \\ &= -4a(-2a - ap^2 + 2a) \\ &= -4a(y + 2a) \end{aligned}$$

So G lies on the parabola $x^2 = -4a(y + 2a)$.

This parabola has vertex $(0, -2a)$, focal length a units, directrix $y = -a$.

So G lies on a parabola with the same focal length and directrix as $x^2 = 4ay$.

[Or the only parabola with the same focal length and directrix as $x^2 = 4ay$ has its vertex at $(0, -2a)$. Its equation is $x^2 = -4a(y + 2a)$. The coordinates of G satisfy this equation.]

(2 marks)

(c) (i) Projectile fired from A:

$$y = ut \sin\alpha - \frac{g}{2}t^2$$

$$\dot{y} = u \sin\alpha - gt$$

$$\ddot{y} = -g$$

The maximum ($\ddot{y} < 0$) height will occur when $\dot{y} = 0$

i.e. $u \sin\alpha - gt = 0$

$$gt = u \sin\alpha$$

$$t = \frac{u}{g}\sin\alpha$$

The projectile fired from A will reach its maximum height after $\frac{u}{g}\sin\alpha$ seconds.

(2 marks)

(ii) Similarly, the projectile fired from B will reach its maximum height after $\frac{w}{g}\sin\beta$ seconds.

But both reach their maximum height at the same time, when they collide.

So $\frac{u}{g}\sin\alpha = \frac{w}{g}\sin\beta$

$u \sin\alpha = w \sin\beta$

(1 mark)

(iii) For the projectile fired from A the horizontal distance from A to the point of impact is given by:

$x = ut\cos\alpha$

When $t = \frac{u}{g}\sin\alpha$,

$$x = \frac{u^2}{g}\sin\alpha\cos\alpha$$

Similarly for the projectile fired from B:

$$x = \frac{w^2}{g}\sin\beta\cos\beta$$

$$\begin{aligned} d &= \frac{u^2}{g}\sin\alpha\cos\alpha + \frac{w^2}{g}\sin\beta\cos\beta \\ &= \frac{1}{g}(u^2\sin\alpha\cos\alpha + w^2\sin\beta\cos\beta) \\ &= \frac{1}{g}(u\sin\alpha\,(u\cos\alpha) + w\sin\beta\,(w\cos\beta)) \end{aligned}$$

But $u\sin\alpha = w\sin\beta$

$$\begin{aligned} d &= \frac{1}{g}(w\sin\beta\,(u\cos\alpha) + u\sin\alpha\,(w\cos\beta)) \\ &= \frac{uw}{g}(\sin\beta\cos\alpha + \sin\alpha\cos\beta) \\ &= \frac{uw}{g}(\sin\alpha\cos\beta + \cos\alpha\sin\beta) \\ &= \frac{uw}{g}\sin(\alpha + \beta) \end{aligned}$$

(2 marks)

(d) Join QT and TP.

The circles have a common tangent at T.

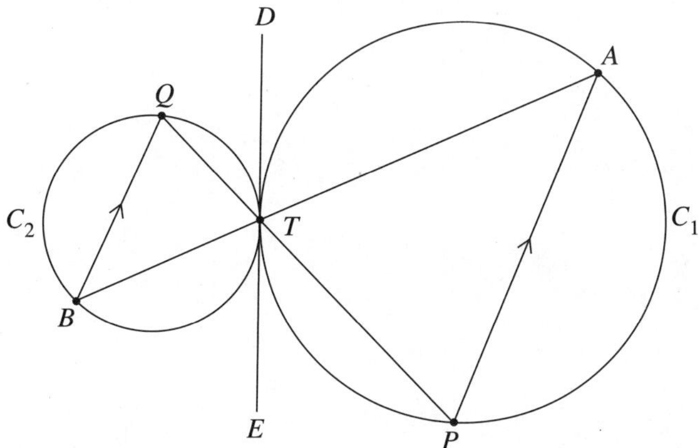

Let D and E be points on the tangent as shown.

The angle between a tangent and chord equals the angle in alternate segment, so

$$\angle DTQ = \angle TBQ$$
$$\text{and } \angle ETP = \angle TAP$$
$$\text{but } \angle TBQ = \angle TAP \text{ (alternate, } QB \parallel AP\text{)}$$
$$\therefore \angle DTQ = \angle ETP$$
$$\angle ETP + \angle PTD = 180^\circ \text{ (angles in straight line)}$$
$$\therefore \angle DTQ + \angle PTD = 180^\circ$$
$$\therefore \angle PTQ = 180^\circ$$

$\therefore Q$, T and P are collinear. *(3 marks)*

QUESTION 14

(a) (i) $$\frac{1}{(k+1)^2} - \frac{1}{k} + \frac{1}{k+1} = \frac{k - (k+1)^2 + k(k+1)}{k(k+1)^2}$$
$$= \frac{k - k^2 - 2k - 1 + k^2 + k}{k(k+1)^2}$$
$$= \frac{-1}{k(k+1)^2}$$

If $k > 0$, $\dfrac{-1}{k(k+1)^2} < 0$

$$\therefore \frac{1}{(k+1)^2} - \frac{1}{k} + \frac{1}{k+1} < 0 \text{ for } k > 0$$

(1 mark)

(ii) Aim to prove

$$\frac{1}{1^2} + \frac{1}{2^2} + \frac{1}{3^2} + \ldots + \frac{1}{n^2} < 2 - \frac{1}{n} \text{ for } n \geq 2$$

If $n = 2$,

$$\text{LHS} = \frac{1}{1^2} + \frac{1}{2^2} = 1\frac{1}{4}$$

$$\text{RHS} = 2 - \frac{1}{2} = 1\frac{1}{2}$$

Now $1\frac{1}{4} < 1\frac{1}{2}$

$\therefore$ the statement is true for $n = 2$.

Assume true for $n = k$:

i.e. $\frac{1}{1^2} + \frac{1}{2^2} + \frac{1}{3^2} + \ldots + \frac{1}{k^2} < 2 - \frac{1}{k}$

Show true for $n = k + 1$:

We wish to show that

$$\frac{1}{1^2} + \frac{1}{2^2} + \frac{1}{3^2} + \ldots + \frac{1}{k^2} + \frac{1}{(k+1)^2} < 2 - \frac{1}{k+1}$$

$$\begin{aligned}\text{LHS} &< 2 - \frac{1}{k} + \frac{1}{(k+1)^2}\\ &= 2 - \frac{1}{k} + \frac{1}{(k+1)^2} + \frac{1}{k+1} - \frac{1}{k+1}\\ &= 2 - \frac{1}{k+1} + \frac{1}{(k+1)^2} - \frac{1}{k} + \frac{1}{k+1}\end{aligned}$$

But $\frac{1}{(k+1)^2} - \frac{1}{k} + \frac{1}{k+1} < 0$ for $k > 0$ from part (i)

$\therefore \text{LHS} < 2 - \frac{1}{k+1}$

So, if true for $n = k$, the statement is also true for $n = k + 1$.

It is true for $n = 2$, so by the process of induction the statement is true for all integers $n \geq 2$.

(3 marks)

(b) (i) $(1 + x)^{4n} = \binom{4n}{0} + \binom{4n}{1}x + \binom{4n}{2}x^2 + \binom{4n}{3}x^3 + \ldots + \binom{4n}{4n}x^{4n}$

The coefficient of x^{2n} is $\binom{4n}{2n}$.

(1 mark)

(ii) $(1 + x^2 + 2x)^{2n}$

$$\begin{aligned}&= ((x^2 + 2x) + 1)^{2n}\\ &= \binom{2n}{0}(x^2 + 2x)^{2n} + \binom{2n}{1}(x^2 + 2x)^{2n-1} + \binom{2n}{2}(x^2 + 2x)^{2n-2} + \ldots + \binom{2n}{2n}\\ &= \sum_{k=0}^{2n}\binom{2n}{k}(x^2 + 2x)^{2n-k}\\ &= \sum_{k=0}^{2n}\binom{2n}{k}(x(x + 2))^{2n-k}\\ &= \sum_{k=0}^{2n}\binom{2n}{k}x^{2n-k}(x + 2)^{2n-k}\end{aligned}$$

(2 marks)

(iii) $(1 + x^2 + 2x)^{2n} = \sum_{k=0}^{2n}\binom{2n}{k}x^{2n-k}(x + 2)^{2n-k}$

Now $x^{2n-k}(x + 2)^{2n-k} = \binom{2n-k}{0}2^{2n-k}x^{2n-k} + \binom{2n-k}{1}2^{2n-k-1}x^{2n-k+1} + \binom{2n-k}{2}2^{2n-k-2}x^{2n-k+2}$

$$+ \ldots + \binom{2n-k}{2n-k}2^0x^{4n-2k}$$

$$\text{So } (1 + x^2 + 2x)^{2n} = \binom{2n}{0}\left[\binom{2n}{0}2^{2n}x^{2n} + \binom{2n}{1}2^{2n-1}x^{2n+1} + \ldots\right]$$
$$+ \binom{2n}{1}\left[\binom{2n-1}{0}2^{2n-1}x^{2n-1} + \binom{2n-1}{1}2^{2n-2}x^{2n} + \ldots\right]$$
$$+ \binom{2n}{2}\left[\binom{2n-2}{0}2^{2n-2}x^{2n-2} + \binom{2n-2}{1}2^{2n-3}x^{2n-1} + \binom{2n-2}{2}2^{2n-4}x^{2n} + \ldots\right]$$
$$+ \ldots$$

The coefficient of x^{2n} in this expansion

$$= \binom{2n}{0}\binom{2n}{0}2^{2n} + \binom{2n}{1}\binom{2n-1}{1}2^{2n-2} + \binom{2n}{2}\binom{2n-2}{2}2^{2n-4} + \ldots \binom{2n}{n}\binom{n}{n}2^0$$
$$= \sum_{k=0}^{n}\binom{2n}{k}\binom{2n-k}{k}2^{2n-2k}$$

Now $(1 + x^2 + 2x)^{2n} = ((1 + x)^2)^{2n}$
$= (1 + x)^{4n}$

The coefficient of x^{2n} in the expansion $(1 + x)^{4n}$ is $\binom{4n}{2n}$ from part (i).

$$\therefore \binom{4n}{2n} = \sum_{k=0}^{n} 2^{2n-2k}\binom{2n}{k}\binom{2n-k}{k}$$ *(3 marks)*

(c) (i) $e^t = \dfrac{1}{t}$

$e^t - \dfrac{1}{t} = 0$

$f(t) = e^t - \dfrac{1}{t}$
$= e^t - t^{-1}$

$f'(t) = e^t + t^{-2}$
$= e^t + \dfrac{1}{t^2}$

By Newton's method:

$$t_1 = t_0 - \frac{f(t_0)}{f'(t_0)}$$
$$= 0.5 - \frac{e^{0.5} - \dfrac{1}{0.5}}{e^{0.5} + \dfrac{1}{(0.5)^2}}$$
$= 0.562\,1873\ldots$
$= 0.56$ [2 d.p.]

$\therefore t_1 = 0.56$ is an approximate solution of $e^t = \dfrac{1}{t}$. *(2 marks)*

(ii) $y = e^{rx}, y = \log_e x$

Let the point of intersection be (x_1, y_1).

$y = e^{rx}$

$\dfrac{dy}{dx} = re^{rx}$

At (x_1, y_1)

$\dfrac{dy}{dx} = re^{rx_1}$

$\therefore$ the gradient of the tangent is re^{rx_1}.

$y = \log_e x$

$\dfrac{dy}{dx} = \dfrac{1}{x}$

At (x_1, y_1)

$\dfrac{dy}{dx} = \dfrac{1}{x_1}$

$\therefore$ the gradient of the tangent is $\dfrac{1}{x_1}$.

But the curves have a common tangent at (x_1, y_1).

$\therefore re^{rx_1} = \dfrac{1}{x_1}$

$\therefore e^{rx_1} = \dfrac{1}{rx_1}$

From part (i) an approximate value of rx_1 is 0.56.

$y_1 = e^{rx_1}$
$\approx e^{0.56}$
≈ 1.75

But $y_1 = \log_e x_1$

$\log_e x_1 \approx 1.75$
$x_1 \approx e^{1.75}$
≈ 5.75
$rx_1 \approx 0.56$
$r \approx 0.56 \div 5.75$
$= 0.0973\ldots$
$r \approx 0.1$

An approximate value of r for which the graphs have a common tangent is 0.1.

(3 marks)

Solutions to replacement questions

QUESTION 3

$$\begin{aligned}\overrightarrow{BC} &= \overrightarrow{BO} + \overrightarrow{OC}\\ &= -(-2\underset{\sim}{i} + 7\underset{\sim}{j}) + (5\underset{\sim}{i} + 6\underset{\sim}{j})\\ &= 2\underset{\sim}{i} - 7\underset{\sim}{j} + 5\underset{\sim}{i} + 6\underset{\sim}{j}\\ &= 7\underset{\sim}{i} - \underset{\sim}{j}\end{aligned}$$

Answer A

(1 mark)

QUESTION 6

$$P(X = 4) = \binom{4}{4} p^4(1-p)^0$$

$$p^4 = 0.105\,56$$

$$\begin{aligned}p &= 0.569\,999\,986\ldots\\ &= 0.57 \text{ (2 dec. pl.)}\end{aligned}$$

Answer C

(1 mark)

QUESTION 11

(e) Consider $a = -1 < 0$ and $\Delta < 0$:

$$\begin{aligned}\Delta = 2^2 - 4(-1)(1-2k) &< 0\\ 4 + 4 - 8k &< 0\\ 8k &> 8\\ k &> 1\end{aligned}$$

(1 mark)

QUESTION 12

(d) (i)

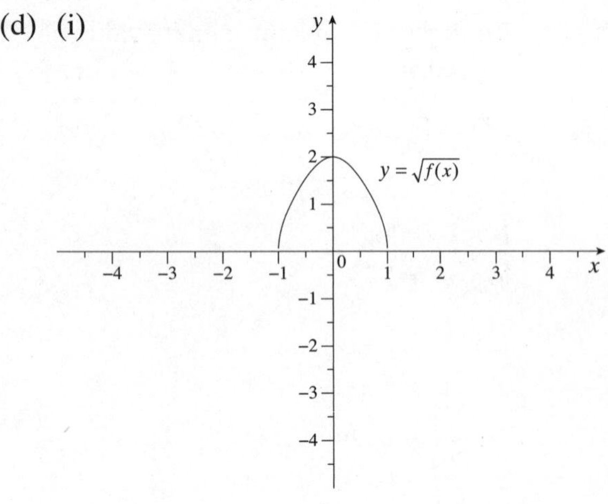

(2 marks)

(ii)

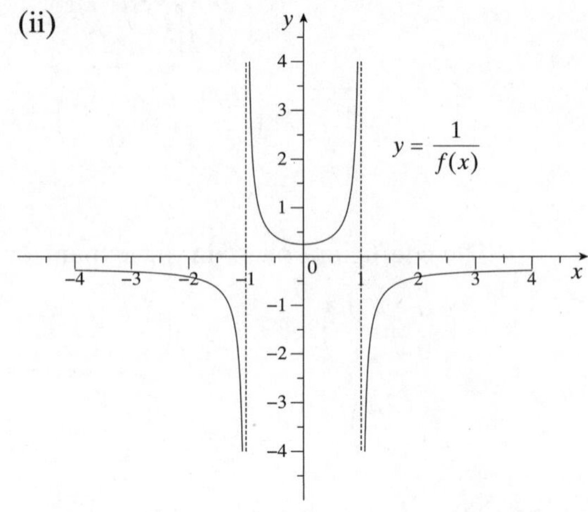

(2 marks)

(e)
$$\begin{aligned}\underset{\sim}{a}\bullet\underset{\sim}{b} &= 4 \times 5 + (-10) \times 2\\ &= 20 - 20\\ &= 0\end{aligned}$$

As $\underset{\sim}{a}\bullet\underset{\sim}{b} = 0$ then the vectors are perpendicular.

(2 marks)

QUESTION 13

(b) $u = \cos 2x \qquad \dfrac{du}{dx} = -2\sin 2x$

$$dx = \frac{du}{-2\sin 2x}$$

Also, $x = \dfrac{\pi}{6}, u = \dfrac{1}{2}; x = 0, u = 1$

$$\begin{aligned}\int_0^{\frac{\pi}{6}} \cos^2 2x \sin 2x\, dx &= \int_1^{\frac{1}{2}} u^2 \sin 2x . \frac{du}{-2\sin 2x}\\ &= \frac{1}{2}\int_{\frac{1}{2}}^{1} u^2\, du\\ &= \frac{1}{2}\left[\frac{u^3}{3}\right]_{\frac{1}{2}}^{1}\\ &= \frac{1}{2}\left[\frac{1^3}{3} - \frac{\left(\frac{1}{2}\right)^3}{3}\right]\\ &= \frac{1}{2}\left[\frac{1}{3} - \frac{1}{24}\right]\\ &= \frac{7}{48}\end{aligned}$$

(4 marks)

(d) $\cos 5\theta + \cos 3\theta + \cos \theta = 0$

$$2\cos 4\theta \cos \theta + \cos \theta = 0$$

$$\cos \theta (2\cos 4\theta + 1) = 0$$

$$\cos \theta = 0 \qquad \cos 4\theta = -\frac{1}{2}$$

$$\theta = \frac{\pi}{2} \qquad 4\theta = \frac{2\pi}{3}, \frac{4\pi}{3}, \frac{8\pi}{3}, \frac{10\pi}{3}$$

$$\theta = \frac{\pi}{6}, \frac{\pi}{3}, \frac{2\pi}{3}, \frac{5\pi}{6}$$

$$\therefore x = \frac{\pi}{6}, \frac{\pi}{3}, \frac{\pi}{2}, \frac{2\pi}{3}, \frac{5\pi}{6}$$

(3 marks)

QUESTION 14

(b)* Let X denote the number of females in the sample.

X has a binomial distribution with $n = 100$ and $p = 0.5$

The distribution of the sample proportion $\hat{p}$ is approximately normal with $E(\hat{p}) = 0.5$

$$Var(\hat{p}) = \frac{p(1-p)}{n}$$

$$= \frac{0.5(1-0.5)}{100}$$

$$= 0.0025$$

$$\sigma = \sqrt{0.025}$$

$$= 0.05$$

$$P(\hat{p} < 0.45) = P\left(z < \frac{0.45 - 0.5}{0.05}\right)$$

$$= P(z < -1)$$

$$= \frac{100\% - 68\%}{2}$$

$$= 16\%$$

(2 marks)

(b)** Let $P(x) = -x^3 + 7x - 6$

Consider $P(1) = -1^3 + 7(1) - 6 = 0$

As $x = 1$ is root, then $(x - 1)$ is factor.

$$\begin{array}{r} -x^2 - x + 6 \\ x-1 \overline{\big) -x^3 + 0x^2 + 7x - 6} \\ \underline{-x^3 + x^2} \qquad\qquad \\ -x^2 + 7x \qquad \\ \underline{-x^2 + x} \qquad \\ 6x - 6 \\ \underline{6x - 6} \\ 0 \end{array}$$

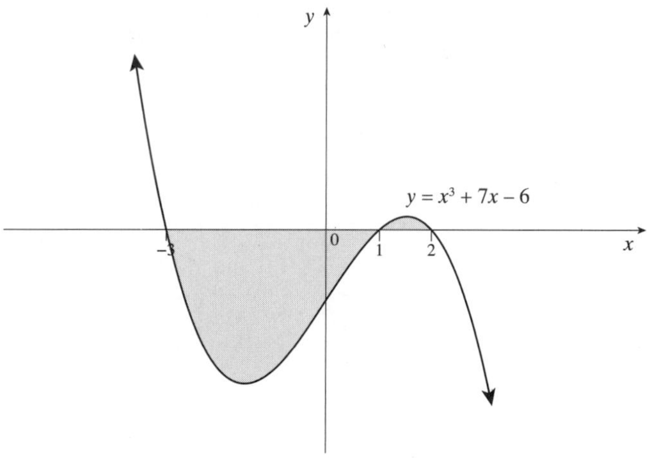

$$P(x) = (x-1)(6 - x - x^2)$$

$$= (x-1)(3+x)(2-x)$$

$\therefore y = (x-1)(3+x)(2-x)$ has zeros of $1, -3, 2$

$$\text{Area} = \left|\int_{-3}^{1} \left(-x^3 + 7x - 6\right)\right| + \int_{1}^{2} \left(-x^3 + 7x - 6\right) dx$$

$$= \left|\left[-\frac{x^4}{4} + \frac{7x^2}{2} - 6x\right]_{-3}^{1}\right| + \left[-\frac{x^4}{4} + \frac{7x^2}{2} - 6x\right]_{1}^{2}$$

$$= \left|\left[-\frac{1^4}{4} + \frac{7(1)^2}{2} - 6(1) - \left(-\frac{(-3)^4}{4} + \frac{7(-3)^2}{2} - 6(-3)\right)\right]\right| + \left[-\frac{2^4}{4} + \frac{7(2)^2}{2} - 6(2) - \left(-\frac{1^4}{4} + \frac{7(1)^2}{2} - 6(1)\right)\right]$$

$$= 32 + 0.75$$

$$= 32.75$$

$\therefore$ area is 32.75 units^2.

(4 marks)

(c)* From $x = \frac{t}{2}$,

$t = 2x$

Substitute into y:

$y = 3(2x)^2$

$= 12x^2$

Also, when $t = 1$, then $x = \frac{1}{2}$.

Now, $\frac{dy}{dx} = 24x$

$\frac{dy}{dx}\left(\frac{1}{2}\right) = 24\left(\frac{1}{2}\right)$

$= 12$

$\therefore$ the gradient is 12.

(2 marks)

(c)** $(1 + x^2)\frac{dy}{dx} = 4xy$

$\int \frac{1}{y} dy = \int \frac{4x}{x^2+1} dx$

$\log_e y = 2 \log_e(x^2 + 1) + c$

Substitute $x = 1, y = 2$:

$\log_e 2 = 2 \log_e(1^2 + 1) + c$

$\log_e 2 = 2 \log_e 2 + c$

$c = -\log_e 2$

$\log_e y = \log_e (x^2 + 1)^2 - \log_e 2$

$\log_e y = \log_e \frac{(x^2+1)^2}{2}$

$y = \frac{(x^2+1)^2}{2}$

(3 marks)

2014 **HIGHER SCHOOL CERTIFICATE EXAMINATION**

Mathematics Extension 1

General Instructions

- Reading time – 5 minutes
- Working time – 2 hours
- Write using black or blue pen
  Black pen is preferred
- Board-approved calculators may be used
- A table of standard integrals is provided at the back of this paper
- In Questions 11–14, show relevant mathematical reasoning and/or calculations

Total marks – 70

Section I

10 marks

- Attempt Questions 1–10
- Allow about 15 minutes for this section

Section II

60 marks

- Attempt Questions 11–14
- Allow about 1 hour and 45 minutes for this section

Section I

10 marks
Attempt Questions 1–10
Allow about 15 minutes for this section

Use the multiple-choice answer sheet for Questions 1–10.

---

1 The points A, B and C lie on a circle with centre O, as shown in the diagram. The size of $\angle ACB$ is 40°.

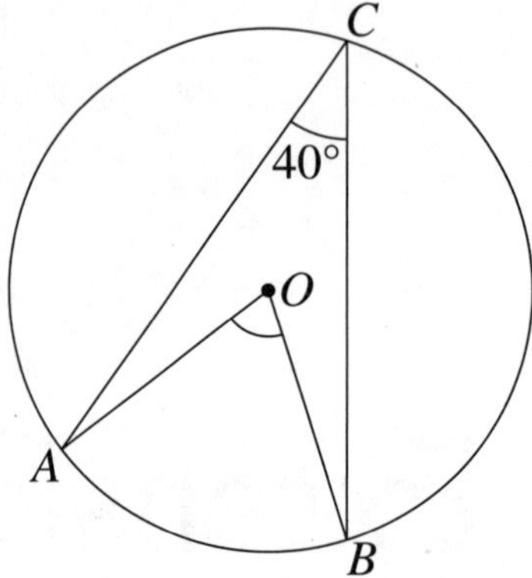

NOT TO SCALE

What is the size of $\angle AOB$?

(A) 20°

(B) 40°

(C) 70°

(D) 80°

2 Which expression is equal to $\cos x - \sin x$?

(A) $\sqrt{2}\cos\left(x + \dfrac{\pi}{4}\right)$

(B) $\sqrt{2}\cos\left(x - \dfrac{\pi}{4}\right)$

(C) $2\cos\left(x + \dfrac{\pi}{4}\right)$

(D) $2\cos\left(x - \dfrac{\pi}{4}\right)$

3 What is the constant term in the binomial expansion of $\left(2x - \frac{5}{x^3}\right)^{12}$?

(A) $\binom{12}{3} 2^9 5^3$

(B) $\binom{12}{9} 2^3 5^9$

(C) $-\binom{12}{3} 2^9 5^3$

(D) $-\binom{12}{9} 2^3 5^9$

✗ **4** The acute angle between the lines $2x + 2y = 5$ and $y = 3x + 1$ is θ.

What is the value of $\tan\theta$?

(A) $\frac{1}{7}$

(B) $\frac{1}{2}$

(C) 1

(D) 2

5 Which group of three numbers could be the roots of the polynomial equation $x^3 + ax^2 - 41x + 42 = 0$?

(A) 2, 3, 7

(B) 1, −6, 7

(C) −1, −2, 21

(D) −1, −3, −14

6 What is the derivative of $3\sin^{-1}\frac{x}{2}$?

(A) $\dfrac{6}{\sqrt{4-x^2}}$

(B) $\dfrac{3}{\sqrt{4-x^2}}$

(C) $\dfrac{3}{2\sqrt{4-x^2}}$

(D) $\dfrac{3}{4\sqrt{4-x^2}}$

✗ **7** A particle is moving in simple harmonic motion with period 6 and amplitude 5.

Which is a possible expression for the velocity, v, of the particle?

(A) $v = \dfrac{5\pi}{3}\cos\left(\dfrac{\pi}{3}t\right)$

(B) $v = 5\cos\left(\dfrac{\pi}{3}t\right)$

(C) $v = \dfrac{5\pi}{6}\cos\left(\dfrac{\pi}{6}t\right)$

(D) $v = 5\cos\left(\dfrac{\pi}{6}t\right)$

8 In how many ways can 6 people from a group of 15 people be chosen and then arranged in a circle?

(A) $\dfrac{14!}{8!}$

(B) $\dfrac{14!}{8!6}$

(C) $\dfrac{15!}{9!}$

(D) $\dfrac{15!}{9!6}$

9 The remainder when the polynomial $P(x) = x^4 - 8x^3 - 7x^2 + 3$ is divided by $x^2 + x$ is $ax + 3$.

What is the value of a?

(A) –14

(B) –11

(C) –2

(D) 5

✗ **10** Which equation describes the locus of points (x, y) which are equidistant from the distinct points $(a + b,\ b - a)$ and $(a - b,\ b + a)$?

(A) $bx + ay = 0$

(B) $bx + ay = 2ab$

(C) $bx - ay = 0$

(D) $bx - ay = 2ab$

Section II

60 marks
Attempt Questions 11–14
Allow about 1 hour and 45 minutes for this section

Answer each question in a SEPARATE writing booklet. Extra writing booklets are available.

In Questions 11–14, your responses should include relevant mathematical reasoning and/or calculations.

Question 11 (15 marks) Use a SEPARATE writing booklet.

(a) Solve $\left(x+\frac{2}{x}\right)^2-6\left(x+\frac{2}{x}\right)+9=0$. **3**

(b) The probability that it rains on any particular day during the 30 days of November is 0.1. **2**

Write an expression for the probability that it rains on fewer than 3 days in November.

(c) Sketch the graph $y=6\tan^{-1}x$, clearly indicating the range. **2**

(d) Evaluate $\int_2^5 \frac{x}{\sqrt{x-1}}\,dx$ using the substitution $x=u^2+1$. **3**

(e) Solve $\frac{x^2+5}{x}>6$. **3**

(f) Differentiate $\frac{e^x \ln x}{x}$. **2**

Question 12 (15 marks) Use a SEPARATE writing booklet.

(a) A particle is moving in simple harmonic motion about the origin, with displacement x metres. The displacement is given by $x = 2\sin 3t$, where t is time in seconds. The motion starts when $t = 0$.

(i) What is the total distance travelled by the particle when it first returns to the origin? **1**

(ii) What is the acceleration of the particle when it is first at rest? **2**

(b) The region bounded by $y = \cos 4x$ and the x-axis, between $x = 0$ and $x = \frac{\pi}{8}$, is rotated about the x-axis to form a solid. **3**

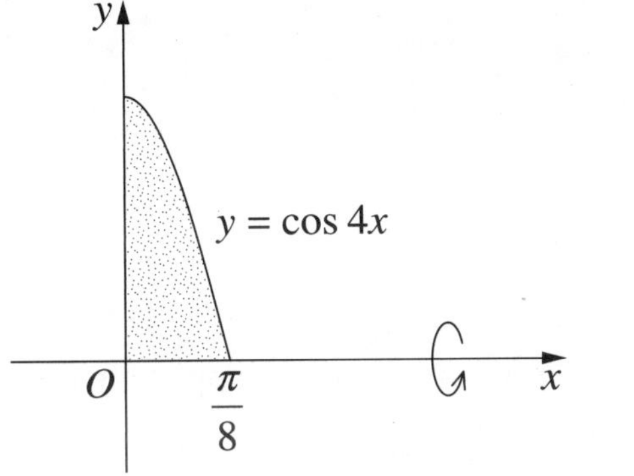

NOT TO SCALE

Find the volume of the solid.

(c) A particle moves along a straight line with displacement x m and velocity v m s^{-1}. The acceleration of the particle is given by **3**

$$\ddot{x} = 2 - e^{-\frac{x}{2}}.$$

Given that $v = 4$ when $x = 0$, express v^2 in terms of x.

Question 12 continues on page 8

Question 12 (continued)

(d) Use the binomial theorem to show that **2**

$$0 = \binom{n}{0} - \binom{n}{1} + \binom{n}{2} - \cdots + (-1)^n \binom{n}{n}.$$

(e) The diagram shows the graph of a function $f(x)$. **1**

The equation $f(x) = 0$ has a root at $x = \alpha$. The value x_1, as shown in the diagram, is chosen as a first approximation of α.

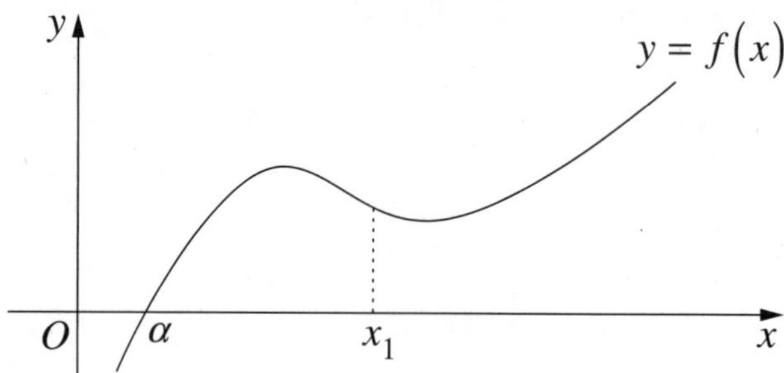

A second approximation, x_2, of α is obtained by applying Newton's method once, using x_1 as the first approximation.

Using a diagram, or otherwise, explain why x_1 is a closer approximation of α than x_2.

(f) Milk taken out of a refrigerator has a temperature of 2°C. It is placed in a room of constant temperature 23°C. After t minutes the temperature, T°C, of the milk is given by **3**

$$T = A - Be^{-0.03t},$$

where A and B are positive constants.

How long does it take for the milk to reach a temperature of 10°C?

End of Question 12

Question 13 (15 marks) Use a SEPARATE writing booklet.

(a) Use mathematical induction to prove that $2^n + (-1)^{n+1}$ is divisible by 3 for all integers $n \geq 1$. **3**

(b) One end of a rope is attached to a truck and the other end to a weight. The rope passes over a small wheel located at a vertical distance of 40 m above the point where the rope is attached to the truck.

The distance from the truck to the small wheel is L m, and the horizontal distance between them is x m. The rope makes an angle θ with the horizontal at the point where it is attached to the truck.

The truck moves to the right at a constant speed of 3 m s^{-1}, as shown in the diagram.

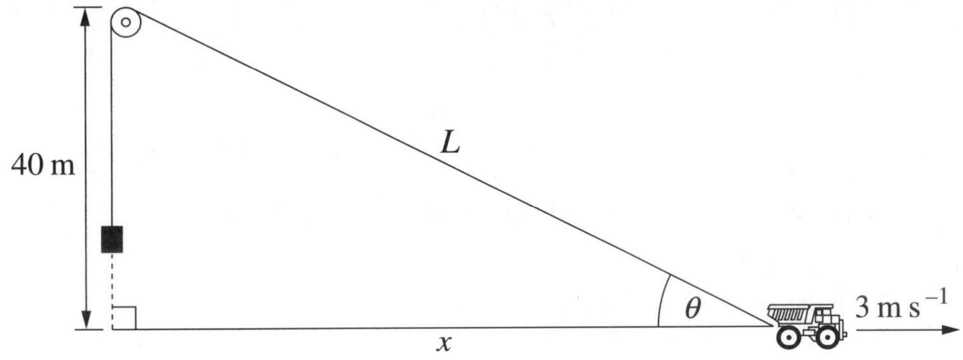

(i) Using Pythagoras' Theorem, or otherwise, show that $\dfrac{dL}{dx} = \cos\theta$. **2**

(ii) Show that $\dfrac{dL}{dt} = 3\cos\theta$. **1**

Question 13 continues on page 10

Question 13 (continued)

(c) The point $P\left(2at, at^2\right)$ lies on the parabola $x^2 = 4ay$ with focus S.

The point Q divides the interval PS internally in the ratio $t^2 : 1$.

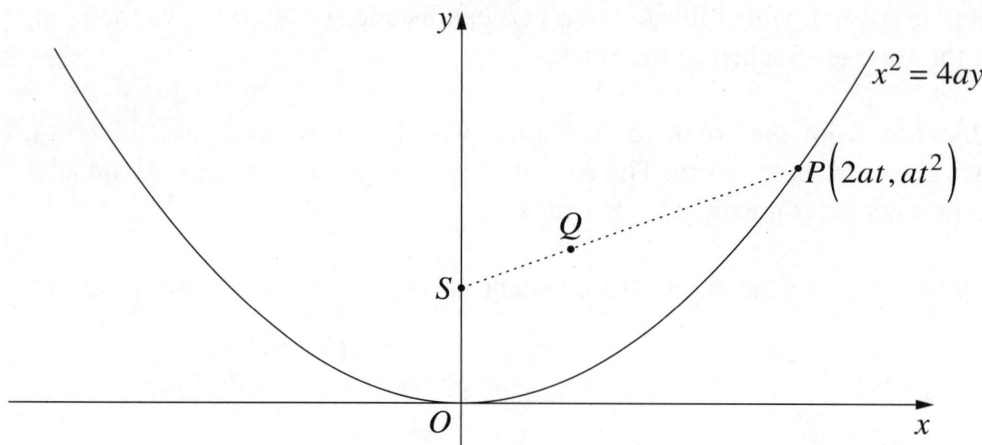

(i) Show that the coordinates of Q are $x = \dfrac{2at}{1+t^2}$ and $y = \dfrac{2at^2}{1+t^2}$. **2**

(ii) Express the slope of OQ in terms of t. **1**

(iii) Using the result from part (ii), or otherwise, show that Q lies on a fixed circle of radius a. **3**

Question 13 continues on page 11

Question 13 (continued)

(d) In the diagram, AB is a diameter of a circle with centre O. The point C is chosen such that $\triangle ABC$ is acute-angled. The circle intersects AC and BC at P and Q respectively.

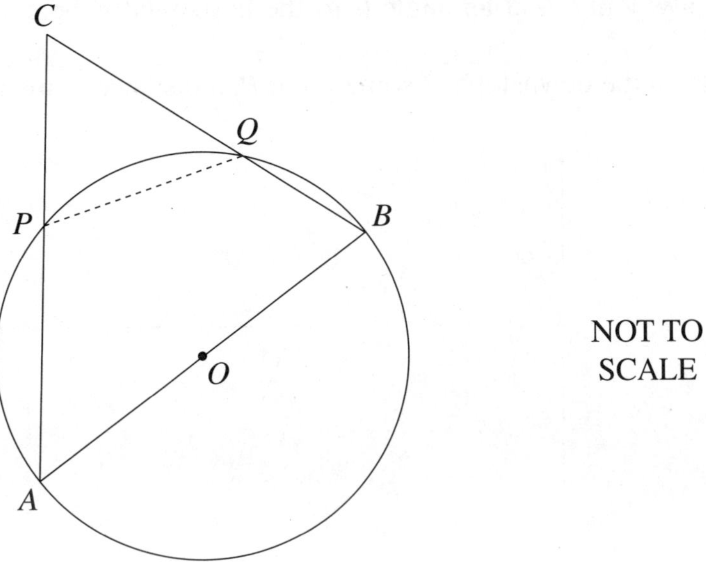

NOT TO SCALE

Copy or trace the diagram into your writing booklet.

(i) Why is $\angle BAC = \angle CQP$? **1**

(ii) Show that the line OP is a tangent to the circle through P, Q and C. **2**

End of Question 13

Question 14 (15 marks) Use a SEPARATE writing booklet.

* (a) The take-off point O on a ski jump is located at the top of a downslope. The angle between the downslope and the horizontal is $\frac{\pi}{4}$. A skier takes off from O with velocity V m s^{-1} at an angle θ to the horizontal, where $0 \le \theta < \frac{\pi}{2}$. The skier lands on the downslope at some point P, a distance D metres from O.

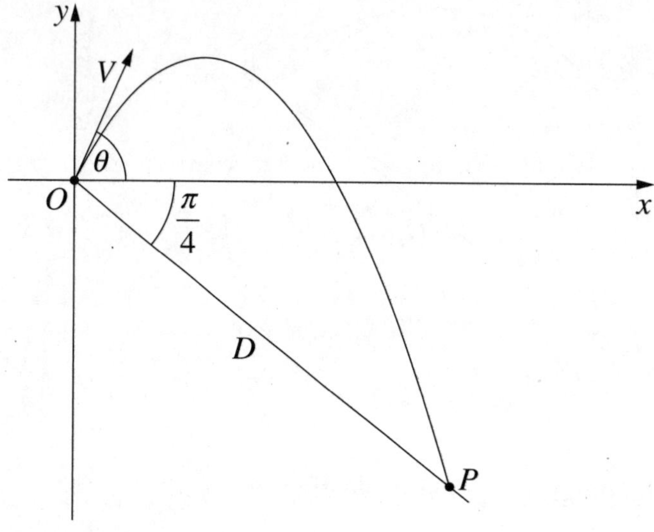

The flight path of the skier is given by

$$x = Vt\cos\theta, \quad y = -\frac{1}{2}gt^2 + Vt\sin\theta, \qquad \text{(Do NOT prove this.)}$$

where t is the time in seconds after take-off.

(i) Show that the cartesian equation of the flight path of the skier is given by **2**

$$y = x\tan\theta - \frac{gx^2}{2V^2}\sec^2\theta.$$

* In the new Mathematics Extension 1 course, Projectile Motion questions will be expressed in vector form:

$x = Vt\cos\theta$ and $y = Vt\sin\theta - \frac{1}{2}gt^2$ is expressed as a position vector:

$$\underset{\sim}{r}(t) = (Vt\cos\theta)\underset{\sim}{i} + \left(Vt\sin\theta - \frac{1}{2}gt^2\right)\underset{\sim}{j}$$

Question 14 (continued)

(ii) Show that $D = 2\sqrt{2}\,\dfrac{V^2}{g}\cos\theta\left(\cos\theta + \sin\theta\right)$. **3**

(iii) Show that $\dfrac{dD}{d\theta} = 2\sqrt{2}\,\dfrac{V^2}{g}\left(\cos 2\theta - \sin 2\theta\right)$. **2**

(iv) Show that D has a maximum value and find the value of θ for which this occurs. **3**

(b) Two players A and B play a game that consists of taking turns until a winner is determined. Each turn consists of spinning the arrow on a spinner once. The spinner has three sectors P, Q and R. The probabilities that the arrow stops in sectors P, Q and R are p, q and r respectively.

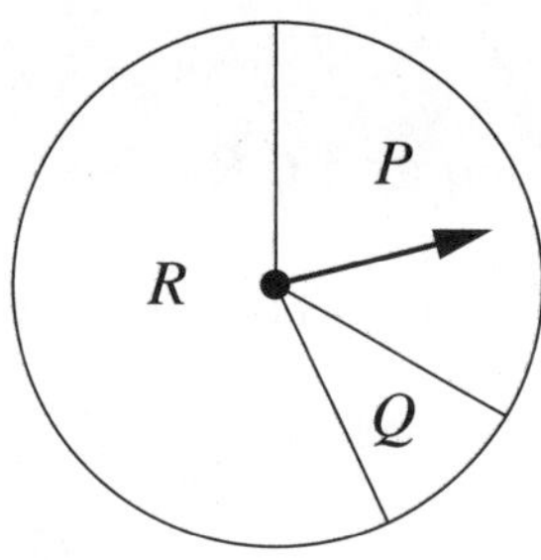

The rules of the game are as follows:

- If the arrow stops in sector P, then the player having the turn wins.
- If the arrow stops in sector Q, then the player having the turn loses and the other player wins.
- If the arrow stops in sector R, then the other player takes a turn.

Player A takes the first turn.

(i) Show that the probability of player A winning on the first or the second turn of the game is $(1 - r)(p + r)$. **2**

(ii) Show that the probability that player A eventually wins the game is **3**

$$\frac{p+r}{1+r}.$$

End of paper

Replacement questions

with content from the most up-to-date syllabus

Marks

Question 1 (1 mark)

What is the smallest group of people that can be assembled where it is guaranteed that at least two people have the same first and last initials in their name? **1**

(A) 27

(B) 53

(C) 675

(D) 677

Question 4 (1 mark)

PQRS is a parallelogram with $\overrightarrow{PQ} = \underset{\sim}{u}$ and $\overrightarrow{QR} = \underset{\sim}{v}$.

If *M* is the midpoint of *RS*, which of these is vector $\overrightarrow{MP}$ in terms of $\underset{\sim}{u}$ and $\underset{\sim}{v}$? **1**

(A) $\frac{1}{2}\underset{\sim}{u} + \underset{\sim}{v}$

(B) $\underset{\sim}{u} + \frac{1}{2}\underset{\sim}{v}$

(C) $\underset{\sim}{u} - \frac{1}{2}\underset{\sim}{v}$

(D) $-\frac{1}{2}\underset{\sim}{u} - \underset{\sim}{v}$

REPLACEMENT QUESTIONS

Marks

Question 7 (1 mark)

Which of these diagrams best represents the direction field of the differential equation $\frac{dy}{dx} = \frac{y-2x}{2y+x}$? **1**

(A)

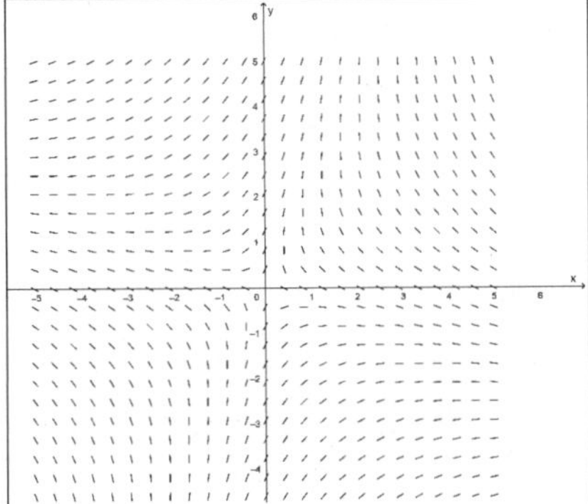

(B)

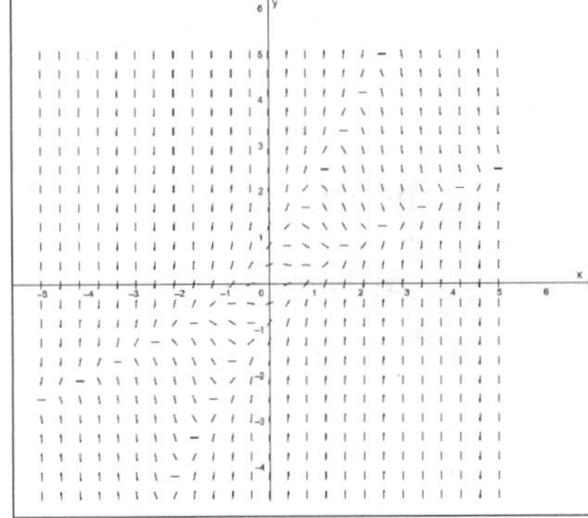

(C)

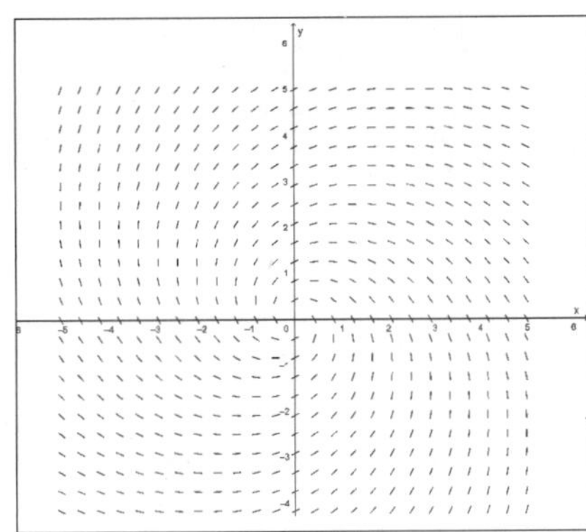

(D)

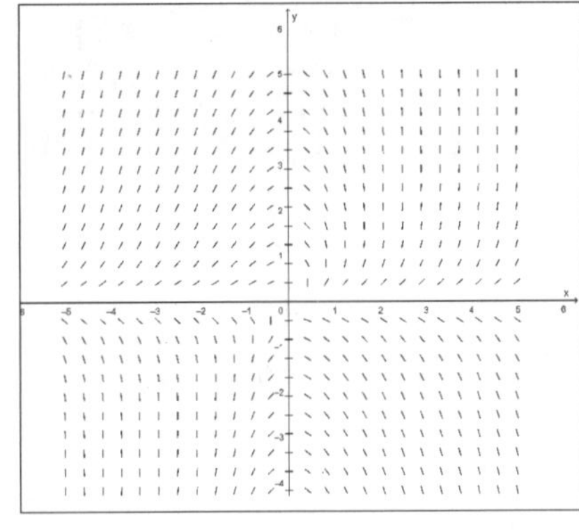

Question 10 (1 mark)

Suppose $X \sim \text{Bin}(n, p)$, where the mean is 20 and the standard deviation is 3. **1**

Which of these is the probability of success p?

(A) 0.35

(B) 0.45

(C) 0.55

(D) 0.65

REPLACEMENT QUESTIONS

Marks

Question 12 (9 marks)

(a) Solve $|x^2 - 4x| \leq x - 4$ **3**

(c) Find the cartesian equation of the curve represented by the equations $x = \frac{t}{2} + 1$ and $y = \frac{t^2}{4} - 1$, and show that a stationary point exists at $x = 1$. **3**

(d) Two particles P and Q move so that their vector equations are given by $\underset{\sim}{r}_P(t) = (3t - 2)\underset{\sim}{i} + (t + 1)\underset{\sim}{j}$ and $\underset{\sim}{r}_Q(t) = t^2\underset{\sim}{i} + (5 - 2t)\underset{\sim}{j}$.
Find the distance between the particles when $t = 3$. **2**

(e) The number plane shows the graph of $y = f(x)$. **1**

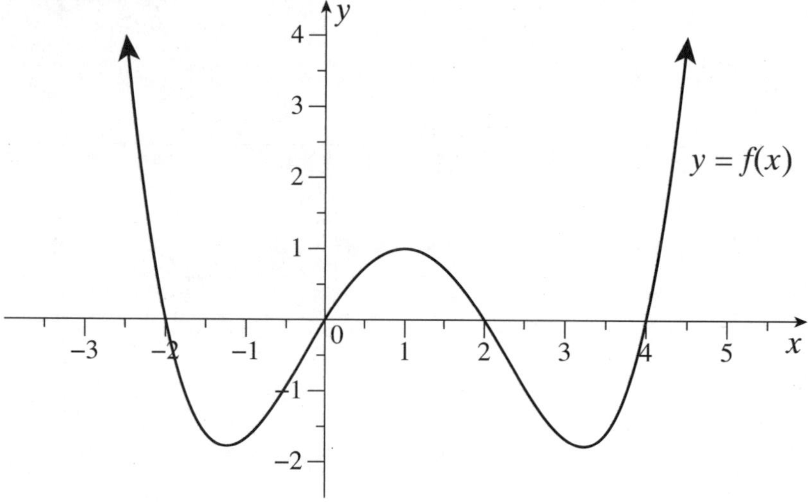

A second graph is drawn based on the graph of $y = f(x)$.

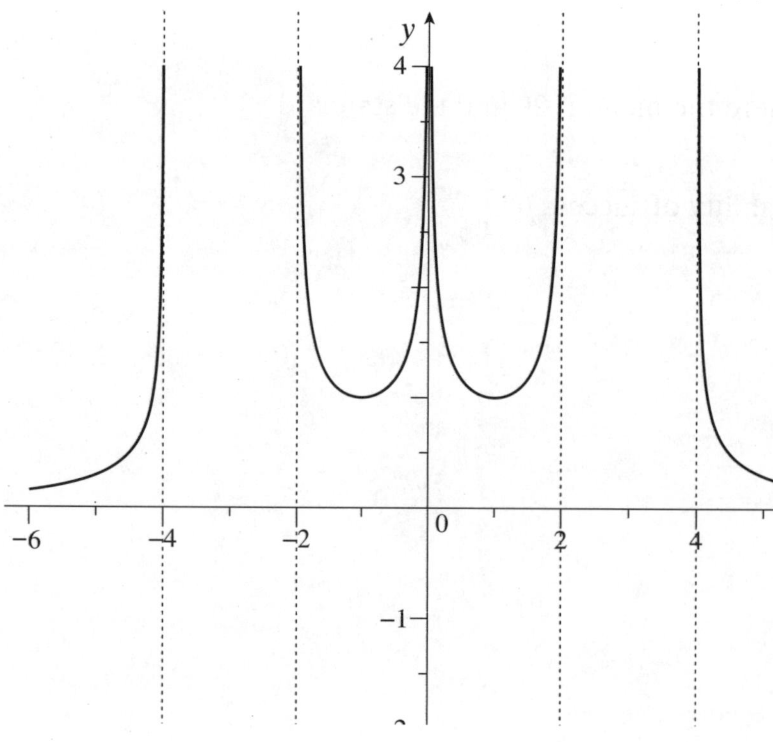

Marks

Question 12 (continued)

Which of these is the equation for the second graph?

A $y = \sqrt{f(x)}$

B $y = \dfrac{1}{\sqrt{f(x)}}$

C $y = \sqrt{f(|x|)}$

D $y = \dfrac{1}{\sqrt{f(|x|)}}$

Question 13 (9 marks)

(c)* Show $\dfrac{\cos 3x + \cos x}{\cos 3x - \cos x} = -\cot 2x \cot x$ **2**

(c)** A population grows according to the logistical differential equation $\dfrac{dP}{dt} = 0.02P\left(1 - \dfrac{P}{1200}\right)$, where P is the population after t months, and the initial population is 800. Also, $0 < P < 1200$.

[Use the result $\dfrac{60\,000}{P(1200-P)} = \dfrac{50}{P} + \dfrac{50}{1200-P}$]
Find the population P at time t. **4**

(d) If $\underset{\sim}{a} = 2\underset{\sim}{i} - 4\underset{\sim}{j}$ and $\underset{\sim}{b} = -3\underset{\sim}{i} + \underset{\sim}{j}$, show that $\underset{\sim}{a} \cdot \underset{\sim}{b} = -10$ and hence find the vector projection of $\underset{\sim}{a}$ onto $\underset{\sim}{b}$. **3**

REPLACEMENT QUESTIONS

2014 Higher School Certificate
Worked answers

Section I

(*Total 10 marks*)

1. D	**2.** A	**3.** C	**4.** D	**5.** B
6. B	**7.** A	**8.** D	**9.** C	**10.** C

1. The angle at the centre is twice the angle at circumference standing on the same arc.

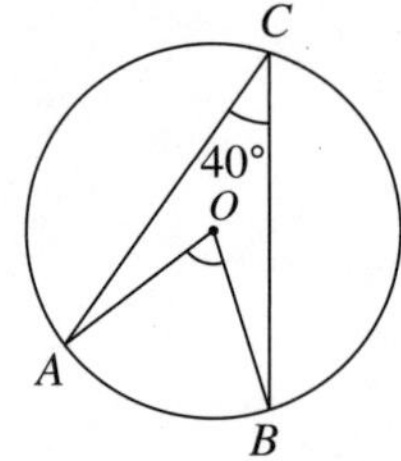

$\angle AOB = 80°$

Answer D

2. $\cos\left(x + \frac{\pi}{4}\right) = \cos x \cos\frac{\pi}{4} - \sin x \sin\frac{\pi}{4}$

$= \frac{1}{\sqrt{2}}(\cos x - \sin x)$

So $\sqrt{2}\cos\left(x + \frac{\pi}{4}\right) = \cos x - \sin x$

Answer A

3. $\left(2x - \frac{5}{x^3}\right)^{12}$

$= \binom{12}{0}(2x)^{12} + \binom{12}{1}(2x)^{11}\left(\frac{-5}{x^3}\right) + \binom{12}{2}(2x)^{10}\left(\frac{-5}{x^3}\right)^2 + \ldots$

General term $= \binom{12}{k}(2x)^{12-k}\left(\frac{-5}{x^3}\right)^k$

$= \binom{12}{k}2^{12-k}x^{12-k}(-5)^k x^{-3k}$

$= \binom{12}{k}2^{12-k}(-5)^k x^{12-4k}$

Constant term when $12 - 4k = 0$

$k = 3$

Constant term $= \binom{12}{3}2^9(-5)^3$

$= -\binom{12}{3}2^9 5^3$

Answer C

4. $2x + 2y = 5$

$m_1 = -1$

$y = 3x + 1$

$m_2 = 3$

$\tan\theta = \frac{m_1 - m_2}{1 + m_1 m_2}$

$= \frac{-1 - 3}{1 + (-1) \times 3}$

$= 2$

Answer D

5. $x^3 + ax^2 - 41x + 42 = 0$

$\alpha\beta\gamma = -42$

[So options A and C are eliminated.]

$\alpha\beta + \alpha\gamma + \beta\gamma = -41$

[So option D is eliminated.]

Of the options only 1, −6 and 7 could be the roots of the equation.

Answer B

6. $y = 3\sin^{-1}\frac{x}{2}$

$\frac{dy}{dx} = 3 \times \frac{1}{\sqrt{2^2 - x^2}}$

$= \frac{3}{\sqrt{4 - x^2}}$

Answer B

7. Amplitude = 5

Period = 6

$\therefore \frac{2\pi}{n} = 6$

$n = \frac{2\pi}{6}$

$= \frac{\pi}{3}$

$x = 5\sin\left(\frac{\pi}{3}t\right)$

$v = \frac{5\pi}{3}\cos\left(\frac{\pi}{3}t\right)$

Answer A

8. Combinations of 6 people from 15

$= \dfrac{15!}{9!6!}$

6 people can be arranged in a circle in 5! ways.

Total arrangements $= \dfrac{15!}{9!6!} \times 5!$

$= \dfrac{15!}{9!6}$

Answer D

9.

$$\begin{array}{r} x^2 - 9x + 2 \\ x^2 + x \overline{)x^4 - 8x^3 - 7x^2 + 0x + 3} \\ \underline{x^4 + x^3} \qquad\qquad\qquad \\ -9x^3 - 7x^2 \qquad\quad \\ \underline{-9x^3 - 9x^2} \qquad\quad \\ 2x^2 + 0x \quad \\ \underline{2x^2 + 2x} \quad \\ -2x + 3 \end{array}$$

The remainder is $-2x + 3$.

So $a = -2$.

Answer C

10. $(a + b, b - a), (a - b, b + a)$

The locus of points equidistant from two points is the perpendicular bisector.

Midpoint: $\left(\dfrac{a + b + a - b}{2}, \dfrac{b - a + b + a}{2}\right)$

$= (a, b)$

Of the options, only C passes through the point (a, b).

Answer C

Section II

QUESTION 11

(a) $\left(x + \dfrac{2}{x}\right)^2 - 6\left(x + \dfrac{2}{x}\right) + 9 = 0$

Let $u = x + \dfrac{2}{x}$

So
$$\begin{aligned} u^2 - 6u + 9 &= 0 \\ (u - 3)^2 &= 0 \\ u &= 3 \\ \therefore x + \frac{2}{x} &= 3 \\ x^2 + 2 &= 3x \\ x^2 - 3x + 2 &= 0 \\ (x - 2)(x - 1) &= 0 \\ x = 2 \quad \text{or} \quad x &= 1 \end{aligned}$$

(3 marks)

(b) $P(\text{rain on any day}) = 0.1$

$\therefore P(\text{no rain on any day}) = 0.9$

$P(\text{rain on fewer than 3 days})$

$= P(\text{rain 0 days}) + P(\text{rain 1 day}) + P(\text{rain 2 days})$

$= \binom{30}{0}(0.9)^{30} + \binom{30}{1}(0.1)(0.9)^{29} + \binom{30}{2}(0.1)^2(0.9)^{28}$

(2 marks)

(c) $y = 6\tan^{-1}x$

Now $-\dfrac{\pi}{2} < \tan^{-1}x < \dfrac{\pi}{2}$

$\therefore -3\pi < 6\tan^{-1}x < 3\pi$

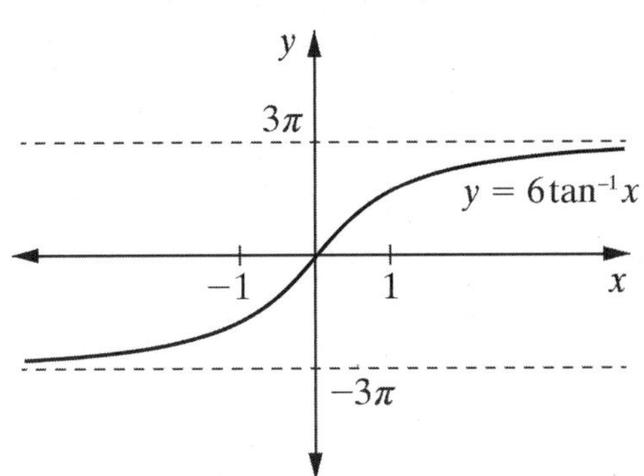

(2 marks)

(d) Let $x = u^2 + 1$

$dx = 2u\,du$

If $u = 1, x = 2$

if $u = 2, x = 5$

$$\int_2^5 \frac{x}{\sqrt{x-1}}dx = \int_1^2 \frac{u^2+1}{u}2u\,du$$

$$= 2\int_1^2 (u^2+1)\,du$$

$$= 2\left[\frac{u^3}{3} + u\right]_1^2$$

$$= 2\left(\frac{2^3}{3} + 2 - \left(\frac{1^3}{3} + 1\right)\right)$$

$$= 6\frac{2}{3}$$

(3 marks)

(e)

$$\frac{x^2+5}{x} > 6$$

$$x(x^2+5) > 6x^2$$

$$x^3 + 5x > 6x^2$$

$$x^3 - 6x^2 + 5x > 0$$

$$x(x^2 - 6x + 5) > 0$$

$$x(x-1)(x-5) > 0$$

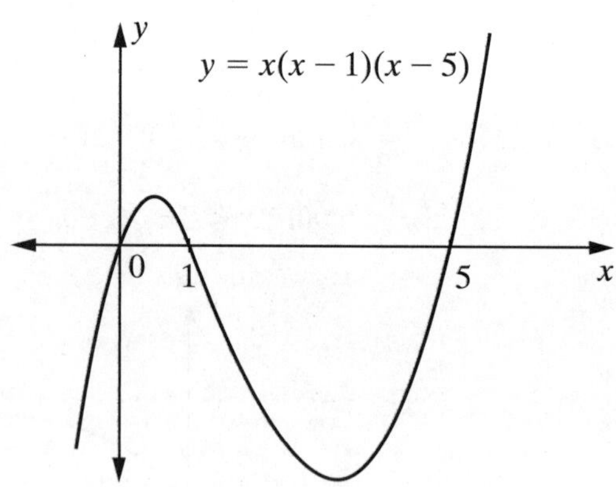

From the graph $x(x-1)(x-5) > 0$ when $0 < x < 1$ or $x > 5$.

$\therefore \frac{x^2+5}{x} > 6$ when $0 < x < 1$ or $x > 5$ *(3 marks)*

(f) $y = \frac{e^x \ln x}{x}$

$$\frac{dy}{dx} = \frac{x(e^x \times \frac{1}{x} + (\ln x) \times e^x) - (e^x \ln x) \times 1}{x^2}$$

$$= \frac{e^x + xe^x \ln x - e^x \ln x}{x^2}$$

$$= \frac{e^x}{x^2}(1 + x\ln x - \ln x)$$

(2 marks)

QUESTION 12

(a) (i) $x = 2\sin 3t$

Amplitude = 2

The particle is at the origin, O, when

$2\sin 3t = 0$

$3t = 0, \pi, 2\pi, \ldots$

The particle first returns to O when

$t = \frac{\pi}{3}$.

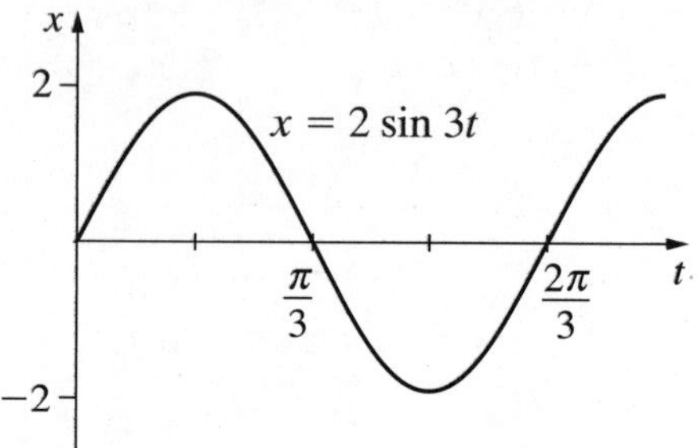

Total distance travelled when it first returns to O is (2 + 2) m or 4 m.

(1 mark)

(ii) $\dot{x} = 6\cos 3t$

The particle is at rest when $\cos 3t = 0$

It is first at rest when $3t = \frac{\pi}{2}$

$t = \frac{\pi}{6}$

$\ddot{x} = -18\sin 3t$

When $t = \frac{\pi}{6}$,

$\ddot{x} = -18\sin\frac{\pi}{2}$

$= -18\text{ m s}^{-2}$

(2 marks)

(b) $V = \int \pi y^2 dx$

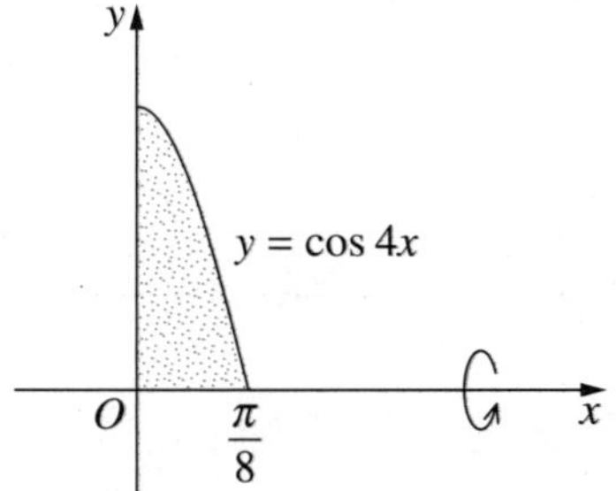

$$V = \pi \int_0^{\frac{\pi}{8}} \cos^2 4x \, dx$$

$$= \frac{\pi}{2} \int_0^{\frac{\pi}{8}} (1 + \cos 8x) dx$$

$$= \frac{\pi}{2}\left[x + \frac{1}{8}\sin 8x\right]_0^{\frac{\pi}{8}}$$

$$= \frac{\pi}{2}\left(\frac{\pi}{8} + \frac{1}{8}\sin \pi - \left(0 + \frac{1}{8}\sin 0\right)\right)$$

$$= \frac{\pi^2}{16} \text{ units}^3$$

(3 marks)

(c) $\ddot{x} = 2 - e^{-\frac{x}{2}}$

$$\frac{1}{2}v^2 = \int (2 - e^{-\frac{x}{2}}) dx$$

$$= 2x + 2e^{-\frac{x}{2}} + c$$

Now $v = 4$ when $x = 0$

$$8 = 0 + 2 + c$$

$$c = 6$$

$$\frac{1}{2}v^2 = 2x + 2e^{-\frac{x}{2}} + 6$$

$$v^2 = 4x + 4e^{-\frac{x}{2}} + 12$$

(3 marks)

(d) $(1 + x)^n = \binom{n}{0} + \binom{n}{1}x + \binom{n}{2}x^2 + \ldots + \binom{n}{n}x^n$

Let $x = -1$,

$$(1 - 1)^n = \binom{n}{0} + \binom{n}{1}(-1) + \binom{n}{2}(-1)^2 + \ldots + \binom{n}{n}(-1)^n$$

$$0 = \binom{n}{0} - \binom{n}{1} + \binom{n}{2} - \ldots + (-1)^n\binom{n}{n}$$

(2 marks)

(e) x_2 is the x-intercept of the tangent to the curve at x_1. So, from the diagram, x_1 is a closer approximation than x_2.

[But the next approximation, x_3, will be closer to the root.]

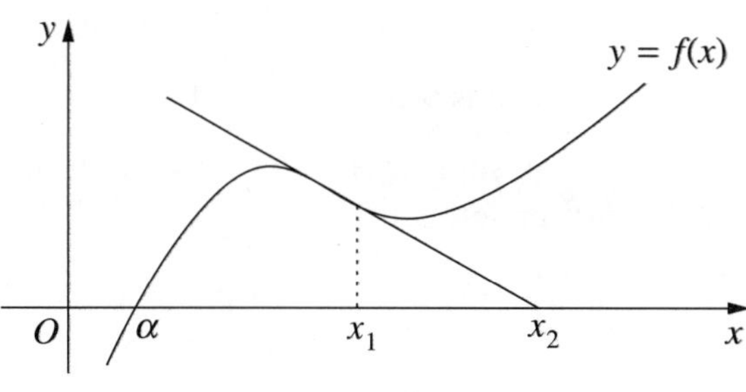

(1 mark)

(f) $T = A - Be^{-0.03t}$

Room has constant temperature of 23 °C.

So $A = 23$

$T = 23 - Be^{-0.03t}$

When $t = 0$, $T = 2$

$$2 = 23 - Be^0$$

$$B = 23 - 2$$

$$= 21$$

$$\therefore T = 23 - 21e^{-0.03t}$$

When $T = 10$,

$$10 = 23 - 21e^{-0.03t}$$

$$21e^{-0.03t} = 13$$

$$e^{-0.03t} = \frac{13}{21}$$

$$-0.03t = \ln\left(\frac{13}{21}\right)$$

$$t = \frac{\ln\left(\frac{13}{21}\right)}{-0.03}$$

$$= 15.985\,769\ldots$$

$$= 16.0 \quad \text{(1 d.p.)}$$

It takes 16 minutes for the milk to reach 10 °C.

(3 marks)

QUESTION 13

(a) To prove $2^n + (-1)^{n+1}$ is divisible by 3 for $n \geq 1$.

When $n = 1$,

$2^1 + (-1)^2 = 3$

$\therefore$ it is true when $n = 1$.

Assume $2^k + (-1)^{k+1} = 3m$ where m is an integer.

If $n = k + 1$,

$$\begin{aligned} &2^{k+1} + (-1)^{k+1+1} \\ &= 2 \times 2^k + (-1) \times (-1)^{k+1} \\ &= 3 \times 2^k - 1 \times 2^k - (-1)^{k+1} \\ &= 3 \times 2^k - (2^k + (-1)^{k+1}) \\ &= 3 \times 2^k - 3m \\ &= 3(2^k - m) \end{aligned}$$

So, if true for $n = k$ it is also true for $n = k + 1$.

It is true for $n = 1$, so it is true for $n = 2$ and hence it is true for $n = 3$ and so on.

By the process of mathematical induction it is true for all integers $n \geq 1$.

(3 marks)

(b) (i) By Pythagoras' theorem:

$$\begin{aligned} L^2 &= x^2 + 40^2 \\ &= x^2 + 1600 \\ L &= \sqrt{x^2 + 1600} \qquad (L > 0) \\ &= (x^2 + 1600)^{\frac{1}{2}} \end{aligned}$$

$$\begin{aligned} \frac{dL}{dx} &= \frac{1}{2}(x^2 + 1600)^{-\frac{1}{2}} \times 2x \\ &= \frac{x}{\sqrt{x^2 + 1600}} \\ &= \frac{x}{L} \\ &= \cos\theta \end{aligned}$$

(2 marks)

(ii)

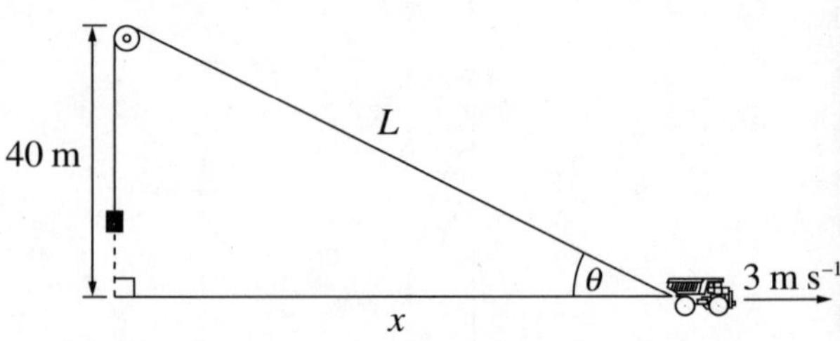

The truck moves right at 3 m s^{-1}

So $\dfrac{dx}{dt} = 3$

$$\begin{aligned} \frac{dL}{dt} &= \frac{dL}{dx} \cdot \frac{dx}{dt} \\ &= (\cos\theta) \times 3 \\ &= 3\cos\theta \end{aligned}$$

(1 mark)

(c) (i) $P(2at, at^2)$ $\quad S(0, a)$ $\quad t^2{:}1$

$$\begin{aligned} x &= \frac{kx_2 + lx_1}{k + l} \\ &= \frac{t^2 \times 0 + 1 \times 2at}{t^2 + 1} \\ &= \frac{2at}{1 + t^2} \end{aligned}$$

$$\begin{aligned} y &= \frac{ky_2 + ly_1}{k + l} \\ &= \frac{t^2 \times a + 1 \times at^2}{t^2 + 1} \\ &= \frac{2at^2}{1 + t^2} \end{aligned}$$

(2 marks)

(ii) $O(0, 0)$, $Q\left(\dfrac{2at}{1 + t^2}, \dfrac{2at^2}{1 + t^2}\right)$

$$\begin{aligned} m &= \frac{y_2 - y_1}{x_2 - x_1} \\ &= \frac{\dfrac{2at^2}{1 + t^2} - 0}{\dfrac{2at}{1 + t^2} - 0} \\ &= \frac{2at^2}{1 + t^2} \times \frac{1 + t^2}{2at} \\ &= t \end{aligned}$$

(1 mark)

(iii) OQ has gradient t

$\therefore$ the equation of OQ is $y = tx$

$$\text{So} \quad y^2 = t^2x^2$$
$$t^2 = \frac{y^2}{x^2}$$
$$\text{At } Q, \quad y = \frac{2at^2}{1 + t^2}$$
$$= \frac{2a\left(\frac{y^2}{x^2}\right)}{1 + \frac{y^2}{x^2}}$$
$$= \frac{2ay^2}{x^2 + y^2}$$
$$y(x^2 + y^2) = 2ay^2$$
$$x^2 + y^2 = 2ay \quad (y \neq 0)$$
$$x^2 + y^2 - 2ay + a^2 = a^2$$
$$x^2 + (y - a)^2 = a^2$$

So Q lies on the circle, centre $(0, a)$, radius a units

[Or: Let θ be the angle that OQ makes with the positive direction of the x-axis.

$\therefore \tan\theta = t$

$\tan(90° - \theta) = \frac{1}{t}$

So $\angle SOQ = \tan^{-1}\left(\frac{1}{t}\right)$

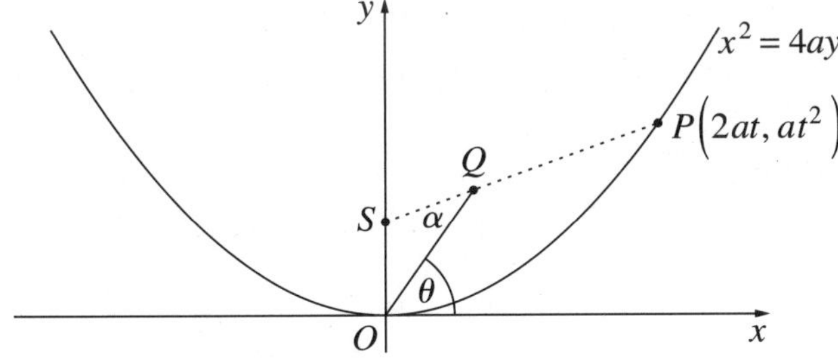

$S(0, a), P(2at, at^2)$

$$m_{SP} = \frac{at^2 - a}{2at - 0}$$
$$= \frac{t^2 - 1}{2t}$$

Let $\angle SQO = \alpha$

$$\tan\alpha = \frac{m_1 - m_2}{1 + m_1m_2}$$
$$= \frac{t - \frac{t^2 - 1}{2t}}{1 + t\left(\frac{t^2 - 1}{2t}\right)}$$
$$= \frac{2t^2 - t^2 + 1}{2t + t^3 - t}$$
$$= \frac{t^2 + 1}{t^3 + t}$$
$$= \frac{t^2 + 1}{t(t^2 + 1)}$$
$$= \frac{1}{t}$$

So $\angle SQO = \tan^{-1}\left(\frac{1}{t}\right)$

$\angle SQO = \angle SOQ$

$\therefore SO = SQ$ (sides opposite equal angles, isosceles triangle)

but $SO = a$ units

$\therefore SQ = a$ units

$\therefore Q$ is always a units from the focus.

So Q lies on a circle, centre S, radius a units.] *(3 marks)*

(d) (i)

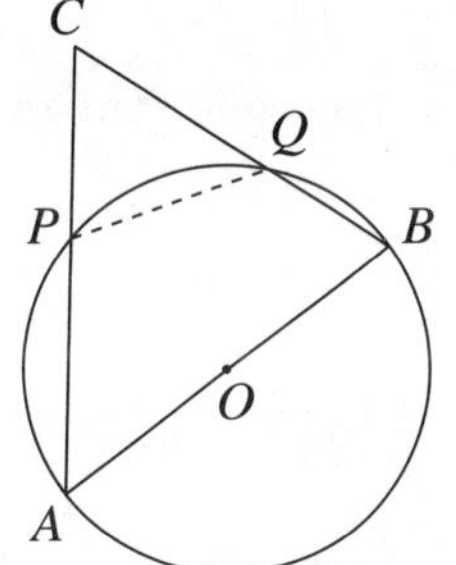

$\angle BAC = \angle CQP$ because $APQB$ is a cyclic quadrilateral and an exterior angle of a cyclic quadrilateral is equal to the interior opposite angle. *(1 mark)*

(ii) $OA = OP$ (radii of circle)

$\therefore \angle APO = \angle PAO$ (base angles isos. $\triangle$)

$\therefore \angle APO = \angle CQP$ (from part (i))

$\angle APQ = \angle CQP + \angle PCQ$ (ext. $\angle$ of $\triangle$)

i.e. $\angle APO + \angle OPQ = \angle CQP + \angle PCQ$

$\therefore \angle OPQ = \angle PCQ$

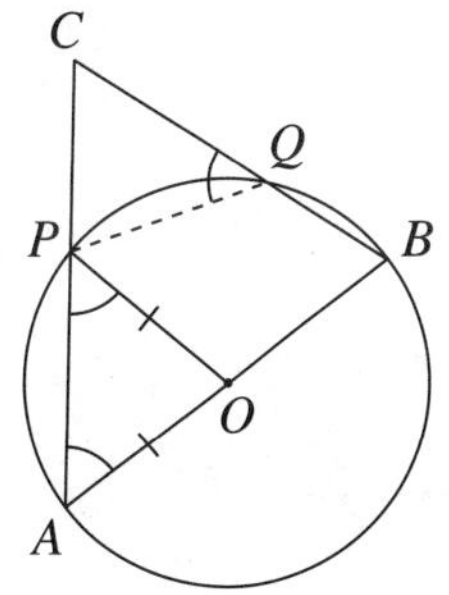

$\therefore OP$ is a tangent to the circle through P, Q and C because the angle between the tangent and the chord is equal to the angle in the alternate segment.

(2 marks)

QUESTION 14

(a) (i) $x = Vt\cos\theta$

$\therefore t = \dfrac{x}{V\cos\theta}$

$$y = -\frac{1}{2}gt^2 + Vt\sin\theta$$

$$= -\frac{1}{2}g\left(\frac{x}{V\cos\theta}\right)^2 + V\frac{x}{V\cos\theta}\sin\theta$$

$$= -\frac{1}{2}g\frac{x^2}{V^2\cos^2\theta} + \frac{x\sin\theta}{\cos\theta}$$

$$= x\tan\theta - \frac{gx^2}{2V^2\cos^2\theta}$$

$$= x\tan\theta - \frac{gx^2}{2V^2}\sec^2\theta$$

(2 marks)

(ii) The angle between the slope and the horizontal is $\frac{\pi}{4}$.

$\therefore$ the slope has gradient -1.

Equation of OP is $y = -x$

At points of intersection:

$$-x = x\tan\theta - \frac{gx^2}{2V^2}\sec^2\theta$$

$$0 = x\tan\theta + x - \frac{gx^2}{2V^2}\sec^2\theta$$

$$= x(\tan\theta + 1 - \frac{gx}{2V^2}\sec^2\theta)$$

$$x = 0 \text{ or } \tan\theta + 1 - \frac{gx}{2V^2}\sec^2\theta = 0$$

At O, $x = 0$

At P, $\tan\theta + 1 - \dfrac{gx}{2V^2}\sec^2\theta = 0$

$$\frac{gx}{2V^2}\sec^2\theta = \tan\theta + 1$$

$$x = \frac{2V^2}{g sec^2\theta}(\tan\theta + 1)$$

$$= \frac{2V^2}{g}\cos^2\theta\,(\tan\theta + 1)$$

$$= \frac{2V^2}{g}\cos\theta(\sin\theta + \cos\theta)$$

$y = -x$

So $y = -\dfrac{2V^2}{g}\cos\theta(\sin\theta + \cos\theta)$

$$D^2 = x^2 + y^2$$

$$= x^2 + (-x)^2$$

$$= 2x^2$$

$$D = \sqrt{2x^2}$$

$$= \sqrt{2}x \qquad (x > 0)$$

$$= \sqrt{2} \times \frac{2V^2}{g}\cos\theta(\sin\theta + \cos\theta)$$

$$= 2\sqrt{2}\frac{V^2}{g}\cos\theta(\cos\theta + \sin\theta)$$

(3 marks)

(iii) $$\frac{dD}{d\theta} = 2\sqrt{2}\frac{V^2}{g}(\cos\theta(-\sin\theta + \cos\theta) + (\cos\theta + \sin\theta)(-\sin\theta))$$

$$= 2\sqrt{2}\frac{V^2}{g}(-\sin\theta\cos\theta + \cos^2\theta - \sin\theta\cos\theta - \sin^2\theta)$$

$$= 2\sqrt{2}\frac{V^2}{g}(\cos^2\theta - \sin^2\theta - 2\sin\theta\cos\theta)$$

$$= 2\sqrt{2}\frac{V^2}{g}(\cos 2\theta - \sin 2\theta)$$

(2 marks)

(iv) $$\frac{dD}{d\theta} = 2\sqrt{2}\frac{V^2}{g}(\cos 2\theta - \sin 2\theta)$$

Stationary points occur when $\dfrac{dD}{d\theta} = 0$

i.e. $2\sqrt{2}\dfrac{V^2}{g}(\cos 2\theta - \sin 2\theta) = 0$

$$\cos 2\theta - \sin 2\theta = 0$$

$$\sin 2\theta = \cos 2\theta$$

$$\tan 2\theta = 1 \qquad (\cos 2\theta \neq 0)$$

$$2\theta = \frac{\pi}{4} \qquad \left(0 \le \theta < \frac{\pi}{2}\right)$$

$$\theta = \frac{\pi}{8}$$

$$\frac{d^2D}{d\theta^2} = 2\sqrt{2}\frac{V^2}{g}(-2\sin 2\theta - 2\cos 2\theta)$$

$$= -4\sqrt{2}\frac{V^2}{g}(\sin 2\theta + \cos 2\theta)$$

When $\theta = \dfrac{\pi}{8}$,

$$\frac{d^2D}{d\theta^2} = -4\sqrt{2}\frac{V^2}{g}\left(\frac{1}{\sqrt{2}} + \frac{1}{\sqrt{2}}\right) \quad (< 0)$$

$\therefore$ D has a maximum value when $\theta = \dfrac{\pi}{8}$

(3 marks)

(b) (i)

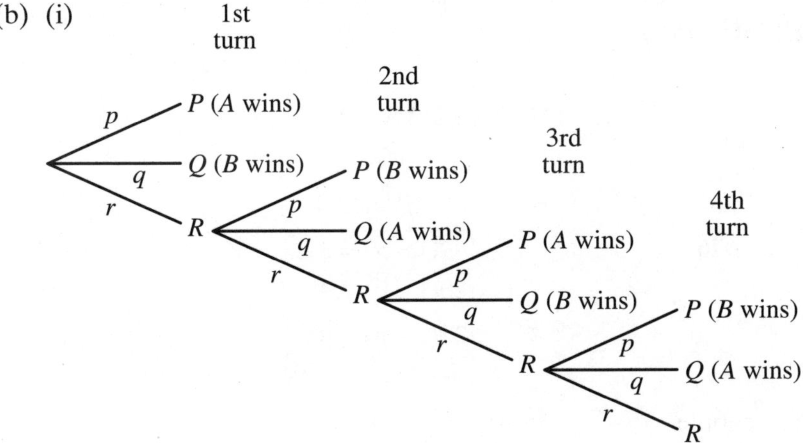

$P(A \text{ wins on 1st or 2nd turn}) = p + rq$

Now $p + q + r = 1$

$$q = 1 - p - r$$

So $P(A \text{ wins on 1st or 2nd turn}) = p + r(1 - p - r)$

$$= p + r - rp - r^2$$
$$= p - pr + r - r^2$$
$$= p(1 - r) + r(1 - r)$$
$$= (1 - r)(p + r)$$

(2 marks)

(ii) $P(A \text{ eventually wins}) = (1 - r)(p + r) + r^2(1 - r)(p + r) + r^4(1 - r)(p + r) + \ldots$

$$= (1 - r)(p + r)(1 + r^2 + r^4 + \ldots)$$

Now $1 + r^2 + r^4 + \ldots$ is an infinite geometric series with first term 1 and common ratio r^2.

$$S = \frac{a}{1 - r}$$
$$= \frac{1}{1 - r^2}$$
$$= \frac{1}{(1 + r)(1 - r)}$$

So $P(A \text{ eventually wins}) = (1 - r)(p + r) \times \dfrac{1}{(1 + r)(1 - r)}$

$$= \frac{p + r}{1 + r}$$

(3 marks)

Solutions to replacement questions

QUESTION 1

As there are 2 initials, then $n = 2$.

The number of initial pairings $(k) = 26^2$

$= 676$

Smallest group $= (n-1)k + 1$

$= 1 \times 676 + 1$

$= 677$

$\therefore$ the group must have at least 677 people.

Answer D

(1 mark)

QUESTION 4

As $\vec{PQ} = \underset{\sim}{u}$ then $\vec{RS} = -\underset{\sim}{u}$.

Hence, $\vec{MS} = -\frac{1}{2}\underset{\sim}{u}$.

Also, as $\vec{QR} = \underset{\sim}{v}$, then $\vec{SP} = -\underset{\sim}{v}$.

Now, $\vec{MP} = \vec{MS} + \vec{SP}$,

$= -\frac{1}{2}\underset{\sim}{u} + -\underset{\sim}{v}$

$= -\frac{1}{2}\underset{\sim}{u} - \underset{\sim}{v}$

Answer D

(1 mark)

QUESTION 7

At the point (2, 4) gradient $= \dfrac{4-2(2)}{2(4)+2} = 0$.

So possible solutions are B and C.

At the point (4, 0) gradient $= \dfrac{0-2(4)}{2(0)+4} = -\text{ve}$.

So the solution is C.

Hence,

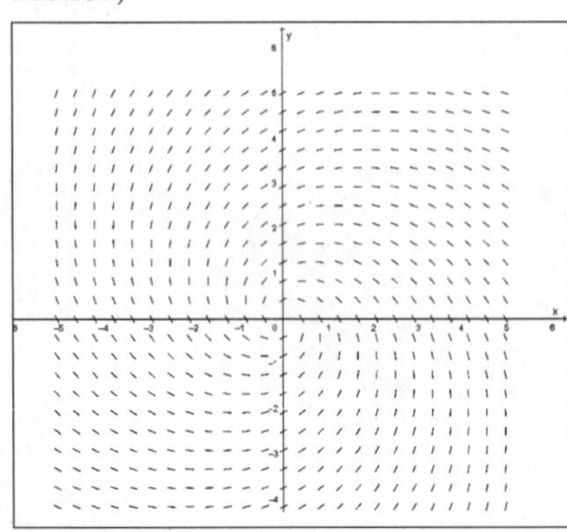

Answer C

(1 mark)

QUESTION 10

$E(X) = np = 20$

$\sigma = \sqrt{\text{Var}(X)}$

$\therefore \sqrt{np(1-p)} = 3$

$\sqrt{20(1-p)} = 3$

$20(1-p) = 9$

$1 - p = 0.45$

$p = 0.55$

Answer C

(3 marks)

QUESTION 12

(a) Solve graphically $y = |x^2 - 4x|$ and $y = x - 4$:

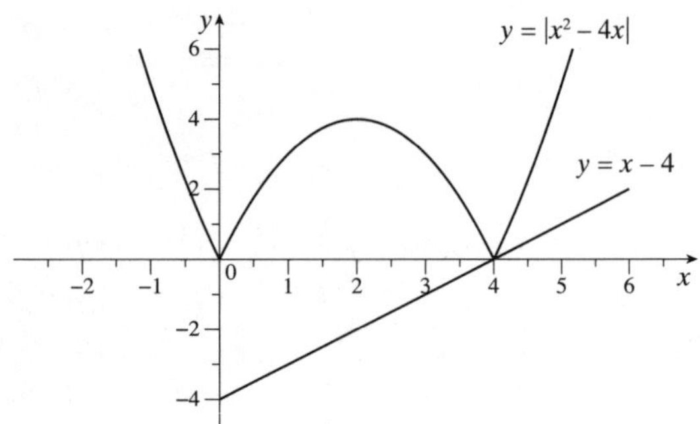

The solution to $|x^2 - 4x| \le x - 4$ is $x = 4$.

Note: Solving $|x^2 - 4x| \le x - 4$ algebraically would provide two cases:

$x^2 - 4x \le x - 4$	$-x^2 + 4x \le x - 4$
$x^2 - 5x + 4 \le 0$	$x^2 - 3x - 4 \ge 0$
$(x-4)(x-1)$	$(x-4)(x+1) \ge 0$
$\therefore 1 \le x \le 4$	$\therefore x \le -1$ or $x \ge 4$

However, testing of solutions would yield $x = 4$ as the only solution.

(3 marks)

(c) From $x = \frac{t}{2} + 1$,

$t = 2(x-1)$

Substitute into y:

$y = \dfrac{[2(x-1)]^2}{4} - 1$

$= (x-1)^2 - 1$

$= x^2 - 2x + 1 - 1$

$= x^2 - 2x$

$\therefore \quad y = x^2 - 2x$

$$\frac{dy}{dx} = 2x - 2$$

$$\frac{dy}{dx}(1) = 2(1) - 2 = 0$$

$\therefore$ a stationary point exists at $x = 0$.

(3 marks)

(d) $\underset{\sim P}{r}(3) = (3(3) - 2)\underset{\sim}{i} + (3 + 1)\underset{\sim}{j} = 7\underset{\sim}{i} + 4\underset{\sim}{j}$

$\underset{\sim Q}{r}(3) = 3^2\underset{\sim}{i} + (5 - 2(3))\underset{\sim}{j}. = 9\underset{\sim}{i} - \underset{\sim}{j}$

$\underset{\sim P}{r}(3) - \underset{\sim Q}{r}(3) = (7\underset{\sim}{i} + 4\underset{\sim}{j}) - (9\underset{\sim}{i} - \underset{\sim}{j}) = -2\underset{\sim}{i} + 5\underset{\sim}{j}$

$$\text{Distance} = \sqrt{(-2)^2 + 5^2} = \sqrt{29}$$

$\therefore$ the particles are $\sqrt{29}$ units apart.

(2 marks)

(e) The second graph is reflected about the y-axis.

Also, there are discontinuities at the values of x where $f(x) = 0$ in the first graph.

$\therefore$ the second graph is $y = \dfrac{1}{\sqrt{f(|x|)}}$.

Answer D *(1 mark)*

QUESTION 13

(c)* $$\text{LHS} = \frac{\cos 3x + \cos x}{\cos 3x - \cos x} = \frac{2\cos 2x \cos x}{-2\sin 2x \sin x} = \frac{\cos 2x \cos x}{-\sin 2x \sin x} = -\cot 2x \cot x = \text{RHS}$$

(2 marks)

(c)** $$\frac{dP}{dt} = 0.02P\left(1 - \frac{P}{1200}\right)$$

$$\frac{dP}{dt} = \frac{P}{50}\left(\frac{1200 - P}{1200}\right)$$

$$\frac{dP}{dt} = \frac{P(1200 - P)}{60000}$$

$$\frac{dP}{dt} = \frac{60000}{P(1200 - P)}$$

$$\frac{dP}{dt} = \frac{50}{P} + \frac{50}{1200 - P}$$

$$t = 50\int\left(\frac{1}{P} + \frac{1}{1200 - P}\right)dP$$

$$t = 50(\log_e|P| - \log_e|1200 - P|) + c$$

$$t = 50\log_e\frac{P}{1200 - P} + c$$

(as $0 < P < 1200$)

Substitute $t = 0$, $P = 800$:

$$0 = 50\log_e\frac{800}{1200 - 800} + c = 50\log_e 2 + c$$

$$c = -50\log_e 2$$

$$t = 50\log_e\frac{P}{1200 - P} - 50\log_e 2$$

$$t = 50\log_e\frac{P}{2400 - 2P}$$

$$\frac{t}{50} = \log_e\frac{P}{2400 - 2P}$$

$$e^{0.02t} = \log_e\frac{P}{2400 - 2P}$$

$$P = 2400e^{0.02t} - 2Pe^{0.02t}$$

$$P(1 + 2e^{0.02t}) = 2400e^{0.02t}$$

$$P = \frac{2400e^{0.02t}}{1 + 2e^{0.02t}}$$

(4 marks)

(d) $$\underset{\sim}{a} \bullet \underset{\sim}{b} = 2(-3) + (-4)1 = -10$$

$$\frac{\underset{\sim}{a} \bullet \underset{\sim}{b}}{|\underset{\sim}{b}|^2}.\underset{\sim}{b} = \frac{2(-3) + (-4)1}{(-3)^2 + 1^2} \times (-3\underset{\sim}{i} + \underset{\sim}{j}) = \frac{-10}{10} \times (-3\underset{\sim}{i} + \underset{\sim}{j}) = -(-3\underset{\sim}{i} + \underset{\sim}{j}) = 3\underset{\sim}{i} - \underset{\sim}{j}$$

(3 marks)

2015 HIGHER SCHOOL CERTIFICATE EXAMINATION

Mathematics Extension 1

General Instructions

- Reading time – 5 minutes
- Working time – 2 hours
- Write using black pen
- Board-approved calculators may be used
- A table of standard integrals is provided at the back of this paper
- In Questions 11–14, show relevant mathematical reasoning and/or calculations

Total marks – 70

Section I

10 marks

- Attempt Questions 1–10
- Allow about 15 minutes for this section

Section II

60 marks

- Attempt Questions 11–14
- Allow about 1 hour and 45 minutes for this section

Section I

10 marks
Attempt Questions 1–10
Allow about 15 minutes for this section

Use the multiple-choice answer sheet for Questions 1–10.

---

1 What is the remainder when $x^3 - 6x$ is divided by $x + 3$?

(A) -9

(B) 9

(C) $x^2 - 2x$

(D) $x^2 - 3x + 3$

2 Given that $N = 100 + 80e^{kt}$, which expression is equal to $\dfrac{dN}{dt}$?

(A) $k(100 - N)$

(B) $k(180 - N)$

(C) $k(N - 100)$

(D) $k(N - 180)$

3 Two secants from the point P intersect a circle as shown in the diagram.

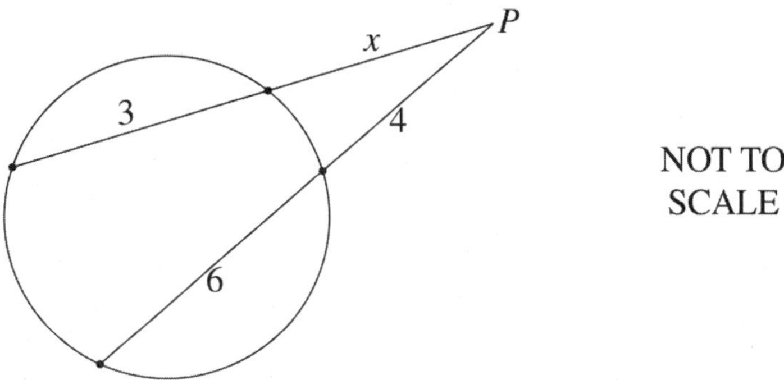

What is the value of x?

(A) 2

(B) 5

(C) 7

(D) 8

4 A rowing team consists of 8 rowers and a coxswain.

The rowers are selected from 12 students in Year 10.

The coxswain is selected from 4 students in Year 9.

In how many ways could the team be selected?

(A) ${}^{12}C_8 + {}^{4}C_1$

(B) ${}^{12}P_8 + {}^{4}P_1$

(C) ${}^{12}C_8 \times {}^{4}C_1$

(D) ${}^{12}P_8 \times {}^{4}P_1$

5 What are the asymptotes of $y = \dfrac{3x}{(x+1)(x+2)}$?

(A) $y = 0,\ \ x = -1,\ \ x = -2$

(B) $y = 0,\ \ x = 1,\ \ x = 2$

(C) $y = 3,\ \ x = -1,\ \ x = -2$

(D) $y = 3,\ \ x = 1,\ \ x = 2$

6 What is the domain of the function $f(x) = \sin^{-1}(2x)$?

(A) $-\pi \le x \le \pi$

(B) $-2 \le x \le 2$

(C) $-\dfrac{\pi}{4} \le x \le \dfrac{\pi}{4}$

(D) $-\dfrac{1}{2} \le x \le \dfrac{1}{2}$

7 What is the value of k such that $\int_0^k \frac{1}{\sqrt{4-x^2}}\,dx = \frac{\pi}{3}$?

(A) 1

(B) $\sqrt{3}$

(C) 2

(D) $2\sqrt{3}$

✗ 8 What is the value of $\lim_{x \to 3} \frac{\sin(x-3)}{(x-3)(x+2)}$?

(A) 0

(B) $\frac{1}{5}$

(C) 5

(D) Undefined

✗ 9 Two particles oscillate horizontally. The displacement of the first is given by $x = 3\sin 4t$ and the displacement of the second is given by $x = a\sin nt$. In one oscillation, the second particle covers twice the distance of the first particle, but in half the time.

What are the values of a and n?

(A) $a = 1.5,\quad n = 2$

(B) $a = 1.5,\quad n = 8$

(C) $a = 6,\quad n = 2$

(D) $a = 6,\quad n = 8$

10 The graph of the function $y = \cos\left(2t - \frac{\pi}{3}\right)$ is shown below.

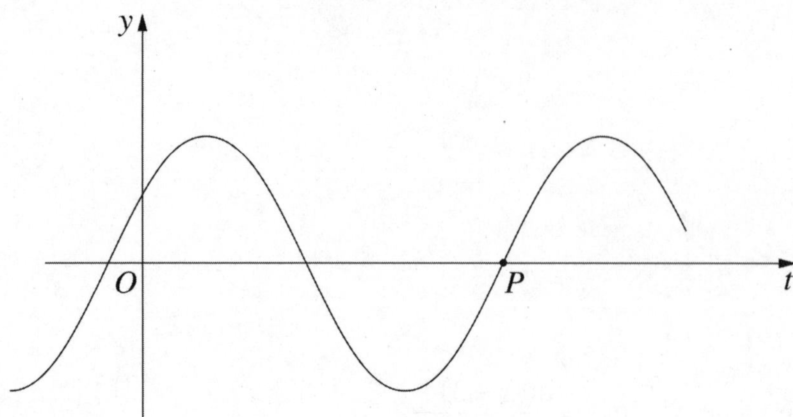

What are the coordinates of the point P?

(A) $\left(\frac{5\pi}{12}, 0\right)$

(B) $\left(\frac{2\pi}{3}, 0\right)$

(C) $\left(\frac{11\pi}{12}, 0\right)$

(D) $\left(\frac{7\pi}{6}, 0\right)$

Section II

60 marks
Attempt Questions 11–14
Allow about 1 hour and 45 minutes for this section

Answer each question in a SEPARATE writing booklet. Extra writing booklets are available.

In Questions 11–14, your responses should include relevant mathematical reasoning and/or calculations.

Question 11 (15 marks) Use a SEPARATE writing booklet.

(a) Find $\int \sin^2 x \, dx$. **2**

(b) Calculate the size of the acute angle between the lines $y = 2x + 5$ and $y = 4 - 3x$. **2**

(c) Solve the inequality $\frac{4}{x+3} \geq 1$. **3**

(d) Express $5\cos x - 12\sin x$ in the form $A\cos(x + \alpha)$, where $0 \leq \alpha \leq \frac{\pi}{2}$. **2**

(e) Use the substitution $u = 2x - 1$ to evaluate $\int_1^2 \frac{x}{(2x-1)^2} dx$. **3**

(f) Consider the polynomials $P(x) = x^3 - kx^2 + 5x + 12$ and $A(x) = x - 3$.

(i) Given that $P(x)$ is divisible by $A(x)$, show that $k = 6$. **1**

(ii) Find all the zeros of $P(x)$ when $k = 6$. **2**

Question 12 (15 marks) Use a SEPARATE writing booklet.

X (a) In the diagram, the points A, B, C and D are on the circumference of a circle, whose centre O lies on BD. The chord AC intersects the diameter BD at Y. The tangent at D passes through the point X.

It is given that $\angle CYB = 100°$ and $\angle DCY = 30°$.

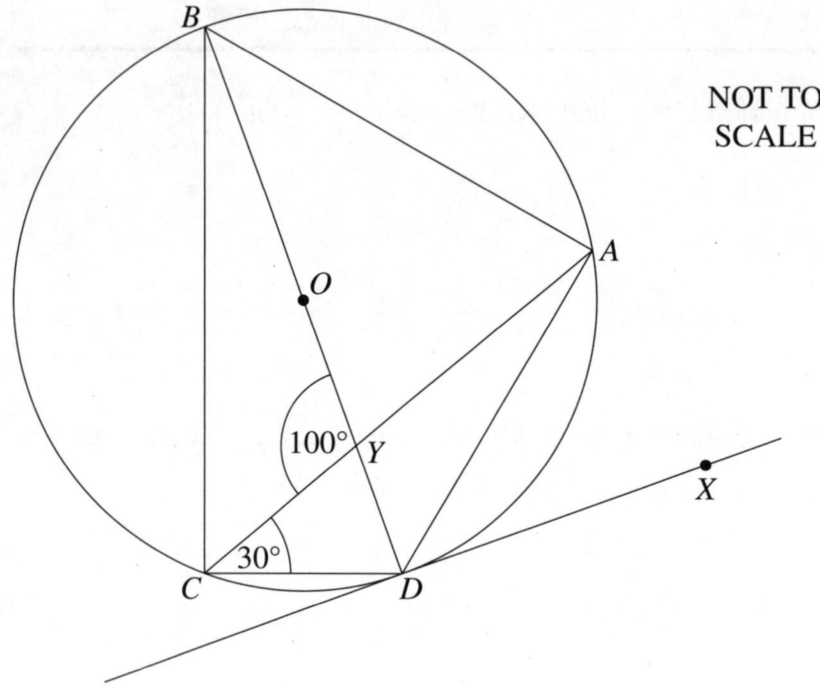

Copy or trace the diagram into your writing booklet.

(i) What is the size of $\angle ACB$? **1**

(ii) What is the size of $\angle ADX$? **1**

(iii) Find, giving reasons, the size of $\angle CAB$. **2**

Question 12 continues on the following page

Question 12 (continued)

(b) The points $P(2ap, ap^2)$ and $Q(2aq, aq^2)$ lie on the parabola $x^2 = 4ay$.

The equation of the chord PQ is given by $(p+q)x - 2y - 2apq = 0$. (Do NOT prove this.)

(i) Show that if PQ is a focal chord then $pq = -1$. **1**

(ii) If PQ is a focal chord and P has coordinates $(8a, 16a)$, what are the coordinates of Q in terms of a? **2**

(c) A person walks 2000 metres due north along a road from point A to point B. The point A is due east of a mountain OM, where M is the top of the mountain. The point O is directly below point M and is on the same horizontal plane as the road. The height of the mountain above point O is h metres.

From point A, the angle of elevation to the top of the mountain is 15°.

From point B, the angle of elevation to the top of the mountain is 13°.

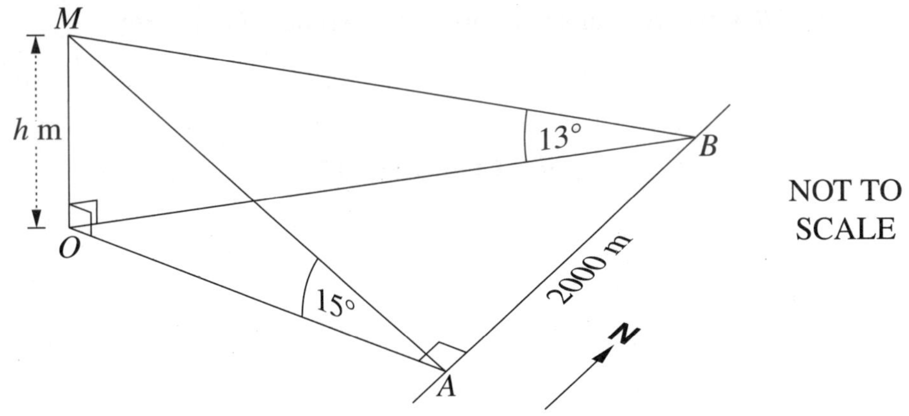

(i) Show that $OA = h\cot 15°$. **1**

(ii) Hence, find the value of h. **2**

Question 12 continues on the following page

Question 12 (continued)

(d) A kitchen bench is in the shape of a segment of a circle. The segment is bounded by an arc of length 200 cm and a chord of length 160 cm. The radius of the circle is r cm and the chord subtends an angle θ at the centre O of the circle.

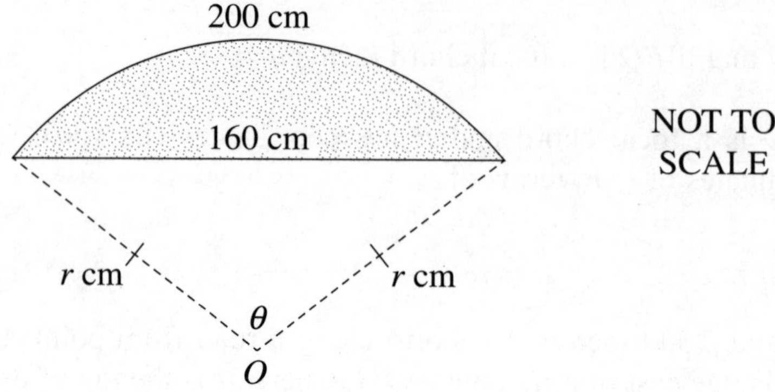

(i) Show that $160^2 = 2r^2(1 - \cos\theta)$. **1**

(ii) Hence, or otherwise, show that $8\theta^2 + 25\cos\theta - 25 = 0$. **2**

(iii) Taking $\theta_1 = \pi$ as a first approximation to the value of θ, use one application of Newton's method to find a second approximation to the value of θ. Give your answer correct to two decimal places. **2**

End of Question 12

Question 13 (15 marks) Use a SEPARATE writing booklet.

(a) A particle is moving along the x-axis in simple harmonic motion. The displacement of the particle is x metres and its velocity is v m s^{-1}. The parabola below shows v^2 as a function of x.

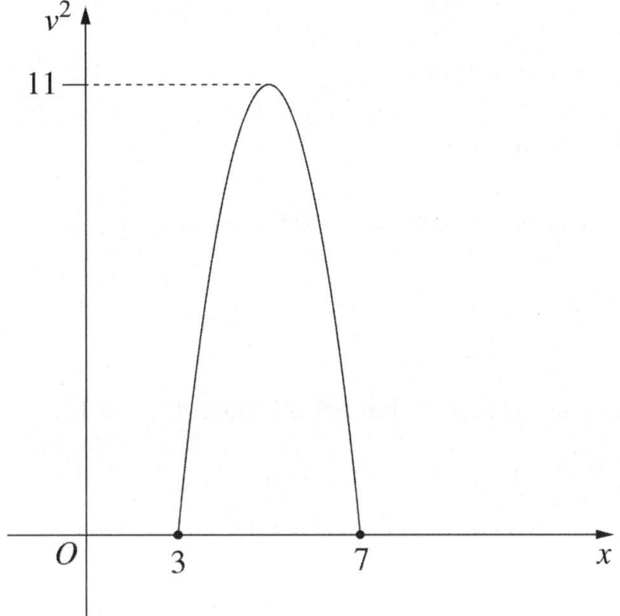

(i) For what value(s) of x is the particle at rest? **1**

(ii) What is the maximum speed of the particle? **1**

(iii) The velocity v of the particle is given by the equation **3**

$$v^2 = n^2\left(a^2 - (x - c)^2\right)$$

where a, c and n are positive constants.

What are the values of a, c and n?

Question 13 continues on the following page

Question 13 (continued)

(b) Consider the binomial expansion

$$\left(2x + \frac{1}{3x}\right)^{18} = a_0 x^{18} + a_1 x^{16} + a_2 x^{14} + \cdots$$

where $a_0, a_1, a_2, \ldots$ are constants.

(i) Find an expression for a_2. **2**

(ii) Find an expression for the term independent of x. **2**

(c) Prove by mathematical induction that for all integers $n \geq 1$, **3**

$$\frac{1}{2!} + \frac{2}{3!} + \frac{3}{4!} + \cdots + \frac{n}{(n+1)!} = 1 - \frac{1}{(n+1)!}.$$

(d) Let $f(x) = \cos^{-1}(x) + \cos^{-1}(-x)$, where $-1 \leq x \leq 1$.

(i) By considering the derivative of $f(x)$, prove that $f(x)$ is constant. **2**

(ii) Hence deduce that $\cos^{-1}(-x) = \pi - \cos^{-1}(x)$. **1**

End of Question 13

Question 14 (15 marks) Use a SEPARATE writing booklet.

* (a) A projectile is fired from the origin O with initial velocity $V\ \text{m s}^{-1}$ at an angle θ to the horizontal. The equations of motion are given by

$$x = Vt\cos\theta, \quad y = Vt\sin\theta - \frac{1}{2}gt^2. \quad \text{(Do NOT prove this.)}$$

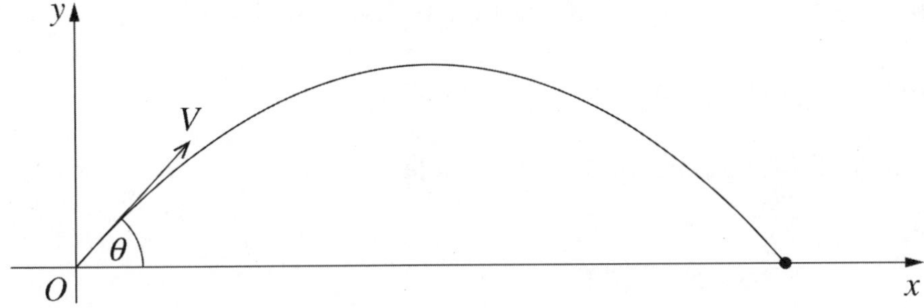

(i) Show that the horizontal range of the projectile is $\dfrac{V^2\sin 2\theta}{g}$. **2**

A particular projectile is fired so that $\theta = \dfrac{\pi}{3}$.

(ii) Find the angle that this projectile makes with the horizontal when $t = \dfrac{2V}{\sqrt{3}\,g}$. **2**

(iii) State whether this projectile is travelling upwards or downwards when $t = \dfrac{2V}{\sqrt{3}\,g}$. Justify your answer. **1**

* In the new Mathematics Extension 1 course, Projectile Motion questions will be expressed in vector form:

$x = Vt\cos\theta$ and $y = Vt\sin\theta - \frac{1}{2}gt^2$ is expressed as a position vector:

$$\underset{\sim}{r}(t) = (Vt\cos\theta)\underset{\sim}{i} + \left(Vt\sin\theta - \frac{1}{2}gt^2\right)\underset{\sim}{j}$$

Question 14 continues on the following page

Question 14 (continued)

✗ (b) A particle is moving horizontally. Initially the particle is at the origin O moving with velocity 1 m s^{-1}.

The acceleration of the particle is given by $\ddot{x} = x - 1$, where x is its displacement at time t.

(i) Show that the velocity of the particle is given by $\dot{x} = 1 - x$. **3**

(ii) Find an expression for x as a function of t. **2**

(iii) Find the limiting position of the particle. **1**

(c) Two players A and B play a series of games against each other to get a prize. In any game, either of the players is equally likely to win.

To begin with, the first player who wins a total of 5 games gets the prize.

(i) Explain why the probability of player A getting the prize in exactly 7 games is $\binom{6}{4}\left(\frac{1}{2}\right)^7$. **1**

(ii) Write an expression for the probability of player A getting the prize in at most 7 games. **1**

(iii) Suppose now that the prize is given to the first player to win a total of $(n + 1)$ games, where n is a positive integer. **2**

By considering the probability that A gets the prize, prove that

$$\binom{n}{n}2^n + \binom{n+1}{n}2^{n-1} + \binom{n+2}{n}2^{n-2} + \cdots + \binom{2n}{n} = 2^{2n}.$$

End of paper

Replacement questions
with content from the most up-to-date syllabus

Marks

Question 3 (1 mark)

The graph of $y = f(x)$ is shown below. **1**

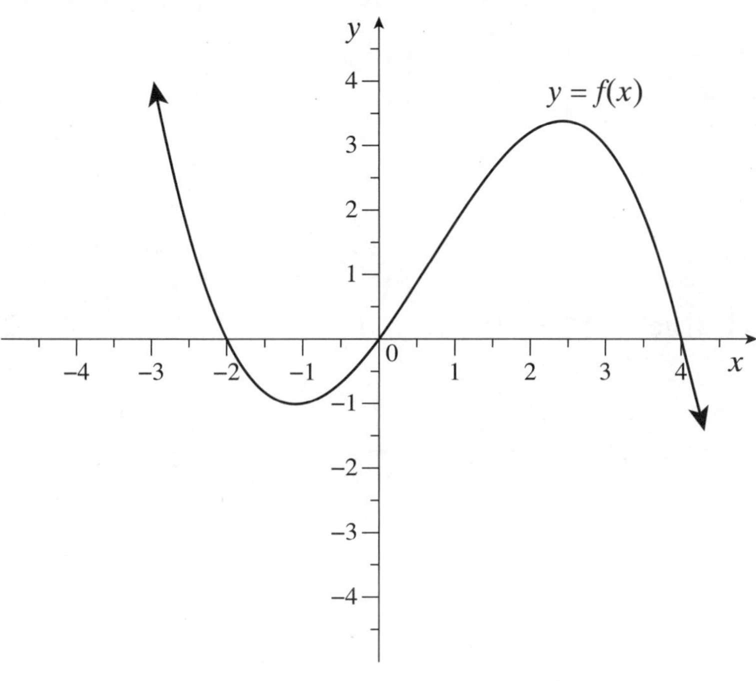

The graph of $y = \dfrac{1}{f(x)}$ is best represented by

(A)

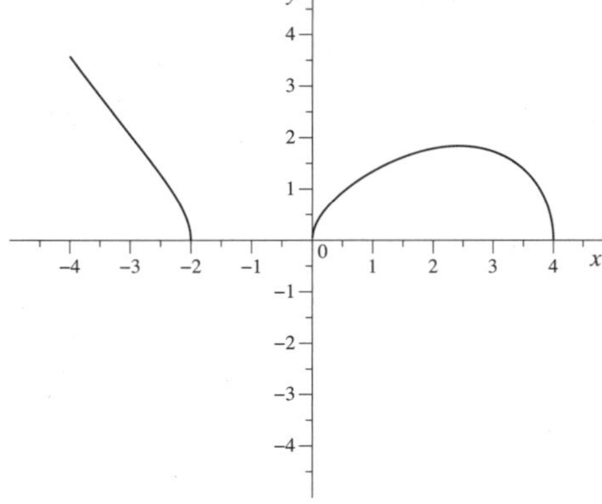

(B)

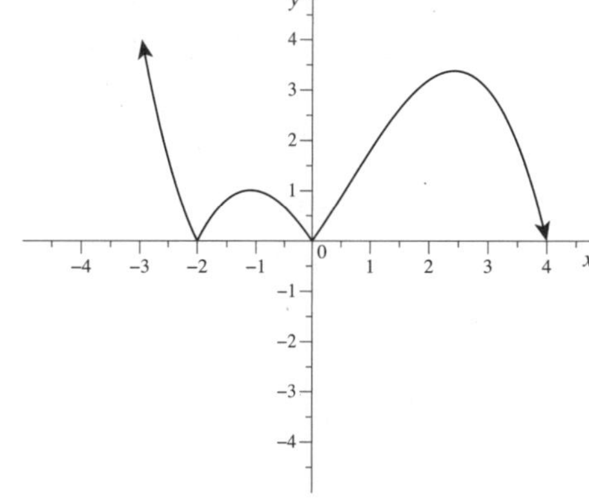

(C)

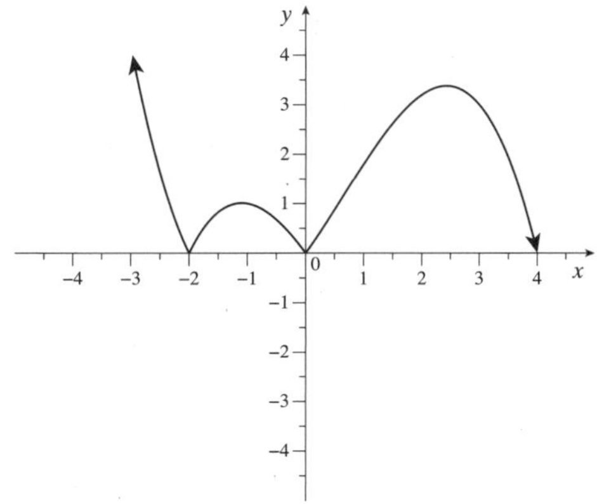

(D)

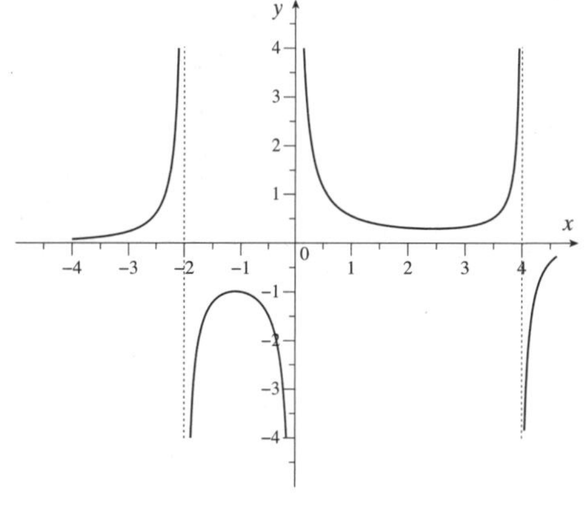

REPLACEMENT QUESTIONS

Marks

Question 8 (1 mark)

In the parallelogram, $|\underset{\sim}{a}| = 2|\underset{\sim}{b}|$.

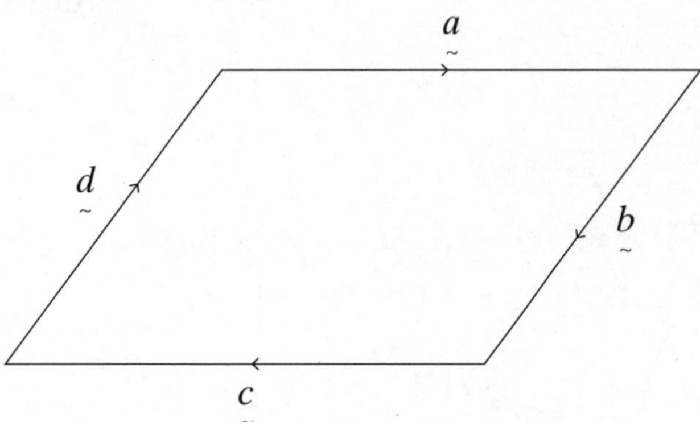

Which of the following statements is true? **1**

(A) $\underset{\sim}{a} = 2\underset{\sim}{b}$

(B) $\underset{\sim}{a} + \underset{\sim}{b} = \underset{\sim}{c} + \underset{\sim}{d}$

(C) $\underset{\sim}{a} - \underset{\sim}{c} = 0$

(D) $\underset{\sim}{b} + \underset{\sim}{d} = 0$

Question 9 (1 mark)

A particle moves so that its position is given by $\underset{\sim}{r}(t) = (t-2)\underset{\sim}{i} + (2t+1)\underset{\sim}{j}$. **1**

How far is the particle from the origin after 3 seconds?

(A) $4\sqrt{2}$ units

(B) 7 units

(C) $5\sqrt{2}$ units

(D) 8 units

Marks

Question 11 (2 marks)

(b) A function is defined as $f(x) = 1 - \sqrt{x-3}$. The graph of $y = f(x)$ is shown below.

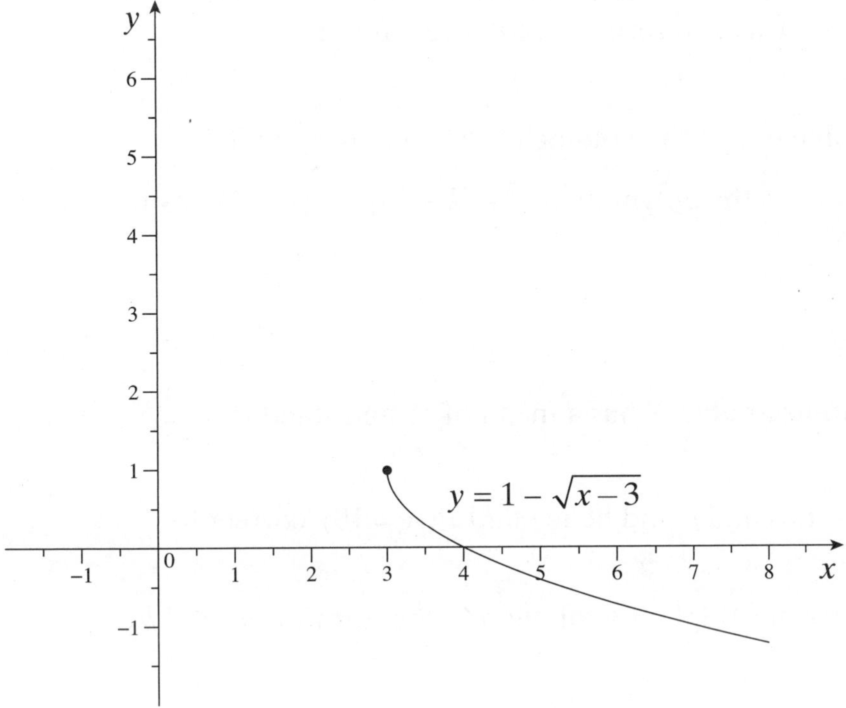

Sketch the graph of $y = f^{-1}(x)$ on the diagram above. **2**

Question 12 (9 marks)

(a) Solve the differential equation $\dfrac{dy}{dx} = \dfrac{x}{1+y^2}$, with the condition $y(4) = 3$ and find the positive value of x when $y = 2$. **4**

(b) Solve $\sin \pi t + \sin 3\pi t = \cos \pi t$, where $0 \le t \le 1$. **3**

(d) (iii) Consider the two vectors $\underset{\sim}{a} = \begin{pmatrix} 3 \\ 2 \end{pmatrix}$ and $\underset{\sim}{b} = \begin{pmatrix} -2 \\ 4 \end{pmatrix}$.

Determine the smallest angle between these two vectors, to the nearest degree. **2**

REPLACEMENT QUESTIONS

Marks

Question 13 (5 marks)

(a)* The region in the first quadrant enclosed by the coordinate axes, the graph with equation $y = e^{-x}$ and the straight line $x = a$, where $a > 0$, is rotated about the x-axis to form a solid of revolution.

Find the exact value of a, if the volume of the solid is $\frac{3\pi}{8}$ units^3. **3**

(a)** For what values of k is the polynomial $x^2 - (k + 2)x + 2(k + 2)$ positive for all values of x? **2**

Question 14 (6 marks)

(b)* A binomial random variable X has a mean of 15 and standard deviation of $\sqrt{6}$.

Find the parameters n and p and hence find $P(X = 10)$, correct to 4 decimal places. **3**

(b)** The diagram shows the graph of a function in the form $y = a \cos^{-1} bx$.

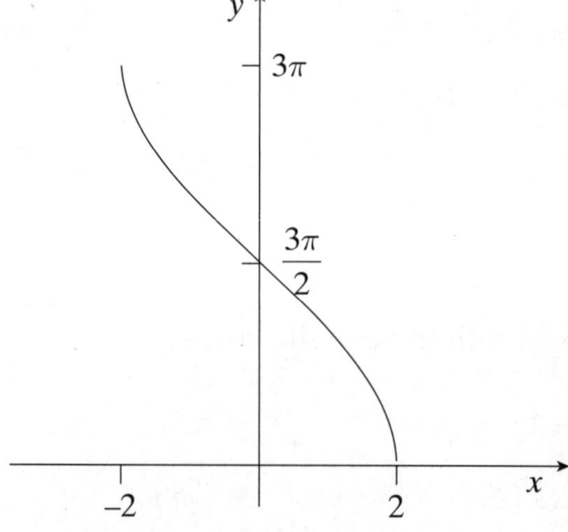

Show that the area between the curve and the y axis between $y = 3\pi$ and $y = \frac{3\pi}{2}$ is 6 units^2. **3**

REPLACEMENT QUESTIONS

2015 Higher School Certificate
Worked answers

Section I (*Total 10 marks*)

1. A	**2.** C	**3.** B	**4.** C	**5.** A
6. D	**7.** B	**8.** B	**9.** D	**10.** C

1. $P(x) = x^3 - 6x$

$P(-3) = (-3)^3 - 6 \times -3$

$= -9$

Answer A

2. $N = 100 + 80e^{kt}$

$\frac{dN}{dt} = k \times 80e^{kt}$

$= k(N - 100)$

Answer C

3. The products of the intercepts of two secants from an external point are equal.

$\therefore x(x + 3) = 4 \times (4 + 6)$

$x^2 + 3x = 40$

$x^2 + 3x - 40 = 0$

$(x + 8)(x - 5) = 0$

$x = -8$ or $x = 5$

But $x > 0$, $\therefore x = 5$

Answer B

4. Unordered selection of 8 rowers from 12 students.

Ways rowers can be selected $= {}^{12}C_8$

For each way they can be selected there are ${}^4C_1$ ways of choosing the coxswain.

So the number of ways is ${}^{12}C_8 \times {}^4C_1$

Answer C

5. $y = \frac{3x}{(x+1)(x+2)}$

$x + 1 \neq 0$

So $x \neq -1$

$x + 2 \neq 0$

So $x \neq -2$

Vertical asymptotes are $x = -1$ and $x = -2$.

As $x \to \pm\infty, y \to 0$.

Horizontal asymptote is $y = 0$.

Answer A

6. $f(x) = \sin^{-1}(2x)$

Now $f(u) = \sin^{-1}u$ has domain $-1 \leq u \leq 1$.

So $\sin^{-1}(2x)$ has domain $-1 \leq 2x \leq 1$.

$-\frac{1}{2} \leq x \leq \frac{1}{2}$

Answer D

7. $\int_0^k \frac{1}{\sqrt{4 - x^2}}\,dx = \left[\sin^{-1}\left(\frac{x}{2}\right)\right]_0^k$

$= \sin^{-1}\left(\frac{k}{2}\right) - \sin^{-1}0$

$= \sin^{-1}\left(\frac{k}{2}\right)$

So $\sin^{-1}\left(\frac{k}{2}\right) = \frac{\pi}{3}$

$\frac{k}{2} = \sin\frac{\pi}{3}$

$= \frac{\sqrt{3}}{2}$

$k = \sqrt{3}$

Answer B

8. $\lim_{x \to 3} \frac{\sin(x-3)}{(x-3)(x+2)}$

$= \lim_{x \to 3} \frac{\sin(x-3)}{(x-3)} \times \lim_{x \to 3} \frac{1}{x+2}$

$= 1 \times \frac{1}{3+2}$

$= \frac{1}{5}$

Answer B

9. $x = 3 \sin 4t$

amplitude = 3, frequency = 4

The second particle covers twice the distance so it will have amplitude 6. So $a = 6$.

The second particle completes an oscillation in half the time so it will have twice the frequency.

So $n = 8$.

Answer D

10. When $\cos\left(2t - \frac{\pi}{3}\right) = 0$

$$2t - \frac{\pi}{3} = \frac{\pi}{2} \text{ or } \frac{3\pi}{2}, \text{ or } \ldots$$

$$2t = \frac{\pi}{2} + \frac{\pi}{3} \text{ or } \frac{3\pi}{2} + \frac{\pi}{3} \text{ or } \ldots$$

$$= \frac{5\pi}{6} \text{ or } \frac{11\pi}{6} \text{ or } \ldots$$

$$t = \frac{5\pi}{12} \text{ or } \frac{11\pi}{12} \text{ or} \ldots$$

So P is the point $\left(\frac{11\pi}{12}, 0\right)$.

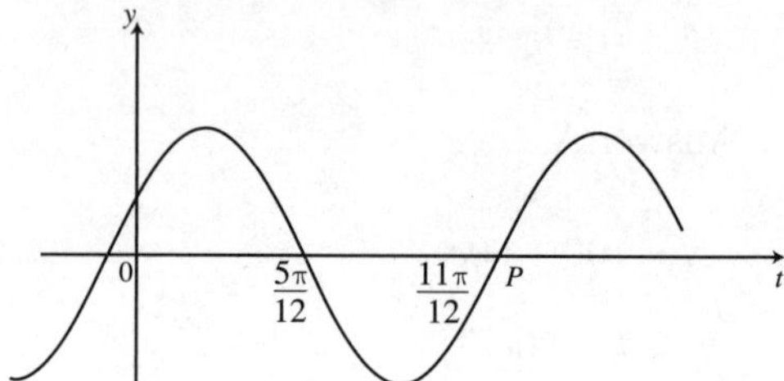

Answer C

Section II

QUESTION 11

(a) $\int sin^2x\, dx = \int \frac{1}{2}(1 - \cos 2x)\, dx$

$$= \frac{1}{2}(x - \frac{1}{2} \sin 2x) + C$$

$$= \frac{x}{2} - \frac{1}{4} \sin 2x + C$$

(2 marks)

(b) $y = 2x + 5$ has gradient $m_1 = 2$

$y = 4 - 3x$ has gradient $m_2 = -3$

If θ is the acute angle between the lines

$$\tan \theta = \left|\frac{m_1 - m_2}{1 + m_1 m_2}\right|$$

$$= \left|\frac{2-(-3)}{1+2\times-3}\right|$$

$$= 1$$

$\therefore \theta = 45°$

(2 marks)

(c) $\frac{4}{x+3} \geq 1 \qquad (x \neq -3)$

$$(x + 3)^2 \times \frac{4}{x+3} \geq (x + 3)^2$$

$$4x + 12 \geq x^2 + 6x + 9$$

$$x^2 + 2x - 3 \leq 0$$

Let $y = x^2 + 2x - 3$

$= (x + 3)(x - 1)$

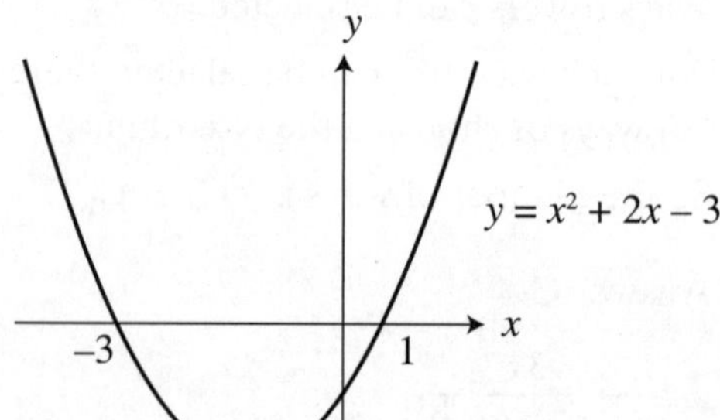

From the graph $x^2 + 2x - 3 \leq 0$ when $-3 \leq x \leq 1$

But $x \neq -3$

So $-3 < x \leq 1$

(3 marks)

(d) $5\cos x - 12\sin x \equiv A\cos(x + \alpha)$

$$\equiv A(\cos x\cos\alpha - \sin x\sin\alpha)$$

$\therefore A\cos\alpha = 5$ and $A\sin\alpha = 12$

$$A^2\cos^2\alpha + A^2\sin^2\alpha = 5^2 + 12^2$$

$$A^2 = 169$$

$$A = 13 \quad (A > 0)$$

Now $\dfrac{13\sin\alpha}{13\cos\alpha} = \dfrac{12}{5}$

$$\tan\alpha = \frac{12}{5}$$

$$\alpha = 1.176 \quad (3 \text{ d.p.}) \quad (0 \le \alpha \le \frac{\pi}{2})$$

So $5\cos x - 12\sin x \equiv 13\cos(x + 1.176)$

(2 marks)

(e) Let $u = 2x - 1$

$$\frac{du}{dx} = 2$$

$$du = 2\,dx$$

$$2x = u + 1$$

$$x = \frac{1}{2}(u + 1)$$

When $x = 1, u = 1$

When $x = 2, u = 3$

$$\therefore \int_1^2 \frac{x}{(2x-1)^2}\,dx$$

$$= \int_1^3 \frac{\frac{1}{2}(u+1)}{u^2}\cdot\frac{1}{2}du$$

$$= \frac{1}{4}\int_1^3 \frac{u+1}{u^2}\,du$$

$$= \frac{1}{4}\int_1^3 \left(\frac{1}{u} + \frac{1}{u^2}\right)du$$

$$= \frac{1}{4}\left[\ln u - \frac{1}{u}\right]_1^3$$

$$= \frac{1}{4}\left(\left(\ln 3 - \frac{1}{3}\right) - (\ln 1 - 1)\right)$$

$$= \frac{1}{4}\left(\ln 3 + \frac{2}{3}\right)$$

$$= \frac{1}{12}(3\ln 3 + 2)$$

(3 marks)

(f) (i) $P(x) = x^3 - kx^2 + 5x + 12$

$P(x)$ is divisible by $(x - 3)$ so $P(3) = 0$

$$\therefore 3^3 - k \times 3^2 + 5 \times 3 + 12 = 0$$

$$54 - 9k = 0$$

$$9k = 54$$

$$k = 6$$

(1 mark)

(ii) $P(x) = x^3 - 6x^2 + 5x + 12$

$$\begin{array}{r} x^2 - 3x - 4 \\ x-3\overline{\smash{)}\,x^3 - 6x^2 + 5x + 12} \\ \underline{x^3 - 3x^2} \\ -3x^2 + 5x \\ \underline{-3x^2 + 9x} \\ -4x + 12 \\ \underline{-4x + 12} \\ 0 \end{array}$$

$$\therefore \quad P(x) = (x - 3)(x^2 - 3x - 4)$$

$$= (x - 3)(x - 4)(x + 1)$$

So the zeros of $P(x)$ are -1, 3 and 4.

(2 marks)

QUESTION 12

(a)

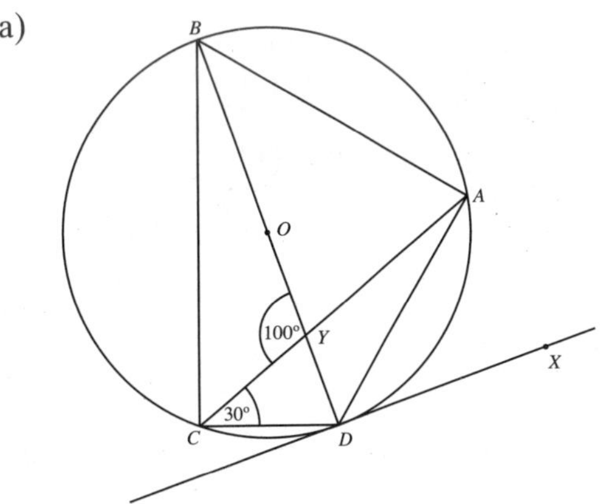

(i) $\angle BCD = 90°$ (angle in a semicircle)

$\therefore \angle ACB = 90° - 30° = 60°$

(1 mark)

(ii) $\angle ADX = \angle ACD$ (the angle between the tangent and chord equals the angle in the alternate segment)

$\therefore \angle ADX = 30°$

(1 mark)

(iii) $\angle ABD = \angle ACD$ (angles in same segment standing on same arc)

So $\angle ABD = 30°$

$\angle CYB = \angle ABD + \angle CAB$

(exterior angle of ΔABY)

So $\angle CAB = 100° - 30° = 70°$

(2 marks)

(b) (i) $x^2 = 4ay$ $P(2ap, ap^2), Q(2aq, aq^2)$

Focus $(0, a)$

$(p + q)x - 2y - 2apq = 0$

If the chord passes through the focus, then:

$$(p + q) \times 0 - 2a - 2apq = 0$$
$$-2apq = 2a$$
$$pq = -1$$

(1 mark)

(ii) P has coordinates $(8a, 16a)$

So $2ap = 8a$ and $ap^2 = 16a$

So $p = 4$

Now PQ is a focal chord so $pq = -1$

$\therefore q = -\dfrac{1}{4}$

$2aq = 2a \times -\dfrac{1}{4} = -\dfrac{a}{2}$

$aq^2 = a \times \left(-\dfrac{1}{4}\right)^2 = \dfrac{a}{16}$

So Q has coordinates $\left(-\dfrac{a}{2}, \dfrac{a}{16}\right)$

(2 marks)

(c)

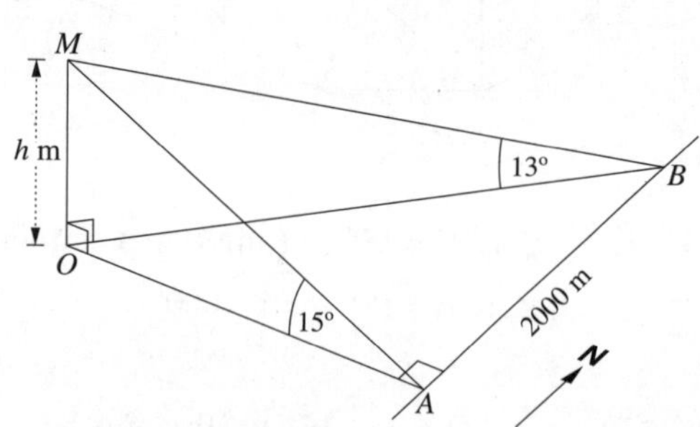

(i) In ΔOMA,

$$\tan 15° = \frac{h}{OA}$$
$$OA = \frac{h}{\tan 15°}$$
$$= h \cot 15°$$

(1 mark)

(ii) In ΔOMB,

$$\tan 13° = \frac{h}{OB}$$
$$OB = \frac{h}{\tan 13°}$$
$$= h \cot 13°$$

In ΔOBA,

$$OB^2 = OA^2 + AB^2$$
$$(h \cot 13°)^2 = (h \cot 15°)^2 + 2000^2$$
$$h^2 \cot^2 13° = h^2 \cot^2 15° + 2000^2$$
$$h^2 \cot^2 13° - h^2 \cot^2 15° = 2000^2$$
$$h^2(\cot^2 13° - \cot^2 15°) = 2000^2$$
$$h^2 = \frac{2000^2}{\cot^2 13° - \cot^2 15°}$$
$$h = \frac{2000}{\sqrt{\cot^2 13° - \cot^2 15°}} \quad (h > 0)$$
$$= 909.703\,84\ldots$$
$$= 910 \quad \text{(nearest unit)}$$

(2 marks)

(d) (i)

$$c^2 = a^2 + b^2 - 2ab \cos C$$
$$160^2 = r^2 + r^2 - 2 \times r \times r \times \cos\theta$$
$$160^2 = 2r^2 - 2r^2 \cos\theta$$
$$= 2r^2(1 - \cos\theta)$$

(1 mark)

(ii) Now $l = r\theta$

$$r\theta = 200$$
$$r = \frac{200}{\theta}$$

So $160^2 = 2\left(\dfrac{200}{\theta}\right)^2(1 - \cos\theta)$

$$25\,600 = \frac{80000}{\theta^2}(1 - \cos\theta)$$
$$256\theta^2 = 800(1 - \cos\theta)$$
$$8\theta^2 = 25(1 - \cos\theta)$$
$$= 25 - 25\cos\theta$$
$$8\theta^2 + 25\cos\theta - 25 = 0$$

(2 marks)

(iii) $f(\theta) = 8\theta^2 + 25\cos\theta - 25$

$f'(\theta) = 16\theta - 25\sin\theta$

$$\theta_2 = \theta_1 - \frac{f(\theta_1)}{f'(\theta_1)}$$

$$= \pi - \frac{8\pi^2 + 25\cos\pi - 25}{16\pi - 25\sin\pi}$$

$= 2.565\ 514\ 72\ldots$

$= 2.57 \quad (2 \text{ d.p.})$

(2 marks)

QUESTION 13

(a) (i) The particle is at rest when $v = 0$.

So it is at rest when $x = 3$ and $x = 7$.

(1 mark)

(ii) The maximum value of v^2 is 11.

So the maximum speed is $\sqrt{11}$ m s^{-1}

(1 mark)

(iii) $v^2 = n^2(a^2 - (x - c)^2)$

The particle is undergoing simple harmonic motion.

The centre of motion is $x = \frac{3+7}{2} = 5$.

So $c = 5$

The amplitude is $7 - 5 = 2$.

So $a = 2$

Now $v^2 = n^2(2^2 - (x - 5)^2)$

When $x = 5$, $v^2 = 11$

$11 = n^2(4 - (5 - 5)^2)$

$4n^2 = 11$

$$n^2 = \frac{11}{4}$$

$$n = \frac{\sqrt{11}}{2} \quad (n > 0)$$

(3 marks)

(b) (i) $\left(2x + \frac{1}{3x}\right)^{18} = a_0x^{18} + a_1x^{16} + a_2x^{14} + \ldots$

But $\left(2x + \frac{1}{3x}\right)^{18} = \binom{18}{0}(2x)^{18} + \binom{18}{1}(2x)^{17}\left(\frac{1}{3x}\right) + \binom{18}{2}(2x)^{16}\left(\frac{1}{3x}\right)^2 + \ldots$

$$\text{So } a_2x^{14} = \binom{18}{2}(2x)^{16}\left(\frac{1}{3x}\right)^2$$

$$= \binom{18}{2}2^{16}\left(\frac{1}{3}\right)^2 x^{14}$$

$$a_2 = \binom{18}{2}\frac{2^{16}}{3^2} \quad [= 17 \times 2^{16} = 1\ 114\ 112]$$

(2 marks)

(ii) Term independent of x

$$= \binom{18}{9}(2x)^9\left(\frac{1}{3x}\right)^9$$

$$= \binom{18}{9}2^9\left(\frac{1}{3}\right)^9$$

$$= \binom{18}{9}\left(\frac{2}{3}\right)^9$$

(2 marks)

(c) To prove for all integers $n \geq 1$:

$$\frac{1}{2!}+\frac{2}{3!}+\frac{3}{4!}+\ldots+\frac{n}{(n+1)!}=1-\frac{1}{(n+1)!}$$

When $n = 1$,

$$\text{LHS} = \frac{1}{2!} = \frac{1}{2}$$

$$\text{RHS} = 1-\frac{1}{2!} = 1-\frac{1}{2} = \frac{1}{2}$$

So the statement is true for $n = 1$.

Assume the statement is true for $n = k$

i.e. assume $\frac{1}{2!}+\frac{2}{3!}+\frac{3}{4!}+\ldots+\frac{k}{(k+1)!}=1-\frac{1}{(k+1)!}$

We need to show that it is true for $n = k + 1$

i.e. that $\frac{1}{2!}+\frac{2}{3!}+\frac{3}{4!}+\ldots+\frac{k+1}{(k+2)!}=1-\frac{1}{(k+2)!}$

When $n = k + 1$

$$\begin{aligned}\text{LHS} &= \frac{1}{2!}+\frac{2}{3!}+\frac{3}{4!}+\ldots+\frac{k}{(k+1)!}+\frac{k+1}{(k+2)!}\\ &= 1-\frac{1}{(k+1)!}+\frac{k+1}{(k+2)!}\\ &= 1-\frac{k+2}{(k+2)(k+1)!}+\frac{k+1}{(k+2)!}\\ &= 1-\left(\frac{k+2}{(k+2)!}-\frac{k+1}{(k+2)!}\right)\\ &= 1-\frac{1}{(k+2)!}\\ &= \text{RHS}\end{aligned}$$

$\therefore$ if it is true for $n = k$ it is also true for $n = k + 1$.

As it is true for $n = 1$, it is true for $n = 2$.

As it is true for $n = 2$, it is true for $n = 3$.

By the process of induction it is true for all integers $n \geq 1$.

(3 marks)

(d) (i) $f(x) = \cos^{-1}(x) + \cos^{-1}(-x)$

$$f'(x) = \frac{-1}{\sqrt{1-x^2}} + \frac{-1}{\sqrt{1-(-x)^2}} \times -1$$

$$= -\frac{1}{\sqrt{1-x^2}} + \frac{1}{\sqrt{1-x^2}}$$

$$= 0$$

So, as the derivative is zero, $f(x)$ is constant.

(2 marks)

(ii) $f(x) = \cos^{-1}(x) + \cos^{-1}(-x)$

$$f(0) = \cos^{-1}(0) + \cos^{-1}(-0)$$

$$= \frac{\pi}{2} + \frac{\pi}{2}$$

$$= \pi$$

But $f(x)$ is constant, so $f(x) = \pi$ for all values of x $(-1 \leq x \leq 1)$

$$\cos^{-1}(x) + \cos^{-1}(-x) = \pi$$

$$\cos^{-1}(-x) = \pi - \cos^{-1}(x)$$

(1 mark)

QUESTION 14

(a) (i) $y = Vt\sin\theta - \frac{1}{2}gt^2$

When $y = 0$,

$$Vt\sin\theta - \frac{1}{2}gt^2 = 0$$

$$t(V\sin\theta - \frac{1}{2}gt) = 0$$

$$t = 0 \text{ or } V\sin\theta - \frac{1}{2}gt = 0$$

$$gt = 2V\sin\theta$$

$$t = \frac{2V\sin\theta}{g}$$

Now $x = Vt\cos\theta$

When $t = \frac{2V\sin\theta}{g}$,

$$x = \frac{2V^2\sin\theta\cos\theta}{g}$$

$$= \frac{V^2\sin 2\theta}{g}$$

So the horizontal range of the projectile is $\frac{V^2\sin 2\theta}{g}$.

(2 marks)

(ii) $\theta = \frac{\pi}{3}$

So $x = Vt\cos\frac{\pi}{3}$

$$= \frac{Vt}{2}$$

$$\dot{x} = \frac{V}{2}$$

$$y = Vt\sin\frac{\pi}{3} - \frac{1}{2}gt^2$$

$$= \frac{\sqrt{3}}{2}Vt - \frac{1}{2}gt^2$$

$$\dot{y} = \frac{\sqrt{3}}{2}V - gt$$

When $t = \frac{2V}{\sqrt{3}g}$,

$$\dot{y} = \frac{\sqrt{3}}{2}V - \frac{2V}{\sqrt{3}}$$

$$= V\left(\frac{\sqrt{3}}{2} - \frac{2\sqrt{3}}{3}\right)$$

$$= \frac{-\sqrt{3}V}{6}$$

Let α be the acute angle the projectile makes with the horizontal.

$$\tan \alpha = \left|\frac{\dot{y}}{\dot{x}}\right|$$

$$= \left|\frac{-\frac{\sqrt{3}V}{6}}{\frac{V}{2}}\right|$$

$$= \frac{\sqrt{3}V}{6} \times \frac{2}{V}$$

$$= \frac{\sqrt{3}}{3}$$

$$= \frac{1}{\sqrt{3}}$$

So $\alpha = \dfrac{\pi}{6}$

(2 marks)

(iii) The projectile is travelling downwards because $\dot{y}$ is negative.

(1 mark)

(b) (i)

$$\ddot{x} = x - 1$$

$$\frac{d}{dx}\left(\frac{1}{2}v^2\right) = x - 1$$

$$\frac{1}{2}v^2 = \frac{x^2}{2} - x + C$$

Now $v = 1$ when $x = 0$

$$\frac{1}{2} = \frac{0^2}{2} - 0 + C$$

So $C = \dfrac{1}{2}$

$$\frac{1}{2}v^2 = \frac{x^2}{2} - x + \frac{1}{2}$$

$$v^2 = x^2 - 2x + 1$$

$$= (x - 1)^2$$

$$v = \pm(x - 1)$$

But, $v = 1$ when $x = 0$

so $v = -(x - 1)$

$$= 1 - x$$

$$\therefore\ \dot{x} = 1 - x$$

(3 marks)

(ii)

$$\frac{dx}{dt} = 1 - x$$

$$\frac{dt}{dx} = \frac{1}{1 - x}$$

$$t = \int \frac{1}{1 - x}\, dx$$

$$= -\ln(1 - x) + c$$

When $t = 0, x = 0$

$$0 = -\ln 1 + c$$

$$c = 0$$

$$\therefore t = -\ln(1 - x)$$

$$-t = \ln(1 - x)$$

$$e^{-t} = 1 - x$$

$$x = 1 - e^{-t}$$

(2 marks)

(iii) As $t \to \infty, e^{-t} \to 0$ so $x \to 1$

The limiting position of the particle is 1 m to the right of O.

(1 mark)

(c) (i) $P(A \text{ wins a game}) = P(B \text{ wins a game}) = \dfrac{1}{2}$

If player A gets the prize in exactly 7 games then A's 5th win must be the 7th game. The other 4 wins can come from any of the previous 6 games. The number of ways this can occur is $\dbinom{6}{4}$.

$$P(A \text{ wins in exactly 7 games}) = \binom{6}{4}\left(\frac{1}{2}\right)^4\left(\frac{1}{2}\right)^2 \times \frac{1}{2}$$
$$= \binom{6}{4}\left(\frac{1}{2}\right)^7$$

(1 mark)

(ii) $P(A \text{ wins in at most 7 games})$

$= P(A \text{ wins in 5 games}) + P(A \text{ wins in 6 games}) + P(A \text{ wins in 7 games})$

$$= \binom{4}{4}\left(\frac{1}{2}\right)^5 + \binom{5}{4}\left(\frac{1}{2}\right)^6 + \binom{6}{4}\left(\frac{1}{2}\right)^7$$

(1 mark)

(iii) The maximum number of games for someone to win $= n + 1 + n = 2n + 1$

$P(A \text{ gets the prize})$

$= P(A \text{ wins in } (n+1) \text{ games}) + P(A \text{ wins in } (n+2) \text{ games} + \ldots. + P(A \text{ wins in } (2n+1)) \text{ games}$

$$= \binom{n}{n}\left(\frac{1}{2}\right)^{n+1} + \binom{n+1}{n}\left(\frac{1}{2}\right)^{n+2} + \binom{n+2}{n}\left(\frac{1}{2}\right)^{n+3} + \ldots + \binom{2n}{n}\left(\frac{1}{2}\right)^{2n+1}$$

But A and B have an equal chance of getting the prize so $P(A \text{ gets the prize}) = \dfrac{1}{2}$

$$\text{So } \binom{n}{n}\left(\frac{1}{2}\right)^{n+1} + \binom{n+1}{n}\left(\frac{1}{2}\right)^{n+2} + \binom{n+2}{n}\left(\frac{1}{2}\right)^{n+3} + \ldots + \binom{2n}{n}\left(\frac{1}{2}\right)^{2n+1} = \frac{1}{2}$$

$$2^{2n+1} \times \left(\binom{n}{n}\left(\frac{1}{2}\right)^{n+1} + \binom{n+1}{n}\left(\frac{1}{2}\right)^{n+2} + \binom{n+2}{n}\left(\frac{1}{2}\right)^{n+3} + \ldots + \binom{2n}{n}\left(\frac{1}{2}\right)^{2n+1}\right) = 2^{2n+1} \times \frac{1}{2}$$

$$\binom{n}{n}\left(\frac{2^{2n+1}}{2^{n+1}}\right) + \binom{n+1}{n}\left(\frac{2^{2n+1}}{2^{n+2}}\right) + \binom{n+2}{n}\left(\frac{2^{2n+1}}{2^{n+3}}\right) + \ldots + \binom{2n}{n}\left(\frac{2^{2n+1}}{2^{2n+1}}\right) = \frac{2^{2n+1}}{2}$$

$$\binom{n}{n}2^n + \binom{n+1}{n}2^{n-1} + \binom{n+2}{n}2^{n-2} + \ldots + \binom{2n}{n} = 2^{2n}$$

(2 marks)

Solutions to replacement questions

QUESTION 3

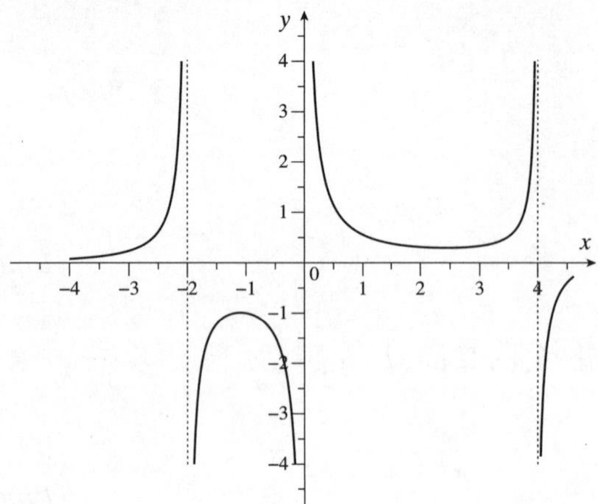

Answer D. *(1 mark)*

QUESTION 8

In a parallelogram, opposite sides equal.

As $\underset{\sim}{b} = -\underset{\sim}{d}$ hence $\underset{\sim}{b} + \underset{\sim}{d} = 0$.

Answer D *(1 mark)*

QUESTION 9

$$\underset{\sim}{r}(3) = (3-2)\underset{\sim}{i} + (2(3)+1)\underset{\sim}{j}$$
$$= \underset{\sim}{i} + 7\underset{\sim}{j}$$
$$|\underset{\sim}{r}(3)| = \sqrt{1^2 + 7^2}$$
$$= \sqrt{50}$$
$$= 5\sqrt{2}$$

$\therefore$ the particle is $5\sqrt{2}$ units from origin.

Answer C *(1 mark)*

QUESTION 11

(b)

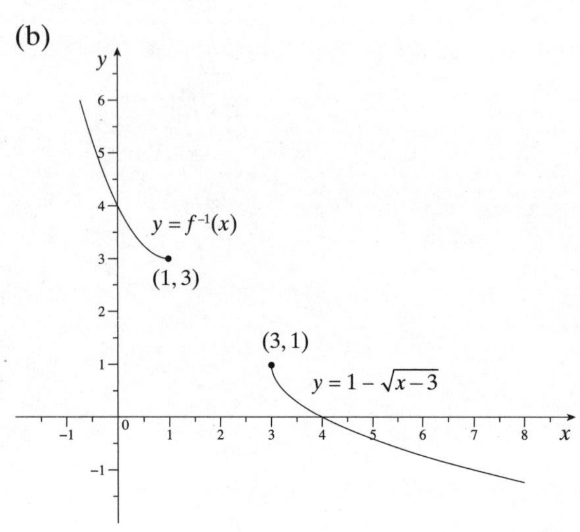

(2 marks)

QUESTION 12

(a)
$$\frac{dy}{dx} = \frac{x}{1+y^2}$$
$$\int (1+y^2)\,dy = \int x\,dx$$
$$y + \frac{y^3}{3} = \frac{x^2}{2} + c$$

Substitute $x = 4, y = 3$:

$$3 + \frac{3^3}{3} = \frac{4^2}{2} + c$$
$$3 + 9 = 8 + c$$
$$c = 4$$
$$y + \frac{y^3}{3} = \frac{x^2}{2} + 4$$

Substitute $y = 2$:

$$2 + \frac{2^3}{3} = \frac{x^2}{2} + 4$$
$$\frac{x^2}{2} = \frac{2}{3}$$
$$x^2 = \frac{4}{3}$$
$$x = \pm\frac{2}{\sqrt{3}}$$

As $x > 0$, then $x = \dfrac{2}{\sqrt{3}}$.

(4 marks)

(b)
$$\sin \pi t + \sin 3\pi t = \cos \pi t$$
$$2(\sin 2\pi t \cos(-\pi t)) = \cos \pi t$$
$$2(\sin 2\pi t \cos \pi t) = \cos \pi t$$
$$2\sin 2\pi t \cos \pi t - \cos \pi t = 0$$
$$\cos \pi t\,(2\sin 2\pi t - 1) = 0$$

$\cos \pi t = 0$ $\qquad$ $2\sin 2\pi t - 1 = 0$

$\cos \pi t = 0$ $\qquad$ $\sin 2\pi t = \dfrac{1}{2}$

$\pi t = \dfrac{\pi}{2}$ $\qquad$ $2\pi t = \dfrac{\pi}{6}, \dfrac{5\pi}{6}$

$t = \dfrac{1}{2}$ $\qquad$ $t = \dfrac{1}{12}, \dfrac{5}{12}$

$$\therefore t = \frac{1}{12}, \frac{5}{12}, \frac{1}{2}$$

(3 marks)

(d) (iii) $\underset{\sim}{a} \cdot \underset{\sim}{b} = 3 \times (-2) + 2 \times 4$

$= 2$

$|\underset{\sim}{a}||\underset{\sim}{b}|\cos\theta = \sqrt{3^2 + 2^2}.\sqrt{(-2)^2 + 4^2}.\cos\theta$

$= \sqrt{260}\cos\theta$

Using $\underset{\sim}{a} \cdot \underset{\sim}{b} = |\underset{\sim}{a}||\underset{\sim}{b}|\cos\theta$,

$2 = \sqrt{260}\cos\theta$

$\cos\theta = \dfrac{2}{\sqrt{260}}$

$= \dfrac{1}{\sqrt{65}}$

$\theta = 82.874\,983\,65\ldots$

$= 83$ (nearest whole)

$\therefore$ the smallest angle is 83°.

(2 marks)

QUESTION 13

(a)* Volume $= \pi\int_0^a (e^{-x})^2\,dx = \dfrac{3\pi}{8}$

$\pi\int_0^a e^{-2x}\,dx = \dfrac{3\pi}{8}$

$\dfrac{-\pi}{2}\left[e^{-2x}\right]_0^a = \dfrac{3\pi}{8}$

$\dfrac{-\pi}{2}\left[e^{-2a} - e^{-2(0)}\right] = \dfrac{3\pi}{8}$

$-\dfrac{1}{2}\left[e^{-2a} - 1\right] = \dfrac{3}{8}$

$e^{-2a} - 1 = -\dfrac{3}{4}$

$e^{-2a} = \dfrac{1}{4}$

$-2a = \log_e \dfrac{1}{4}$

$-2a = -\log_e 4$

$a = \dfrac{1}{2}\log_e 4$

$a = \log_e 2$

(3 marks)

(a)** $\Delta = -(k + 2)^2 - 4(1)(2k + 4) < 0$

$k^2 + 4k + 4 - 8k - 16 < 0$

$k^2 - 4k - 12 < 0$

$(k - 6)(k + 2) < 0$

Consider $(k - 6)(k + 2) = 0$

$k = 6, -2$

$\therefore -2 < k < 6$

(2 marks)

QUESTION 14

(b)* $\mu = np$

$15 = np$

$\sigma^2 = np(1 - p)$

$6 = 15(1 - p)$

$1 - p = 0.4$

$p = 0.6$

Hence, $15 = n(0.6)$

$n = 25$

$\therefore n = 25, p = 0.6$

$\therefore P(X = 10) = \binom{25}{10}(0.6)^{10}(1 - 0.6)^{15}$

$= 0.021\,222\,444\ldots$

$= 0.0212$ (4 dec. pl.)

(3 marks)

(b)** Range is $0 \le y \le 3\pi$, then $a = 3$.

Domain is $-2 \le x \le 2$, and hence $b = \dfrac{1}{2}$.

$\therefore a = 3$ and $b = \dfrac{1}{2}$

$y = 3\cos^{-1}\dfrac{x}{2}$.

Now, $\dfrac{y}{3} = \cos^{-1}\dfrac{x}{2}$

$\cos\dfrac{y}{3} = \dfrac{x}{2}$

$x = 2\cos\dfrac{y}{3}$

Area $= \int_{\frac{3\pi}{2}}^{3\pi} 2\cos\dfrac{y}{3}\,dy$

$= \int_0^{\frac{3\pi}{2}} 2\cos\dfrac{y}{3}\,dy$

$= 2\left[3\sin\dfrac{y}{3}\right]_0^{\frac{3\pi}{2}}$

$= 6\left[\sin\dfrac{\frac{3\pi}{2}}{3} - \sin 0\right]$

$= 6\left[\sin\dfrac{\pi}{2} - \sin 0\right]$

$= 6$

$\therefore$ the area is 6 units^2.

(3 marks)

2016 HSC Mathematics
Extension 1 Examination paper

2016 HIGHER SCHOOL CERTIFICATE EXAMINATION

Mathematics Extension 1

General Instructions

- Reading time – 5 minutes
- Working time – 2 hours
- Write using black pen
- Board-approved calculators may be used
- A reference sheet is provided at the back of this paper
- In Questions 11–14, show relevant mathematical reasoning and/or calculations

Total marks – 70

Section I

10 marks

- Attempt Questions 1–10
- Allow about 15 minutes for this section

Section II

60 marks

- Attempt Questions 11–14
- Allow about 1 hour and 45 minutes for this section

Section I

10 marks
Attempt Questions 1–10
Allow about 15 minutes for this section

Use the multiple-choice answer sheet for Questions 1–10.

---

1 Which sum is equal to $\sum_{k=1}^{20}(2k+1)$?

(A) $1 + 2 + 3 + 4 + \cdots + 20$

(B) $1 + 3 + 5 + 7 + \cdots + 41$

(C) $3 + 4 + 5 + 6 + \cdots + 20$

(D) $3 + 5 + 7 + 9 + \cdots + 41$

2 What is the remainder when $2x^3 - 10x^2 + 6x + 2$ is divided by $x - 2$?

(A) -66

(B) -10

(C) $-x^3 + 5x^2 - 3x - 1$

(D) $x^3 - 5x^2 + 3x + 1$

3 Which expression is equivalent to $\dfrac{\tan 2x - \tan x}{1 + \tan 2x \tan x}$?

(A) $\tan x$

(B) $\tan 3x$

(C) $\dfrac{\tan 2x - 1}{1 + \tan 2x}$

(D) $\dfrac{\tan x}{1 + \tan 2x \tan x}$

✗ **4** In the diagram, O is the centre of the circle ABC, D is the midpoint of BC, AT is the tangent at A and $\angle ATB = 40°$.

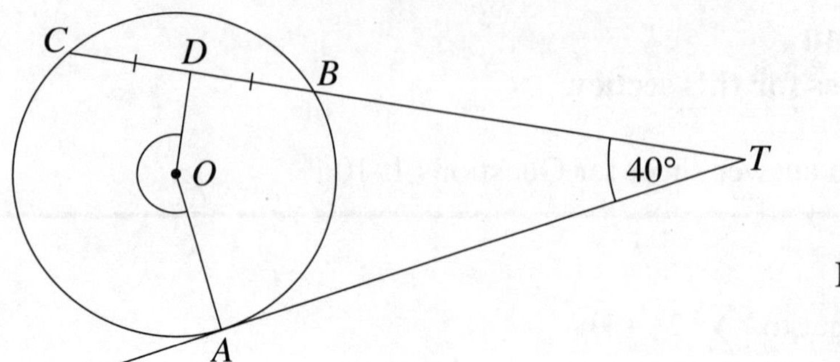

NOT TO SCALE

What is the size of the reflex angle DOA?

(A) 80°

(B) 140°

(C) 220°

(D) 280°

5 Which expression is equal to $\int \sin^2 2x\,dx$?

(A) $\frac{1}{2}\left(x - \frac{1}{4}\sin 4x\right) + c$

(B) $\frac{1}{2}\left(x + \frac{1}{4}\sin 4x\right) + c$

(C) $\frac{\sin^3 2x}{6} + c$

(D) $\frac{-\cos^3 2x}{6} + c$

✗ **6** What is the general solution of the equation $2\sin^2 x - 7\sin x + 3 = 0$?

(A) $n\pi - (-1)^n \frac{\pi}{3}$

(B) $n\pi + (-1)^n \frac{\pi}{3}$

(C) $n\pi - (-1)^n \frac{\pi}{6}$

(D) $n\pi + (-1)^n \frac{\pi}{6}$

7 The displacement x of a particle at time t is given by

$$x = 5\sin 4t + 12\cos 4t.$$

What is the maximum velocity of the particle?

(A) 13

(B) 28

(C) 52

(D) 68

8 A team of 11 students is to be formed from a group of 18 students. Among the 18 students are 3 students who are left-handed.

What is the number of possible teams containing at least 1 student who is left-handed?

(A) 19 448

(B) 30 459

(C) 31 824

(D) 58 344

9 The diagram shows the graph of $y = f(x)$.

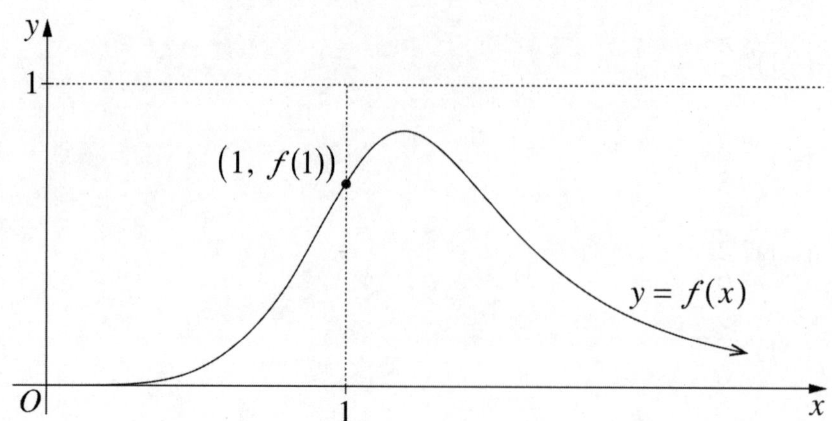

Which of the following is a correct statement?

(A) $f''(1) < f(1) < 1 < f'(1)$

(B) $f''(1) < f'(1) < f(1) < 1$

(C) $f(1) < 1 < f'(1) < f''(1)$

(D) $f'(1) < f(1) < 1 < f''(1)$

10 Consider the polynomial $p(x) = ax^3 + bx^2 + cx - 6$ with a and b positive.

Which graph could represent $p(x)$?

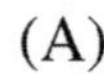
(A)

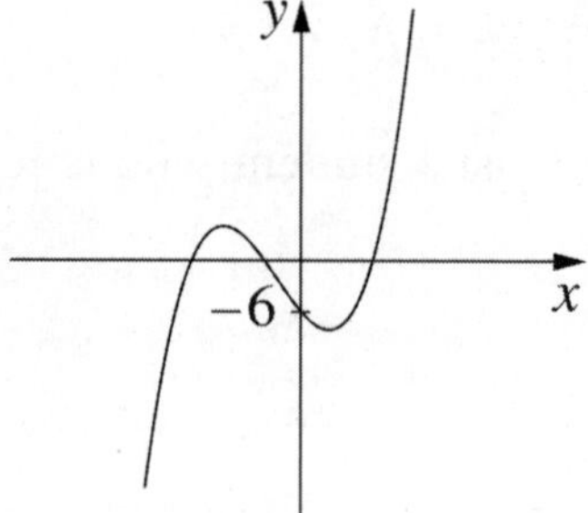

(B)

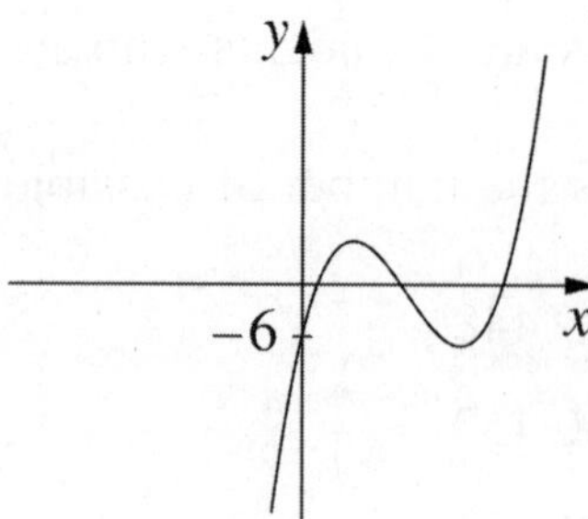

(C)

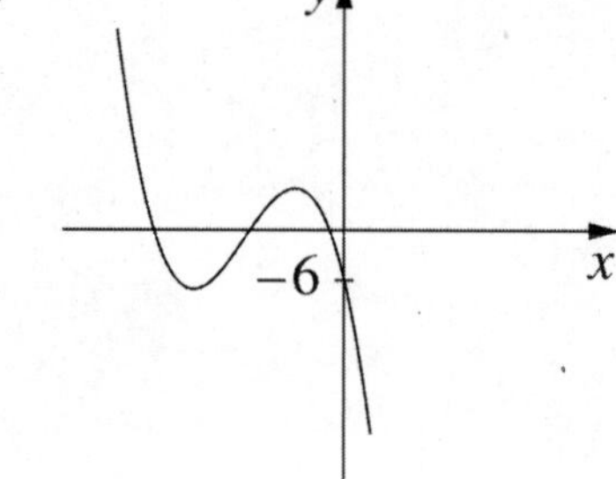

(D)

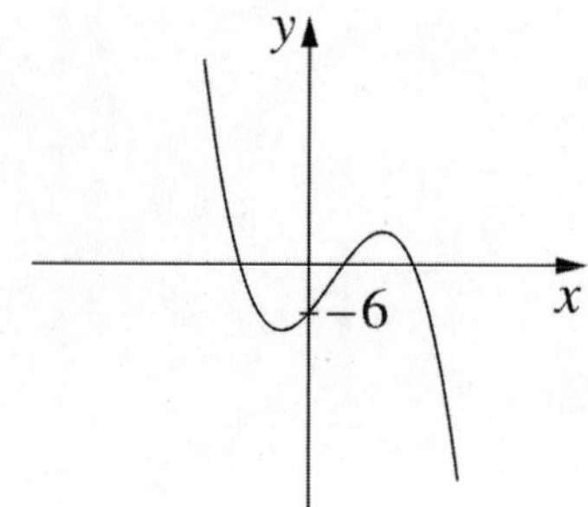

Section II

60 marks
Attempt Questions 11–14
Allow about 1 hour and 45 minutes for this section

Answer each question in a SEPARATE writing booklet. Extra writing booklets are available.

In Questions 11–14, your responses should include relevant mathematical reasoning and/or calculations.

Question 11 (15 marks) Use a SEPARATE writing booklet.

(a) Find the inverse of the function $y = x^3 - 2$. **2**

(b) Use the substitution $u = x - 4$ to find $\int x\sqrt{x-4}\,dx$. **3**

(c) Differentiate $3\tan^{-1}(2x)$. **2**

(d) Evaluate $\lim\limits_{x \to 0}\left(\dfrac{2\sin x\cos x}{3x}\right)$. **2**

(e) Solve $\dfrac{3}{2x+5} - x > 0$. **3**

(f) A darts player calculates that when she aims for the bullseye the probability of her hitting the bullseye is $\dfrac{3}{5}$ with each throw.

(i) Find the probability that she hits the bullseye with exactly one of her first three throws. **1**

(ii) Find the probability that she hits the bullseye with at least two of her first six throws. **2**

Question 12 (15 marks) Use a SEPARATE writing booklet.

(a) The diagram shows a conical soap dispenser of radius 5 cm and height 20 cm.

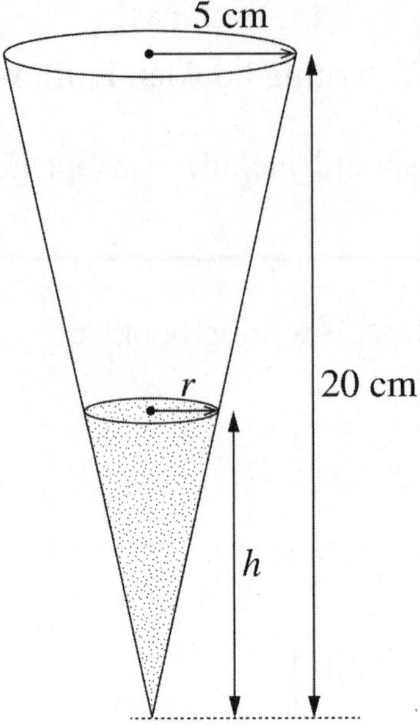

At any time t seconds, the top surface of the soap in the container is a circle of radius r cm and its height is h cm.

The volume of the soap is given by $v = \frac{1}{3}\pi r^2 h$.

(i) Explain why $r = \frac{h}{4}$. **1**

(ii) Show that $\frac{dv}{dh} = \frac{\pi}{16}h^2$. **1**

The dispenser has a leak which causes soap to drip from the container. The area of the circle formed by the top surface of the soap is decreasing at a constant rate of $0.04\ \text{cm}^2\ \text{s}^{-1}$.

(iii) Show that $\frac{dh}{dt} = \frac{-0.32}{\pi h}$. **2**

(iv) What is the rate of change of the volume of the soap, with respect to time, when $h = 10$? **2**

Question 12 continues on the following page

Question 12 (continued)

(b) In a chemical reaction, a compound X is formed from a compound Y. The mass in grams of X and Y are $x(t)$ and $y(t)$ respectively, where t is the time in seconds after the start of the chemical reaction.

Throughout the reaction the sum of the two masses is 500 g.

At any time t, the rate at which the mass of compound X is increasing is proportional to the mass of compound Y.

At the start of the chemical reaction, $x = 0$ and $\frac{dx}{dt} = 2$.

(i) Show that $\frac{dx}{dt} = 0.004(500 - x)$. **3**

(ii) Show that $x = 500 - Ae^{-0.004t}$ satisfies the equation in part (i), and find the value of A. **2**

(c) The graphs of $y = \tan x$ and $y = \cos x$ meet at the point where $x = \alpha$, as shown.

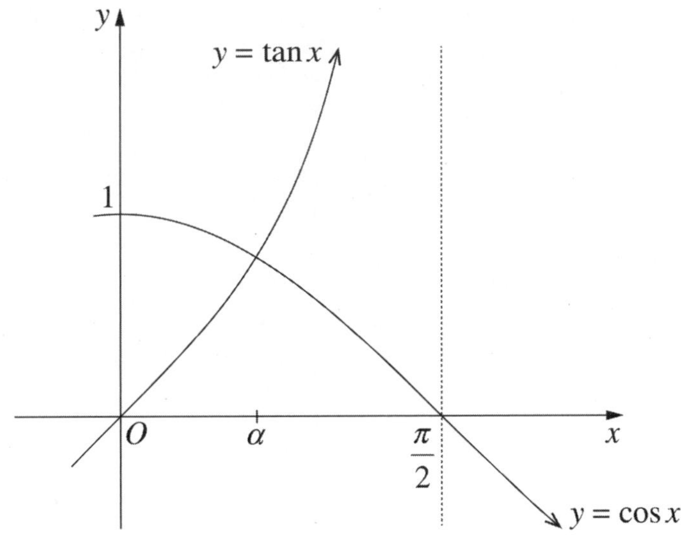

(i) Show that the tangents to the curves at $x = \alpha$ are perpendicular. **2**

(ii) Use one application of Newton's method with $x_1 = 1$ to find an approximate value for α. Give your answer correct to two decimal places. **2**

End of Question 12

Question 13 (15 marks) Use a SEPARATE writing booklet.

(a) The tide can be modelled using simple harmonic motion.

At a particular location, the high tide is 9 metres and the low tide is 1 metre.

At this location the tide completes 2 full periods every 25 hours.

Let t be the time in hours after the first high tide today.

(i) Explain why the tide can be modelled by the function $x = 5 + 4\cos\left(\frac{4\pi}{25}t\right)$. **2**

(ii) The first high tide tomorrow is at 2 am. **2**

What is the earliest time tomorrow at which the tide is increasing at the fastest rate?

Question 13 continues on the following page

Question 13 (continued)

* (b) The trajectory of a projectile fired with speed $u \text{ m s}^{-1}$ at an angle θ to the horizontal is represented by the parametric equations

$$x = ut\cos\theta \quad \text{and} \quad y = ut\sin\theta - 5t^2,$$

where t is the time in seconds.

(i) Prove that the greatest height reached by the projectile is $\dfrac{u^2\sin^2\theta}{20}$. **2**

A ball is thrown from a point 20 m above the horizontal ground. It is thrown with speed 30 m s^{-1} at an angle of 30° to the horizontal. At its highest point the ball hits a wall, as shown in the diagram.

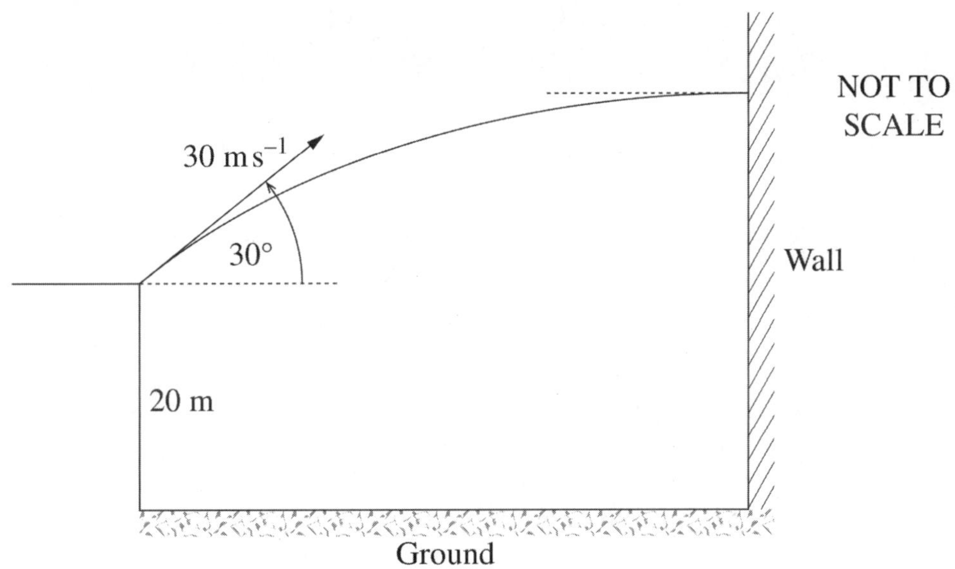

(ii) Show that the ball hits the wall at a height of $\dfrac{125}{4}$ m above the ground. **2**

* In the new Mathematics Extension 1 course, Projectile Motion questions will be expressed in vector form:

$x = Vt\cos\theta$ and $y = Vt\sin\theta - \frac{1}{2}gt^2$ is expressed as a position vector:

$$\underset{\sim}{r}(t) = (Vt\cos\theta)\underset{\sim}{i} + \left(Vt\sin\theta - \frac{1}{2}gt^2\right)\underset{\sim}{j}$$

Question 13 continues on the following page

Question 13 (continued)

The ball then rebounds horizontally from the wall with speed 10 m s^{-1}. You may assume that the acceleration due to gravity is 10 m s^{-2}.

(iii) How long does it take the ball to reach the ground after it rebounds from the wall? **2**

(iv) How far from the wall is the ball when it hits the ground? **1**

(c) The circle centred at O has a diameter AB. From the point M outside the circle the line segments MA and MB are drawn meeting the circle at C and D respectively, as shown in the diagram. The chords AD and BC meet at E. The line segment ME produced meets the diameter AB at F.

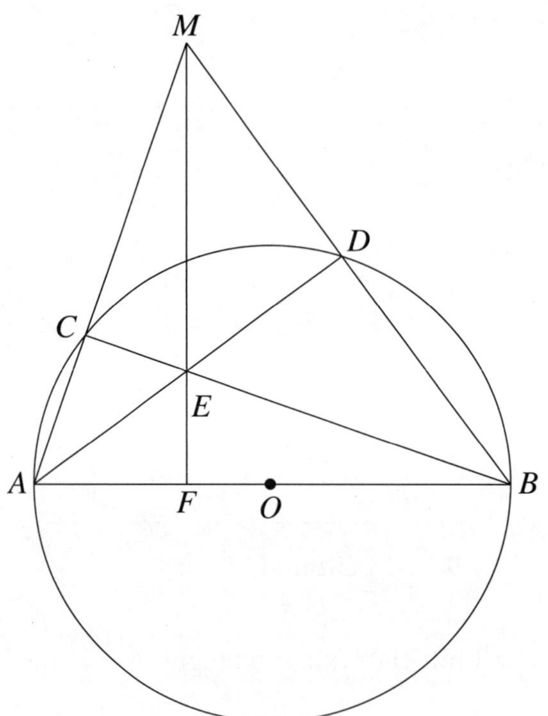

Copy or trace the diagram into your writing booklet.

(i) Show that $CMDE$ is a cyclic quadrilateral. **2**

(ii) Hence, or otherwise, prove that MF is perpendicular to AB. **2**

End of Question 13

Question 14 (15 marks) Use a SEPARATE writing booklet.

(a) (i) Show that $4n^3 + 18n^2 + 23n + 9$ can be written as **1**

$$(n+1)(4n^2+14n+9).$$

(ii) Using the result in part (i), or otherwise, prove by mathematical induction that, for $n \geq 1$, **3**

$$1\times 3+3\times 5+5\times 7+\cdots+(2n-1)(2n+1)=\frac{1}{3}n(4n^2+6n-1).$$

(b) Consider the expansion of $(1+x)^n$, where n is a positive integer.

(i) Show that $2^n = \binom{n}{0}+\binom{n}{1}+\binom{n}{2}+\binom{n}{3}+\cdots+\binom{n}{n}$. **1**

(ii) Show that $n2^{n-1} = \binom{n}{1}+2\binom{n}{2}+3\binom{n}{3}+\cdots+n\binom{n}{n}$. **1**

(iii) Hence, or otherwise, show that $\sum_{r=1}^{n}\binom{n}{r}(2r-n)=n$. **2**

Question 14 continues on the following page

Question 14 (continued)

(c) The point $T(2at, at^2)$ lies on the parabola $\mathcal{P}_1$ with equation $x^2 = 4ay$.

The tangent to the parabola $\mathcal{P}_1$ at T meets the directrix at D.

The normal to the parabola $\mathcal{P}_1$ at T meets the vertical line through D at the point R, as shown in the diagram.

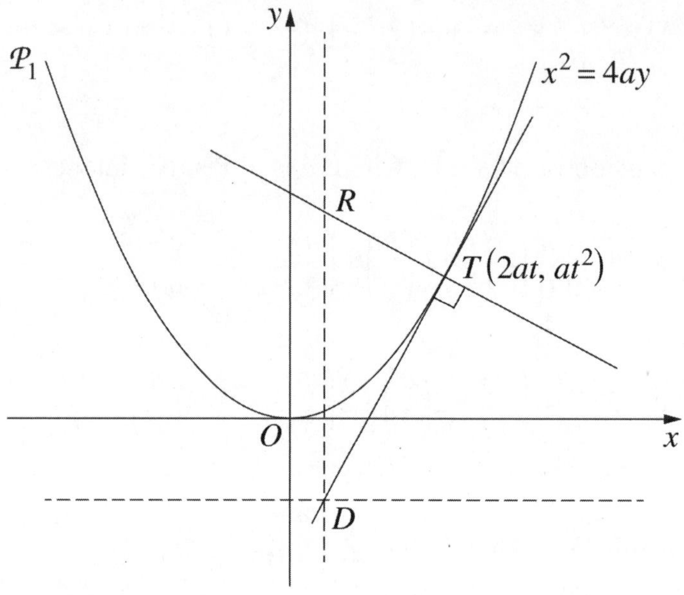

(i) Show that the point D has coordinates $\left(at - \frac{a}{t}, -a\right)$. **1**

(ii) Show that the locus of R lies on another parabola $\mathcal{P}_2$. **3**

(iii) State the focal length of the parabola $\mathcal{P}_2$. **1**

It can be shown that the minimum distance between R and T occurs when the normal to $\mathcal{P}_1$ at T is also the normal to $\mathcal{P}_2$ at R. (Do NOT prove this.)

(iv) Find the values of t so that the distance between R and T is a minimum. **2**

End of paper

Replacement questions

with content from the most up-to-date syllabus

Marks

Question 4 (1 mark)

Find the value of m if the vectors $\underset{\sim}{u} = -2\underset{\sim}{i} + 4\underset{\sim}{j}$ and $\underset{\sim}{v} = (1 - 5m)\underset{\sim}{i} + \underset{\sim}{j}$ are perpendicular. **1**

(A) $-\frac{7}{5}$

(B) $-\frac{1}{5}$

(C) $\frac{1}{5}$

(D) $\frac{7}{5}$

Question 6 (1 mark)

ABC is a triangle. P, Q and R are the midpoints of AB, BC and AC respectively.

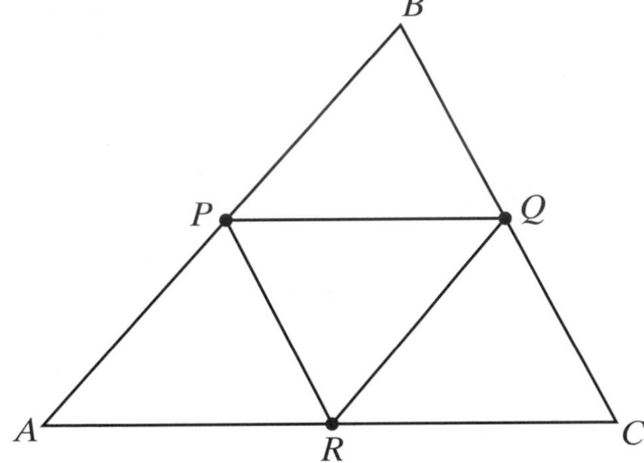

Let $\vec{AB} = \underset{\sim}{a}$, $\vec{BC} = \underset{\sim}{b}$ and $\vec{AC} = \underset{\sim}{c}$.

Which of these is equal to $\vec{QR}$? **1**

(A) $\frac{1}{2}(\underset{\sim}{c} - \underset{\sim}{b})$

(B) $\underset{\sim}{a} - \underset{\sim}{b} + \frac{1}{2}\underset{\sim}{c}$

(C) $\underset{\sim}{a} + \underset{\sim}{b} - \frac{1}{2}\underset{\sim}{c}$

(D) $\frac{1}{2}(\underset{\sim}{b} - \underset{\sim}{c})$

Marks

Question 11 (2 marks)

(d) Show $\cos 3A \cos A = \cos^2 A - \sin^2 2A$ **2**

Question 12 (2 marks)

(c) (ii) The diagram shows the graph of $y = f(x)$.

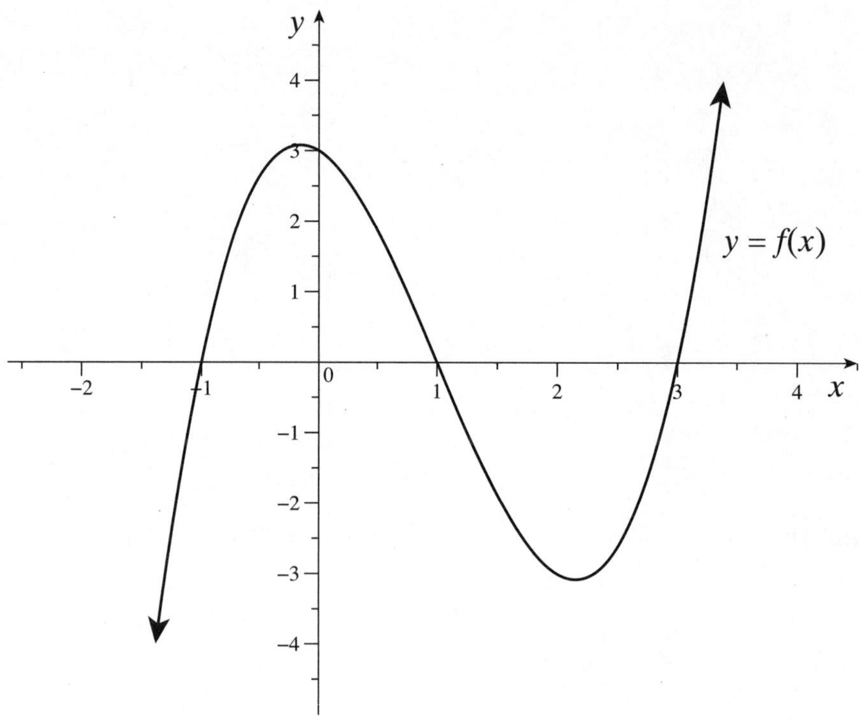

On the number plane below draw the graph of $y = f(\sqrt{x+1})$. **2**

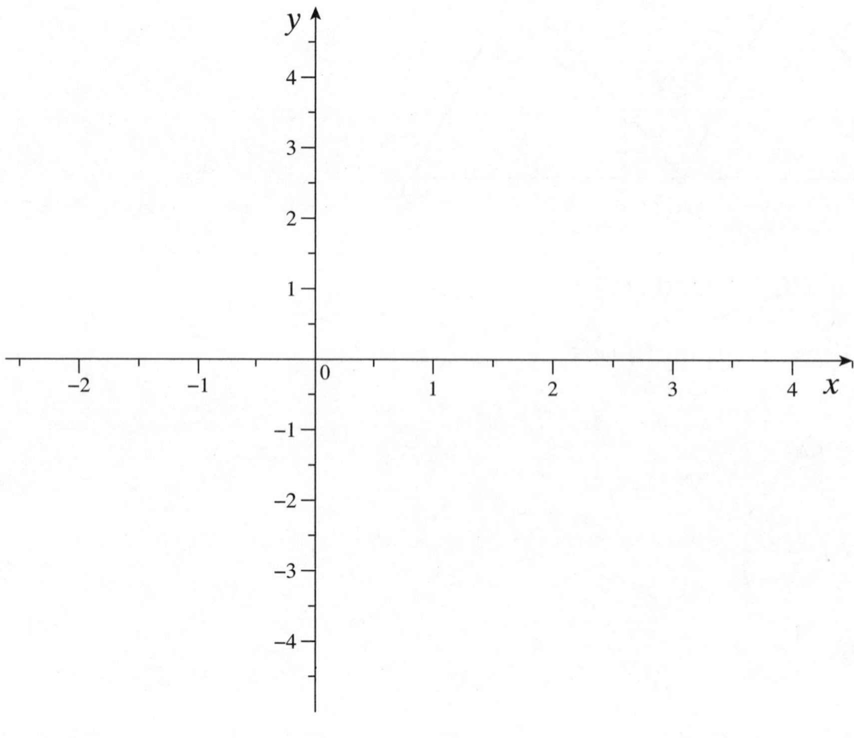

Marks

Question 13 (8 marks)

(a) Find the equation of the normal to the curve represented by the parametric equations $x = \frac{t}{2}$ and $y = 3t^2$ at the point where $t = -1$. **4**

(c) A binomial random variable X has a probability of success of p after 3 trials.

(i) Use the table for the binomial distribution to find the value of p. **1**

x	0	1	2	3
$P(X = x)$	0.064	a	b	0.216

Two values are missing from the table.

(ii) Show that $b = 0.432$. **1**

(iii) Find $P(X \geq 1)$ **1**

(iv) Find the $Var(X)$. **1**

Question 14 (11 marks)

(b) Find the volume of the solid of revolution when $y = \cos 2x \sin x$ between $x = 0$ and $x = \frac{\pi}{6}$ is rotated around the x-axis. **4**

(c)* Consider the differential equation $\frac{dy}{dx} = xy$

If $y = e^2$ when $x = \sqrt{2}$, solve the differential equation, leaving your answer with y in terms of x. **3**

(c)** (i) If $y = 2x \tan^{-1} 2x$, find $\frac{dy}{dx}$. **1**

(ii) Hence, find the area under the curve $y = \tan^{-1} 2x$ between $x = \frac{\sqrt{3}}{2}$ and $x = \frac{1}{2}$. **3**

2016 Higher School Certificate
Worked answers

Section I

(*Total 10 marks*)

1. D	**2.** B	**3.** A	**4.** C	**5.** A
6. D	**7.** C	**8.** B	**9.** A	**10.** A

1. $\sum_{k=1}^{20}(2k+1)$

$= (2 \times 1 + 1) + (2 \times 2 + 1) + (2 \times 3 + 1) + \ldots + (2 \times 20 + 1)$

$= 3 + 5 + 7 + \ldots + 41$

Answer D

2. $P(x) = 2x^3 - 10x^2 + 6x + 2$

$P(2) = 2 \times 2^3 - 10 \times 2^2 + 6 \times 2 + 2$

$= -10$

Answer B

3. $\dfrac{\tan 2x - \tan x}{1 + \tan 2x \tan x} = \tan(2x - x)$

$= \tan x$

Answer A

4. $\angle OAT = 90°$ (angle between tangent and radius)

$\angle ODT = 90°$ (line from the centre of circle to midpoint of chord ⊥ the chord)

So $\angle DOA = 140°$ (angle sum of quadrilateral is 360°)

Reflex angle $DOA = 360° - 140°$

$= 220°$

Answer C

5. $\cos 4x = \cos^2 2x - \sin^2 2x$

$= 1 - 2\sin^2 2x$

So $\sin^2 2x = \frac{1}{2}(1 - \cos 4x)$

$\int \sin^2 2x \, dx = \frac{1}{2}\int (1 - \cos 4x)\, dx$

$= \frac{1}{2}(x - \frac{1}{4}\sin 4x) + c$

Answer A

6. $2\sin^2 x - 7\sin x + 3 = 0$

$(2\sin x - 1)(\sin x - 3) = 0$

$2\sin x = 1$ or $\sin x = 3$ (no solution)

So $\sin x = \frac{1}{2}$

$x = n\pi + (-1)^n \sin^{-1}\left(\frac{1}{2}\right)$

$= n\pi + (-1)^n \frac{\pi}{6}$

Answer D

7. $x = 5\sin 4t + 12\cos 4t$

$= 13\left(\frac{5}{13}\sin 4t + \frac{12}{13}\cos 4t\right)$

$= 13(\cos\alpha \sin 4t + \sin\alpha \cos 4t)$

$= 13\sin(4t + \alpha)$

$\dot{x} = 52\cos(4t + \alpha)$

Now $-1 \le \cos(4t + \alpha) \le 1$

So the maximum velocity is 52.

13
12
α
5

Answer C

8. Total number of possible teams

$= {}^{18}C_{11}$

$= 31\,824$

Number with no left-handed students

$= {}^{15}C_{11}$

$= 1365$

Number of teams with at least 1 left-handed student

$= 31\,824 - 1365$

$= 30\,459$

Answer B

9. When $x = 1$, the gradient is greater than 1.

So $f'(1) > 1$

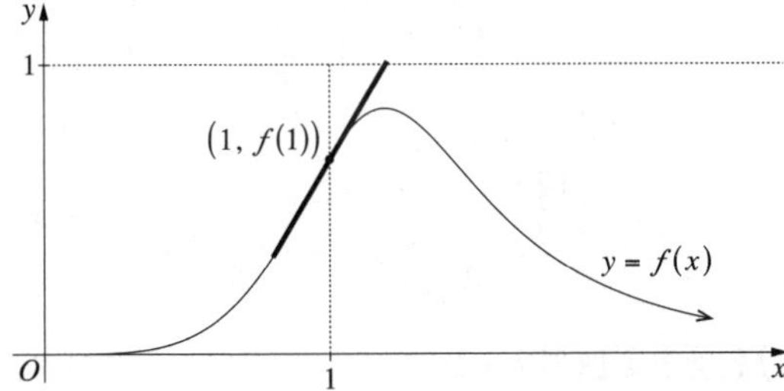

At $x = 1$, the curve changes concavity,

so $f''(1) = 0$

$\therefore\ f''(1) < f(1) < 1 < f'(1)$

Answer A

10. $p(x) = ax^3 + bx^2 + cx - 6$

$a > 0$ so as $x \to \infty$, $p(x) \to \infty$

So options C and D are not correct.

Sum of roots $= -\dfrac{b}{a}$

So the sum of the roots is negative as both a and b are positive.

So option B is not correct.

Answer A

Section II

QUESTION 11

(a) $y = x^3 - 2$

Inverse function is $x = y^3 - 2$

$$y^3 = x + 2$$

$$y = \sqrt[3]{x+2}$$

(2 marks)

(b) $u = x - 4$

$du = dx$

$$\begin{aligned}\int x\sqrt{x-4}\,dx &= \int (u+4)\sqrt{u}\,du \\ &= \int \left(u^{\frac{3}{2}} + 4u^{\frac{1}{2}}\right)du \\ &= \frac{2u^{\frac{5}{2}}}{5} + \frac{8u^{\frac{3}{2}}}{3} + c \\ &= \frac{2(x-4)^2\sqrt{x-4}}{5} + \frac{8(x-4)\sqrt{x-4}}{3} + c \\ &= \frac{2(x-4)\sqrt{x-4}\,(3(x-4)+20)}{15} + c \\ &= \frac{2(3x+8)(x-4)\sqrt{x-4}}{15} + c\end{aligned}$$

(3 marks)

(c) $y = 3\tan^{-1}(2x)$

$$\frac{dy}{dx} = 3 \times \frac{1}{1+(2x)^2} \times 2$$

$$= \frac{6}{1+4x^2}$$

(2 marks)

(d) $$\lim_{x\to 0}\left(\frac{2\sin x\cos x}{3x}\right) = \lim_{x\to 0}\left(\frac{\sin 2x}{3x}\right)$$

$$= \frac{2}{3}\lim_{x\to 0}\left(\frac{\sin 2x}{2x}\right)$$

$$= \frac{2}{3} \times 1$$

$$= \frac{2}{3}$$

(2 marks)

(e) $$\frac{3}{2x+5} - x > 0$$

$$\frac{3}{2x+5} > x$$

If $2x + 5 > 0$

$2x > -5$

$x > -2\frac{1}{2}$

Then $3 > x(2x + 5)$

$3 > 2x^2 + 5x$

$2x^2 + 5x - 3 < 0$

$(2x - 1)(x + 3) < 0$

$-3 < x < \frac{1}{2}$

But $x > -2\frac{1}{2}$

So $-2\frac{1}{2} < x < \frac{1}{2}$

If $2x + 5 < 0$

$x < -2\frac{1}{2}$

and $2x^2 + 5x - 3 > 0$

So $x < -3$

$\therefore x < -3$ or $-2\frac{1}{2} < x < \frac{1}{2}$

(3 marks)

(f) Assuming she aims for the bullseye

(i) $P(\text{bullseye}) = \frac{3}{5}$, $P(\text{no bullseye}) = \frac{2}{5}$

$$P(\text{1 bullseye in 3 throws}) = 3 \times \frac{3}{5} \times \left(\frac{2}{5}\right)^2$$

$$= \frac{36}{125}$$

(1 mark)

(ii) $P(\text{at least 2 bullseyes in 6 throws})$

$= 1 - P(\text{0 bullseyes}) - P(\text{1 bullseye})$

$$= 1 - \left(\frac{2}{5}\right)^6 - 6\left(\frac{2}{5}\right)^5\left(\frac{3}{5}\right)$$

$$= \frac{2997}{3125}$$

(2 marks)

QUESTION 12

(a) (i) The right-angled triangles formed by the radii and height also have a common angle, so are equiangular and similar.

So $\frac{r}{5} = \frac{h}{20}$ (sides are in proportion)

$r = \frac{h}{4}$

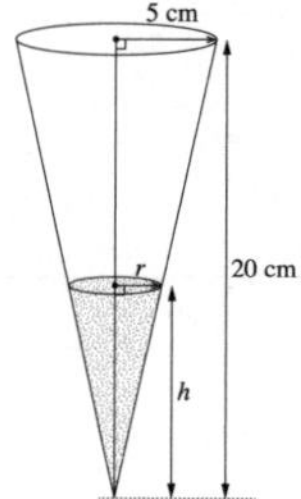

(1 mark)

(ii) $$v = \frac{1}{3}\pi r^2 h$$

$$= \frac{1}{3}\pi\left(\frac{h}{4}\right)^2 h$$

$$= \frac{\pi h^3}{48}$$

So $$\frac{dv}{dh} = \frac{3\pi h^2}{48}$$

$$= \frac{\pi}{16}h^2$$

(1 mark)

(iii) $\frac{dA}{dt} = -0.04$

$$A = \pi r^2$$
$$= \pi\left(\frac{h}{4}\right)^2$$
$$= \frac{\pi h^2}{16}$$
$$\frac{dA}{dh} = \frac{2\pi h}{16}$$
$$= \frac{\pi h}{8}$$
$$\frac{dA}{dt} = \frac{dA}{dh} \cdot \frac{dh}{dt}$$
$$-0.04 = \frac{\pi h}{8} \cdot \frac{dh}{dt}$$

So $\frac{dh}{dt} = -0.04 \times \frac{8}{\pi h}$

$$= \frac{-0.32}{\pi h}$$

(2 marks)

(iv) $\frac{dv}{dt} = \frac{dv}{dh} \cdot \frac{dh}{dt}$

$$= \frac{\pi}{16} h^2 \cdot \frac{-0.32}{\pi h}$$
$$= -0.02h$$

When $h = 10$,

$$\frac{dv}{dt} = -0.02 \times 10 = -0.2$$

The rate of change of the volume of the soap is -0.2 cm^{3}s^{-1}.

(2 marks)

(b) (i) $x + y = 500$

$$y = 500 - x$$

Now $\frac{dx}{dt} = ky$ (k is a constant)

$$= k(500 - x)$$

When $t = 0, x = 0$ and $\frac{dx}{dt} = 2$

So $2 = k(500 - 0)$

$$k = \frac{2}{500}$$
$$= 0.004$$
$$\therefore \frac{dx}{dt} = 0.004(500 - x)$$

(3 marks)

(ii) $x = 500 - Ae^{-0.004t}$

$$\frac{dx}{dt} = 0.004Ae^{-0.004t}$$

But $x = 500 - Ae^{-0.004t}$

So $Ae^{-0.004t} = 500 - x$

$\therefore \frac{dx}{dt} = 0.004(500 - x)$

$\therefore x = 500 - Ae^{-0.004t}$ satisfies the equation in (i)

$x = 500 - Ae^{-0.004t}$

When $t = 0, x = 0$

$0 = 500 - A \times 1$

$A = 500$

(2 marks)

(c) (i) $0 < \alpha < \frac{\pi}{2}$

So $\tan\alpha$, $\cos\alpha$ and $\sin\alpha$ are all positive.

At the point of intersection

$$\tan\alpha = \cos\alpha$$
$$\frac{\sin\alpha}{\cos\alpha} = \cos\alpha$$
$$\sin\alpha = \cos^2\alpha \qquad \text{...(I)}$$
$$y = \tan x$$
$$\frac{dy}{dx} = \sec^2 x$$

At $x = \alpha$, $m_1 = \sec^2\alpha$

$$= \frac{1}{\cos^2\alpha}$$
$$= \frac{1}{\sin\alpha} \qquad \text{from (I)}$$
$$y = \cos x$$
$$\frac{dy}{dx} = -\sin x$$

At $x = \alpha$, $m_2 = -\sin\alpha$

So $m_1 m_2 = \frac{1}{\sin\alpha} \times -\sin\alpha$

$$= -1$$

$\therefore$ the tangents are perpendicular at $x = \alpha$.

(2 marks)

(ii) $f(x) = \tan x - \cos x$

$$f'(x) = \sec^2 x + \sin x$$
$$x_2 = x_1 - \frac{f(x_1)}{f'(x_1)}$$
$$= 1 - \frac{\tan 1 - \cos 1}{\sec^2 1 + \sin 1}$$
$$= 0.761\,633\,97\ldots$$
$$= 0.76 \qquad \text{(2 d.p.)}$$

(2 marks)

QUESTION 13

(a) (i) SHM is of the form $x = b + a \cos nt$.

The high tide is 9 m and low tide 1 m so the centre of motion is $x = 5$.

So $b = 5$

Amplitude $= 9 - 5 = 4$

So $a = 4$

Period = 12.5 h

$$\text{So } \frac{2\pi}{n} = 12.5$$

$$n = \frac{2\pi}{12.5}$$

$$= \frac{4\pi}{25}$$

So $x = 5 + 4\cos\left(\frac{4\pi}{25}t\right)$

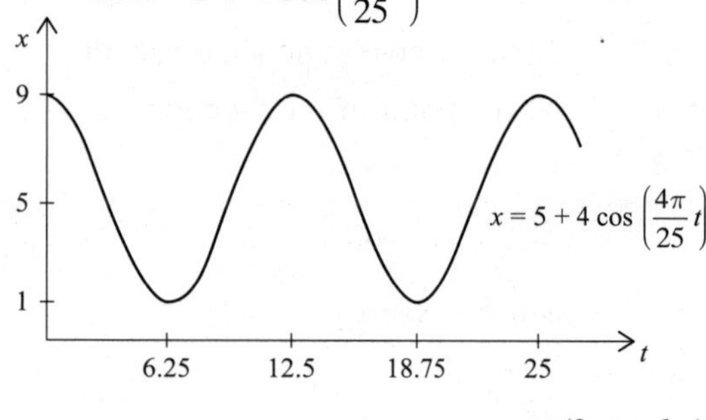

(2 marks)

(ii) The tide increases (or decreases) at the fastest rate at the centre of motion.

So the tide will increase at fastest rate $(12.5 \div 4)$ h or 3.125 h before each high tide.

The first high tide is at 2 am so the earliest time that the tide will increase at the fastest rate will be 3.125 h before the next high tide or $(12.5 - 3.125)$ h $= 9.375$ h after 2 am.

9.375 h = 9 h 22 min 30 s

9 h after 2 am is 11 am so the tide will first increase at its fastest rate at $11.22\frac{1}{2}$ am.

(2 marks)

(b) (i) $y = ut\sin\theta - 5t^2$

$\dot{y} = u\sin\theta - 10t$

The maximum height occurs when $\dot{y} = 0$

i.e. $0 = u\sin\theta - 10t$

$$10t = u\sin\theta$$

$$t = \frac{u\sin\theta}{10}$$

Now $\ddot{y} = -10$ (maximum)

So the maximum height occurs when

$$t = \frac{u\sin\theta}{10}$$

When $t = \frac{u\sin\theta}{10}$,

$$y = \frac{u^2\sin^2\theta}{10} - \frac{5u^2\sin^2\theta}{100}$$

$$= \frac{u^2\sin^2\theta}{20}$$

The greatest height is $\frac{u^2\sin^2\theta}{20}$ m.

(2 marks)

(ii) $u = 30, \theta = 30°$

$$\text{greatest height} = \frac{u^2\sin^2\theta}{20}$$

$$= \frac{30^2\sin^2 30°}{20}$$

$$= 11.25$$

So the ball hits the wall at a point 11.25 m above the starting point.

Height above the ground $= (20 + 11\frac{1}{4})$ m

$$= \frac{125}{4} \text{ m}$$

(2 marks)

(iii) $u = 10, \theta = 0°$

$y = ut\sin\theta - 5t^2$

Ball hits the ground when $y = -\dfrac{125}{4}$

$-\dfrac{125}{4} = 10t \times 0 - 5t^2$

$5t^2 = \dfrac{125}{4}$

$t^2 = \dfrac{25}{4}$

$t = \pm\dfrac{5}{2}$

But $t > 0$,

So the ball reaches the ground 2.5 seconds after it rebounds from the wall.

(2 marks)

(iv) $x = ut\cos\theta$

When $u = 10, t = 2.5$ and $\theta = 0°$

$x = 10 \times 2.5 \times \cos 0°$

$= 25$

So the ball is 25 m from the wall when it hits the ground.

(1 mark)

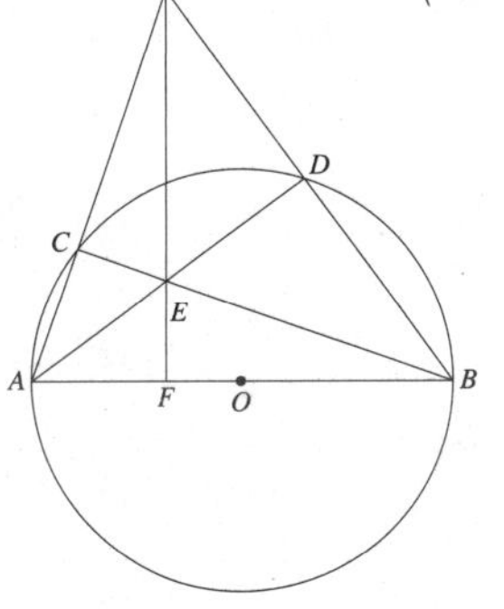

(c) (i) $\angle ADB = 90°$ (angle in a semicircle)

So $\angle MDE = 90°$ (MDB is straight)

$\angle ACB = 90°$ (angle in a semicircle)

$\angle MCE = 90°$ (MCA is straight)

So $\angle MDE + \angle MCE = 180°$

and $\angle CMD + \angle CED = 180°$ (angle sum of quad. is 360°)

$\therefore$ $CMDE$ is a cyclic quadrilateral (opposite angles supplementary).

(2 marks)

(ii) Join CD.

$\angle EMD = \angle ECD$ (angles in same segment on same arc)

$\angle BAD = \angle BCD$ (angles in same segment on same arc)

$\therefore \angle BAD = \angle EMD$

$\angle MED = \angle AEF$ (vert. opp.)

So $\angle AFE = \angle MDE$ (remaining angles equiangular triangles)

But $\angle MDE = 90°$

So $\angle AFE = 90°$

$\therefore MF \perp AB$

(2 marks)

QUESTION 14

(a) (i) $(n+1)(4n^2+14n+9)$

$= n(4n^2+14n+9) + 1(4n^2+14n+9)$

$= 4n^3+14n^2+9n+4n^2+14n+9$

$= 4n^3+18n^2+23n+9$

So $4n^3+18n^2+23n+9$ can be written as $(n+1)(4n^2+14n+9)$.

(1 mark)

(ii) We want to prove, for $n \geq 1$, that

$1\times 3+3\times 5+5\times 7+\ldots+(2n-1)(2n+1)=\frac{1}{3}n(4n^2+6n-1)$

If $n = 1$,

LHS $= 1\times 3 = 3$

RHS $= \frac{1}{3}\times 1\times(4\times 1^2+6\times 1-1)=3$

$\therefore$ it is true for $n = 1$.

Assume true for $n = k$,

i.e. assume $1\times 3+3\times 5+5\times 7+\ldots+(2k-1)(2k+1)=\frac{1}{3}k(4k^2+6k-1)$

When $n = k + 1$,

$$\begin{aligned}\text{LHS} &= 1\times 3+3\times 5+\ldots+(2k-1)(2k+1)+(2(k+1)-1)(2(k+1)+1)\\ &= \frac{1}{3}k(4k^2+6k-1)+(2k+1)(2k+3)\\ &= \frac{1}{3}(4k^3+6k^2-k)+4k^2+8k+3\\ &= \frac{1}{3}(4k^3+6k^2-k+12k^2+24k+9)\\ &= \frac{1}{3}(4k^3+18k^2+23k+9)\\ &= \frac{1}{3}(k+1)(4k^2+14k+9) \quad \text{(from part (i))}\end{aligned}$$

$$\begin{aligned}\text{RHS} &= \frac{1}{3}(k+1)(4(k+1)^2+6(k+1)-1)\\ &= \frac{1}{3}(k+1)(4k^2+8k+4+6k+6-1)\\ &= \frac{1}{3}(k+1)(4k^2+14k+9)\\ &= \text{LHS}\end{aligned}$$

So, if true for $n = k$ it is also true for $n = k + 1$.

It is true for $n = 1$, so it is true for $n = 2$ and so for $n = 3$ and so on.

By the process of induction it is true for all $n \geq 1$.

(3 marks)

(b) (i) $(1+x)^n = \binom{n}{0}+\binom{n}{1}x+\binom{n}{2}x^2+\binom{n}{3}x^3+\ldots+\binom{n}{n}x^n$

When $x = 1$,

$(1+1)^n = \binom{n}{0}+\binom{n}{1}1+\binom{n}{2}1^2+\binom{n}{3}1^3+\ldots+\binom{n}{n}1^n$

$2^n = \binom{n}{0}+\binom{n}{1}+\binom{n}{2}+\binom{n}{3}+\ldots+\binom{n}{n}$

(1 mark)

(ii) $(1+x)^n = \binom{n}{0} + \binom{n}{1}x + \binom{n}{2}x^2 + \binom{n}{3}x^3 + \ldots + \binom{n}{n}x^n$

Differentiating:

$$n(1+x)^{n-1} = \binom{n}{1} + \binom{n}{2}2x + \binom{n}{3}3x^2 + \ldots + \binom{n}{n}nx^{n-1}$$

When $x = 1$,

$$n(1+1)^{n-1} = \binom{n}{1} + \binom{n}{2}2\times1 + \binom{n}{3}3\times1^2 + \ldots + \binom{n}{n}n\times1^{n-1}$$

$$n2^{n-1} = \binom{n}{1} + 2\binom{n}{2} + 3\binom{n}{3} + \ldots + n\binom{n}{n}$$

(1 mark)

(iii) $\sum_{r=1}^{n}\binom{n}{r}(2r-n)$

$$= \binom{n}{1}(2-n) + \binom{n}{2}(4-n) + \binom{n}{3}(6-n) + \ldots + \binom{n}{n}(2n-n)$$

$$= 2\binom{n}{1} - n\binom{n}{1} + 4\binom{n}{2} - n\binom{n}{2} + 6\binom{n}{3} - n\binom{n}{3} + \ldots + 2n\binom{n}{n} - n\binom{n}{n}$$

$$= 2\left(\binom{n}{1} + 2\binom{n}{2} + 3\binom{n}{3} + \ldots + n\binom{n}{n}\right) - n\left(\binom{n}{1} + \binom{n}{2} + \binom{n}{3} + \ldots + \binom{n}{n}\right)$$

$$= 2\left(n2^{n-1}\right) - n\left(2^n - \binom{n}{0}\right)$$

$$= n2^n - n2^n + n \times 1$$

$$= n$$

(2 marks)

(c) (i) $x^2 = 4ay$

Directrix: $y = -a$

Tangent at T: $y = tx - at^2$

When $y = -a$

$$-a = tx - at^2$$

$$tx = at^2 - a$$

$$x = at - \frac{a}{t} \qquad (t \neq 0)$$

So D is the point $\left(at - \frac{a}{t}, -a\right)$

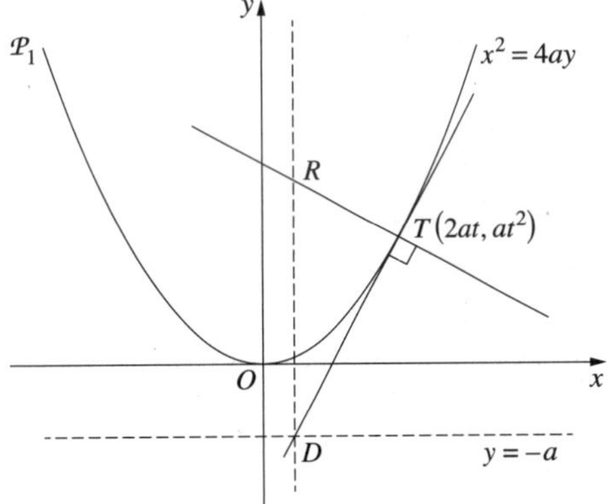

(1 mark)

(ii) Normal at T: $x + ty = at^3 + 2at$

At R, $x = at - \frac{a}{t}$

$$at - \frac{a}{t} + ty = at^3 + 2at$$
$$ty = at^3 + at + \frac{a}{t}$$
$$y = at^2 + a + \frac{a}{t^2}$$

Now
$$x^2 = \left(at - \frac{a}{t}\right)^2$$
$$= a^2t^2 - 2a^2 + \frac{a^2}{t^2}$$
$$= a\left(at^2 - 2a + \frac{a}{t^2}\right)$$
$$= a\left(at^2 + a + \frac{a}{t^2} - 3a\right)$$
$$= a(y - 3a)$$

This is the equation of a parabola.

The locus of R lies on the parabola P_2: $x^2 = a(y - 3a)$.

(3 marks)

(iii) $x^2 = a(y - 3a)$
$$= 4\frac{a}{4}(y - 3a)$$

The focal length is $\frac{a}{4}$ units.

(1 mark)

(iv)
$$x^2 = a(y - 3a)$$
$$y - 3a = \frac{x^2}{a}$$
$$y = \frac{x^2}{a} + 3a$$
$$\frac{dy}{dx} = \frac{2x}{a}$$

At R, $\frac{dy}{dx} = \frac{2}{a}\left(at - \frac{a}{t}\right)$
$$= 2t - \frac{2}{t}$$

So the gradient of the tangent to P_2 at R is $2t - \frac{2}{t}$

The gradient of the normal is $\frac{-1}{2t - \frac{2}{t}}$

The gradient of the tangent to P_1 at T is t.

So the gradient of the normal is $\frac{-1}{t}$

But the normal to P_1 at T is also the normal to P_2 at R

So
$$\frac{-1}{t} = \frac{-1}{2t - \frac{2}{t}}$$
$$2t - \frac{2}{t} = t$$
$$t = \frac{2}{t}$$
$$t^2 = 2$$
$$t = \pm\sqrt{2}$$

The distance between R and T will be a minimum when $t = \pm\sqrt{2}$.

(2 marks)

Solutions to replacement questions

QUESTION 4

$$-2(1-5m)+4\times 1=0$$
$$-2+10m+4=0$$
$$10m+2=0$$
$$m=-\frac{1}{5}$$

Answer B *(1 mark)*

QUESTION 6

As $\vec{BC}=\underset{\sim}{b}$, then $\vec{QC}=\frac{1}{2}\underset{\sim}{b}$.

As $\vec{AC}=\underset{\sim}{c}$, then $\vec{CA}=-\underset{\sim}{c}$, and

hence, $\vec{CR}=-\frac{1}{2}\underset{\sim}{c}$.

Using $\vec{QR}=\vec{QC}+\vec{CR}$
$$=\frac{1}{2}\underset{\sim}{b}+\left(-\frac{1}{2}\underset{\sim}{c}\right)$$
$$=\frac{1}{2}(\underset{\sim}{b}-\underset{\sim}{c})$$

Answer D *(1 mark)*

QUESTION 11

(d) LHS $=\cos 3A\cos A$
$$=\frac{1}{2}[\cos 2A+\cos 4A]$$
$$=\frac{1}{2}[2\cos^2 A-1+1-2\sin^2 2A]$$
$$=\frac{1}{2}[2\cos^2 A-2\sin^2 2A]$$
$$=\cos^2 A-2\sin^2 2A$$
$=$ RHS *(2 marks)*

QUESTION 12

(c) (ii)

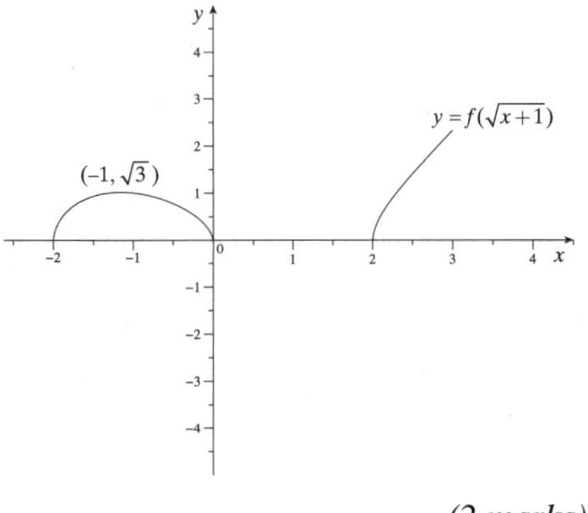

$(-1, \sqrt{3})$

$y=f(\sqrt{x+1})$

(2 marks)

QUESTION 13

(a) $x=\frac{t}{2}$

$\therefore t=2x$

Substitute into y:

$y=3(2x)^2$

$y=12x^2$

Also, when $t=-1$, the point is $\left(-\frac{1}{2},3\right)$.

Now, $\frac{dy}{dx}=24x$
$$\frac{dy}{dx}\left(-\frac{1}{2}\right)=24\left(-\frac{1}{2}\right)$$
$$=-12$$

$\therefore$ gradient of normal is $\frac{1}{12}$.

Equation of normal: $y-3=\frac{1}{12}\left(x+\frac{1}{2}\right)$
$$12y-36=x+\frac{1}{2}$$
$$24y-72=2x+1$$
$$2x-24y-73=0$$

$\therefore$ the equation of the normal

$2x-24y+73=0$.

(4 marks)

(c) (i) $P(X=3)=\binom{3}{3}p^3(1-p)^0=0.216$
$$p^3=0.216$$
$$p=0.6$$

(1 mark)

(ii) $P(X=2)=\binom{3}{2}0.6^2(1-0.6)$
$$=0.432$$

(1 mark)

(iii) $P(X>1)=P(X=2)+P(X=3)$
$$=0.432+0.216$$
$$=0.648$$

(1 mark)

(iv) $E(X)=\mu=np$
$$=3\times 0.6$$
$$=1.8$$

$Var(X)=np(1-p)$
$$=1.8(1-0.6)$$
$$=0.72$$

(1 mark)

QUESTION 14

(b)
$$y = \cos 2x \sin x$$
$$= \frac{1}{2}(\sin 3x - \sin x)$$
$$\therefore y^2 = \frac{1}{4}(\sin^2 3x - 2\sin 3x \sin x + \sin^2 x)$$
$$= \frac{1}{4}(\frac{1}{2}(1 - \cos 6x) + \cos 4x - \cos 2x + \frac{1}{2}(1 - \cos 2x))$$
$$= \frac{1}{4}(\frac{1}{2} - \frac{1}{2}\cos 6x + \cos 4x - \cos 2x + \frac{1}{2} - \frac{1}{2}\cos 2x)$$
$$= \frac{1}{4}(1 - \frac{1}{2}\cos 6x + \cos 4x - \frac{3}{2}\cos 2x)$$

$$\text{Volume} = \pi \int_a^b y^2\, dx$$
$$= \frac{\pi}{4}\int_0^{\frac{\pi}{6}} \left(1 - \frac{1}{2}\cos 6x + \cos 4x - \frac{3}{2}\cos 2x\right) dx$$
$$= \frac{\pi}{4}\left[x - \frac{1}{12}\sin 6x + \frac{1}{4}\sin 4x - \frac{3}{4}\sin 2x\right]_0^{\frac{\pi}{6}}$$
$$= \frac{\pi}{4}\left[\frac{\pi}{6} - \frac{1}{12}\sin 6\left(\frac{\pi}{6}\right) + \frac{1}{4}\sin 4\left(\frac{\pi}{6}\right) - \frac{3}{4}\sin 2\left(\frac{\pi}{6}\right) - 0\right]$$
$$= \frac{\pi}{4}\left[\frac{\pi}{6} - 0 + \frac{1}{4}\left(\frac{\sqrt{3}}{2}\right) - \frac{3}{4}\left(\frac{\sqrt{3}}{2}\right)\right]$$
$$= \frac{\pi}{4}\left[\frac{\pi}{6} + \frac{\sqrt{3}}{8} - \frac{3\sqrt{3}}{8}\right]$$
$$= \frac{\pi}{4}\left[\frac{\pi}{6} - \frac{\sqrt{3}}{4}\right]$$
$$= \frac{\pi}{48}\left[2\pi - 3\sqrt{3}\right]$$

(4 marks)

(c)*
$$\frac{dy}{dx} = xy$$
$$\int \frac{1}{y}\, dy = \int x\, dx$$
$$\ln|y| = \frac{x^2}{2} + c$$
$$y = e^{\frac{x^2}{2} + c}$$

Substitute $y = e^2, x = \sqrt{2}$:
$$e^2 = e^{\frac{(\sqrt{2})^2}{2} + c}$$
$$e^2 = e^{1+c}$$
$$c + 1 = 2$$
$$\therefore c = 1$$
$$\therefore y = e^{\frac{x^2}{2} + 1}$$

(3 marks)

(c)** (i)
$$y = 2x \tan^{-1} 2x$$
$$\frac{dy}{dx} = 2\tan^{-1} 2x + 2x.\frac{1}{1+4x^2}.2$$
$$= 2\tan^{-1} 2x + \frac{4x}{1+4x^2}$$

(1 mark)

(ii) $$\int_{\frac{1}{2}}^{\frac{\sqrt{3}}{2}} \left(2\tan^{-1}2x + \frac{4x}{1+4x^2}\right)dx = \left[2x\tan^{-1}2x\right]_{\frac{1}{2}}^{\frac{\sqrt{3}}{2}}$$

$$\int_{\frac{1}{2}}^{\frac{\sqrt{3}}{2}} \left(\tan^{-1}2x + \frac{2x}{1+4x^2}\right)dx = \left[x\tan^{-1}2x\right]_{\frac{1}{2}}^{\frac{\sqrt{3}}{2}}$$

$$\therefore \int_{\frac{1}{2}}^{\frac{\sqrt{3}}{2}} \tan^{-1}2x\,dx = \left[x\tan^{-1}2x\right]_{\frac{1}{2}}^{\frac{\sqrt{3}}{2}} - \int_{\frac{1}{2}}^{\frac{\sqrt{3}}{2}} \frac{2x}{1+4x^2}dx$$

$$= \left[\frac{\sqrt{3}}{2}\tan^{-1}2\left(\frac{\sqrt{3}}{2}\right) - \frac{1}{2}\tan^{-1}2\left(\frac{1}{2}\right)\right] - \frac{1}{4}\left[\log_e(1+4x^2)\right]_{\frac{1}{2}}^{\frac{\sqrt{3}}{2}}$$

$$= \left[\frac{\sqrt{3}}{2}\tan^{-1}\sqrt{3} - \frac{1}{2}\tan^{-1}1\right] - \frac{1}{4}\left[\log_e\left(1+4\left(\frac{\sqrt{3}}{2}\right)^2\right) - \log_e\left(1+4\left(\frac{1}{2}\right)^2\right)\right]$$

$$= \left[\frac{\sqrt{3}}{2}\left(\frac{\pi}{3}\right) - \frac{1}{2}\left(\frac{\pi}{4}\right)\right] - \frac{1}{4}\left[\log_e(1+3) - \log_e(1+1)\right]$$

$$= \frac{\pi\sqrt{3}}{6} - \frac{\pi}{8} - \frac{1}{4}\log\frac{4}{2}$$

$$= \frac{\pi\sqrt{3}}{6} - \frac{\pi}{8} - \frac{1}{4}\log_e 2$$

$\therefore$ an area of $\left(\frac{\pi\sqrt{3}}{6} - \frac{\pi}{8} - \frac{1}{4}\log_e 2\right)$ units^2.

(3 marks)

2017 HIGHER SCHOOL CERTIFICATE EXAMINATION

Mathematics Extension 1

General Instructions

- Reading time – 5 minutes
- Working time – 2 hours
- Write using black pen
- NESA approved calculators may be used
- A reference sheet is provided at the back of this paper
- In Questions 11–14, show relevant mathematical reasoning and/or calculations

Total marks: 70

Section I – 10 marks

- Attempt Questions 1–10
- Allow about 15 minutes for this section

Section II – 60 marks

- Attempt Questions 11–14
- Allow about 1 hour and 45 minutes for this section

Section I

10 marks
Attempt Questions 1–10
Allow about 15 minutes for this section

Use the multiple-choice answer sheet for Questions 1–10.

---

1 Which polynomial is a factor of $x^3 - 5x^2 + 11x - 10$?

A. $x - 2$

B. $x + 2$

C. $11x - 10$

D. $x^2 - 5x + 11$

2 It is given that $\log_a 8 = 1.893$, correct to 3 decimal places.

What is the value of $\log_a 4$, correct to 2 decimal places?

A. 0.95

B. 1.26

C. 1.53

D. 2.84

✗ **3** The points A, B, C and D lie on a circle and the tangents at A and B meet at T, as shown in the diagram.

The angles BDA and BCD are 65° and 110° respectively.

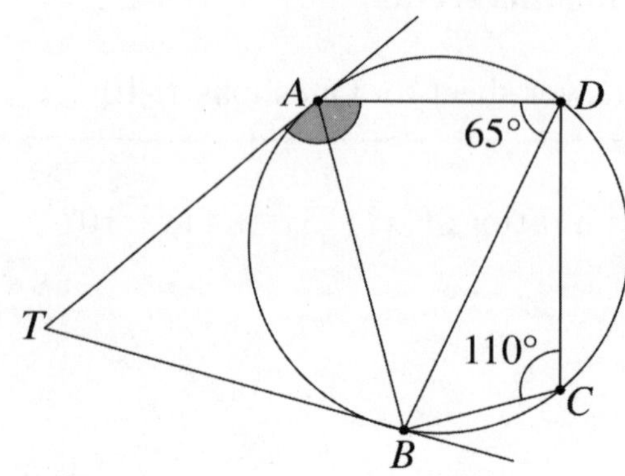

What is the value of $\angle TAD$?

A. 130°

B. 135°

C. 155°

D. 175°

4 What is the value of $\tan\alpha$ when the expression $2\sin x - \cos x$ is written in the form $\sqrt{5}\sin(x-\alpha)$?

A. -2

B. $-\dfrac{1}{2}$

C. $\dfrac{1}{2}$

D. 2

5 Which graph best represents the function $y = \dfrac{2x^2}{1-x^2}$?

A.

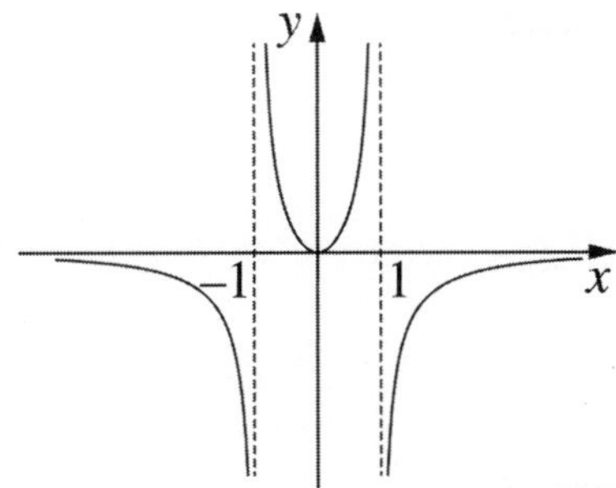

B.

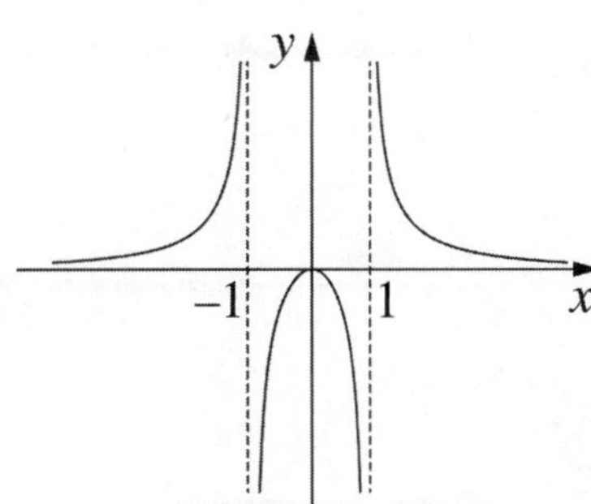

C.

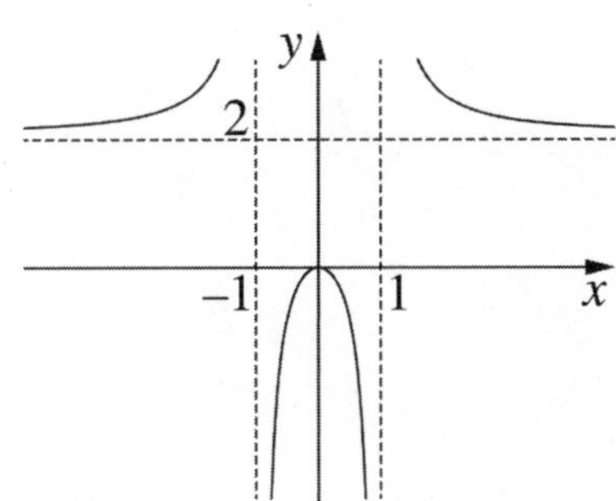

D.

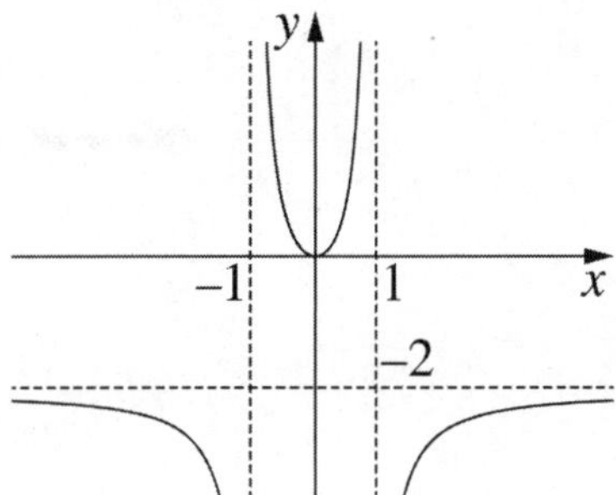

6 The point $P\left(\dfrac{2}{p}, \dfrac{1}{p^2}\right)$, where $p \neq 0$, lies on the parabola $x^2 = 4y$.

What is the equation of the normal at P?

A. $py - x = -p$

B. $p^2y + px = -1$

C. $p^2y - p^3x = 1 - 2p^2$

D. $p^2y + p^3x = 1 + 2p^2$

7 Which diagram represents the domain of the function $f(x) = \sin^{-1}\left(\frac{3}{x}\right)$?

A. 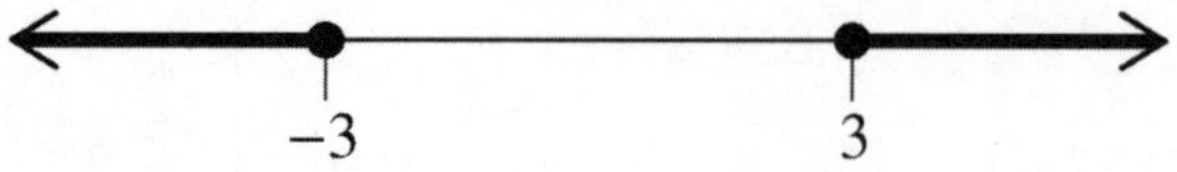

B. 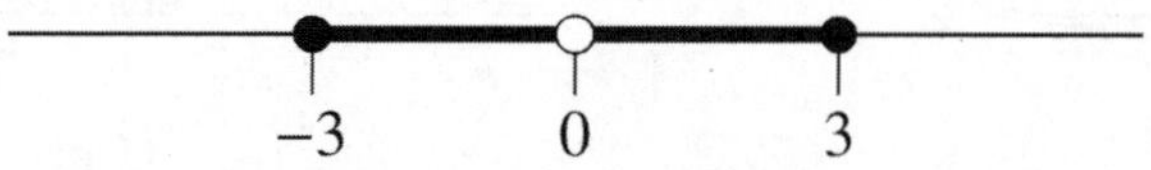

C. 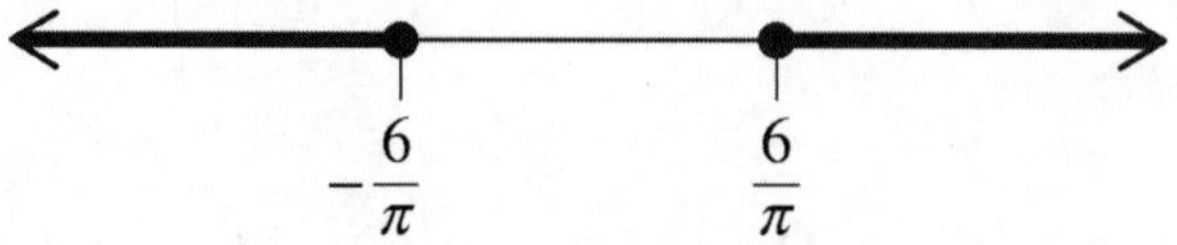

D. 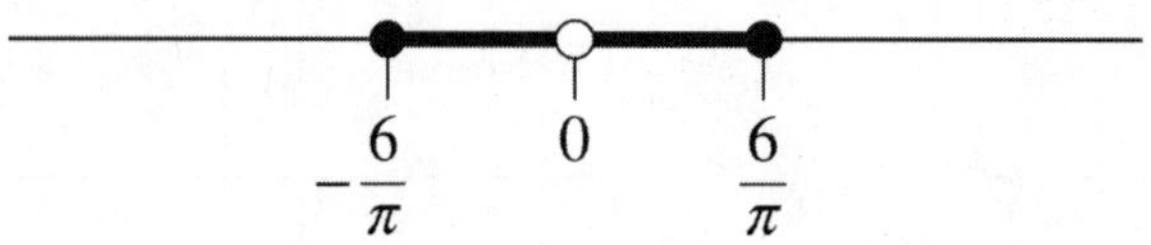

8 A stone drops into a pond, creating a circular ripple. The radius of the ripple increases from 0 cm, at a constant rate of 5 cm s^{-1}.

At what rate is the area enclosed within the ripple increasing when the radius is 15 cm?

A. 25π cm^2 s^{-1}

B. 30π cm^2 s^{-1}

C. 150π cm^2 s^{-1}

D. 225π cm^2 s^{-1}

9 When expanded, which expression has a non-zero constant term?

A. $\left(x + \frac{1}{x^2}\right)^7$

B. $\left(x^2 + \frac{1}{x^3}\right)^7$

C. $\left(x^3 + \frac{1}{x^4}\right)^7$

D. $\left(x^4 + \frac{1}{x^5}\right)^7$

10 Three squares are chosen at random from the 3×3 grid below, and a cross is placed in each chosen square.

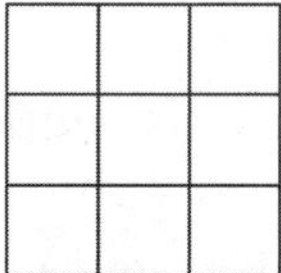

What is the probability that all three crosses lie in the same row, column or diagonal?

A. $\frac{1}{28}$

B. $\frac{2}{21}$

C. $\frac{1}{3}$

D. $\frac{8}{9}$

Section II

60 marks
Attempt Questions 11–14
Allow about 1 hour and 45 minutes for this section

Answer each question in a SEPARATE writing booklet. Extra writing booklets are available.

In Questions 11–14, your responses should include relevant mathematical reasoning and/or calculations.

Question 11 (15 marks) Use a SEPARATE writing booklet.

(a) The point P divides the interval from $A(-4, -4)$ to $B(1, 6)$ internally in the ratio $2:3$. **1**

Find the x-coordinate of P.

(b) Differentiate $\tan^{-1}(x^3)$. **2**

(c) Solve $\dfrac{2x}{x+1} > 1$. **3**

(d) Sketch the graph of the function $y = 2\cos^{-1}x$. **2**

(e) Evaluate $\displaystyle\int_0^3 \frac{x}{\sqrt{x+1}}\,dx$, using the substitution $x = u^2 - 1$. **3**

(f) Find $\displaystyle\int \sin^2 x \cos x \, dx$. **1**

Question 11 continues on the following page

Question 11 (continued)

(g) The probability that a particular type of seedling produces red flowers is $\frac{1}{5}$. Eight of these seedlings are planted.

(i) Write an expression for the probability that exactly three of the eight seedlings produce red flowers. **1**

(ii) Write an expression for the probability that none of the eight seedlings produces red flowers. **1**

(iii) Write an expression for the probability that at least one of the eight seedlings produces red flowers. **1**

End of Question 11

Question 12 (15 marks) Use a SEPARATE writing booklet.

(a) The points A, B and C lie on a circle with centre O, as shown in the diagram. The size of $\angle AOC$ is 100°. **2**

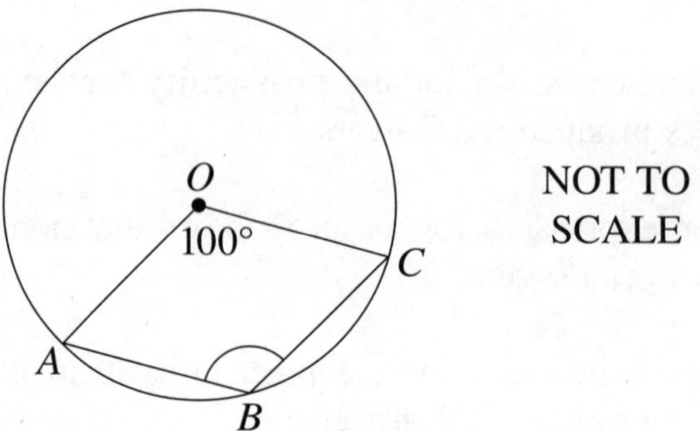

Find the size of $\angle ABC$, giving reasons.

(b) (i) Carefully sketch the graphs of $y = |x+1|$ and $y = 3 - |x-2|$ on the same axes, showing all intercepts. **3**

(ii) Using the graphs from part (i), or otherwise, find the range of values of x for which **1**

$$|x+1| + |x-2| = 3.$$

Question 12 continues on the following page

Question 12 (continued)

(c) The region enclosed by the semicircle $y = \sqrt{1 - x^2}$ and the x-axis is to be divided into two pieces by the line $x = h$, where $0 \le h < 1$.

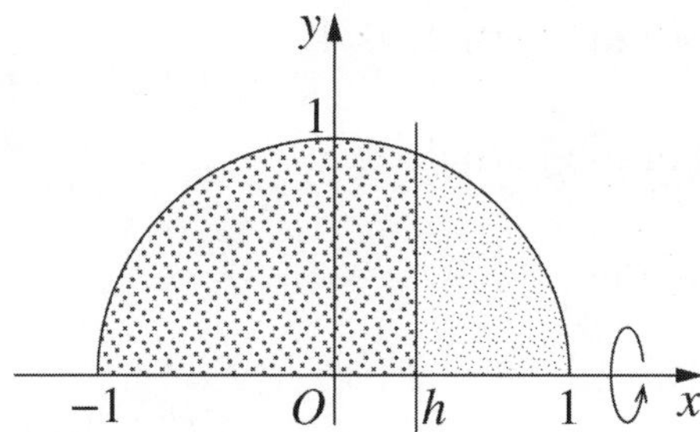

The two pieces are rotated about the x-axis to form solids of revolution. The value of h is chosen so that the volumes of the solids are in the ratio 2:1.

(i) Show that h satisfies the equation $3h^3 - 9h + 2 = 0$. **3**

(ii) Given $h_1 = 0$ as the first approximation for h, use one application of Newton's method to find a second approximation for h. **1**

(d) At time t the displacement, x, of a particle satisfies $t = 4 - e^{-2x}$. **3**

Find the acceleration of the particle as a function of x.

(e) Evaluate $\lim_{x \to 0} \frac{1 - \cos 2x}{x^2}$. **2**

End of Question 12

Question 13 (15 marks) Use a SEPARATE writing booklet.

(a) A particle is moving along the x-axis in simple harmonic motion centred at the origin. **3**

When $x = 2$ the velocity of the particle is 4.

When $x = 5$ the velocity of the particle is 3.

Find the period of the motion.

(b) Let n be a positive EVEN integer.

(i) Show that $(1+x)^n + (1-x)^n = 2\left[\binom{n}{0} + \binom{n}{2}x^2 + \cdots + \binom{n}{n}x^n\right]$. **2**

(ii) Hence show that **1**

$$n\left[(1+x)^{n-1} - (1-x)^{n-1}\right] = 2\left[2\binom{n}{2}x + 4\binom{n}{4}x^3 + \cdots + n\binom{n}{n}x^{n-1}\right].$$

(iii) Hence show that $\binom{n}{2} + 2\binom{n}{4} + 3\binom{n}{6} + \cdots + \frac{n}{2}\binom{n}{n} = n2^{n-3}$. **2**

Question 13 continues on the following page

Question 13 (continued)

* (c) A golfer hits a golf ball with initial speed V m s^{-1} at an angle θ to the horizontal. The golf ball is hit from one side of a lake and must have a horizontal range of 100 m or more to avoid landing in the lake.

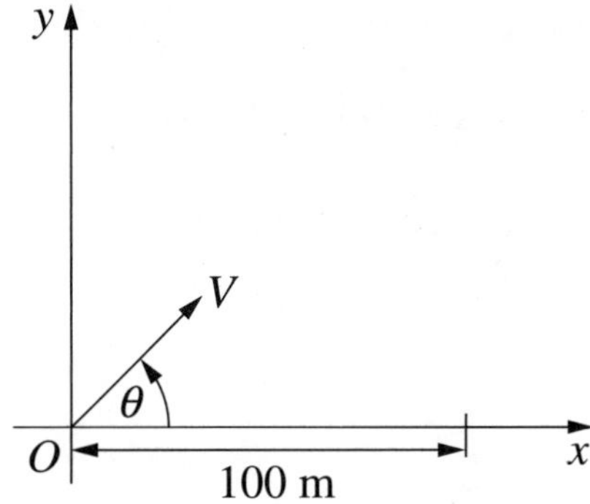

Neglecting the effects of air resistance, the equations describing the motion of the ball are

$$x = Vt\cos\theta$$

$$y = Vt\sin\theta - \frac{1}{2}gt^2,$$

where t is the time in seconds after the ball is hit and g is the acceleration due to gravity in m s^{-2}. Do NOT prove these equations.

(i) Show that the horizontal range of the golf ball is $\dfrac{V^2 \sin 2\theta}{g}$ metres. **2**

(ii) Show that if $V^2 < 100g$ then the horizontal range of the ball is less than 100 m. **1**

It is now given that $V^2 = 200g$ and that the horizontal range of the ball is 100 m or more.

(iii) Show that $\dfrac{\pi}{12} \le \theta \le \dfrac{5\pi}{12}$. **2**

(iv) Find the greatest height the ball can achieve. **2**

* In the new Mathematics Extension 1 course, Projectile Motion questions will be expressed in vector form:

$x = Vt\cos\theta$ and $y = Vt\sin\theta - \frac{1}{2}gt^2$ is expressed as a position vector:

$$\underset{\sim}{r}(t) = (Vt\cos\theta)\underset{\sim}{i} + \left(Vt\sin\theta - \frac{1}{2}gt^2\right)\underset{\sim}{j}$$

End of Question 13

Question 14 (15 marks) Use a SEPARATE writing booklet.

(a) Prove by mathematical induction that $8^{2n+1} + 6^{2n-1}$ is divisible by 7, for any integer $n \geq 1$. **3**

(b) Let $P(2p, p^2)$ be a point on the parabola $x^2 = 4y$.

The tangent to the parabola at P meets the parabola $x^2 = -4ay$, $a > 0$, at Q and R. Let M be the midpoint of QR.

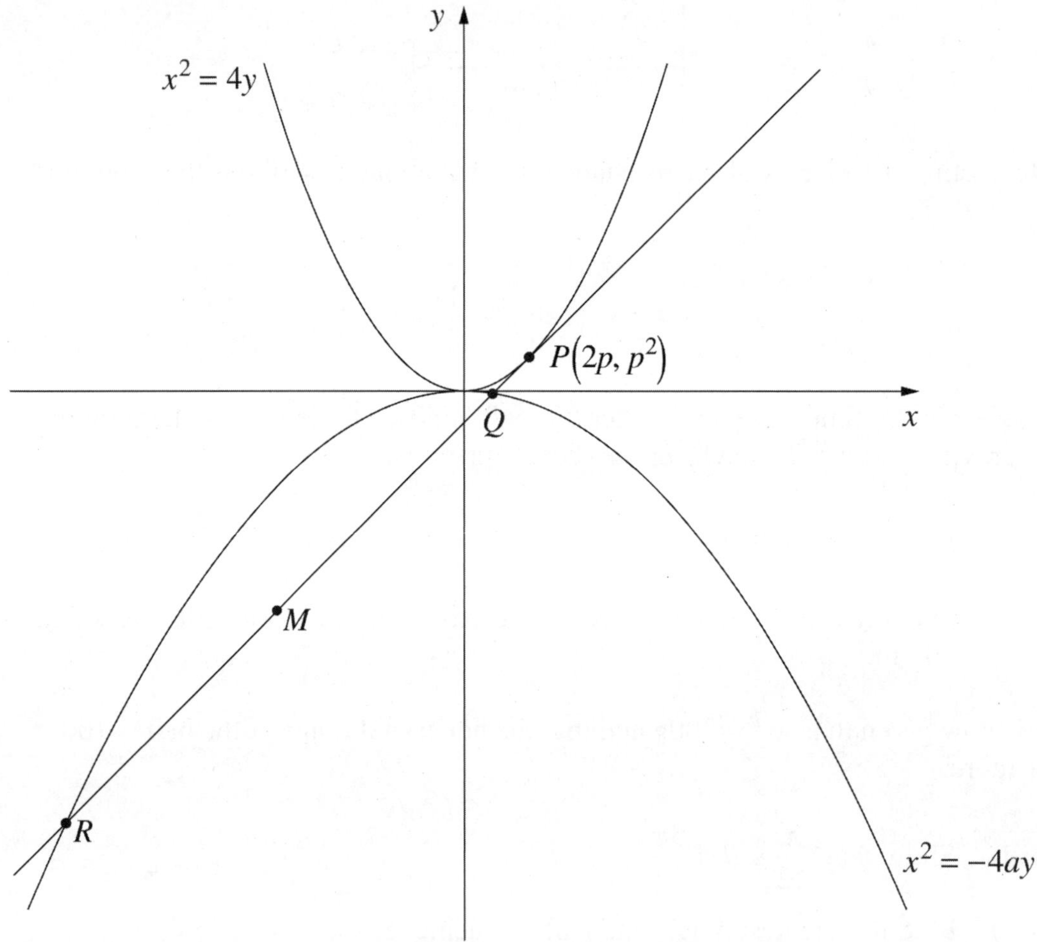

(i) Show that the x coordinates of R and Q satisfy **2**

$$x^2 + 4apx - 4ap^2 = 0.$$

(ii) Show that the coordinates of M are $\left(-2ap, -p^2(2a+1)\right)$. **2**

(iii) Find the value of a so that the point M always lies on the parabola $x^2 = -4y$. **2**

Question 14 continues on the following page

Question 14 (continued)

(c) The concentration of a drug in a body is $F(t)$, where t is the time in hours after the drug is taken.

Initially the concentration of the drug is zero. The rate of change of concentration of the drug is given by

$$F'(t) = 50e^{-0.5t} - 0.4F(t).$$

(i) By differentiating the product $F(t)e^{0.4t}$ show that **2**

$$\frac{d}{dt}\left(F(t)e^{0.4t}\right) = 50e^{-0.1t}.$$

(ii) Hence, or otherwise, show that $F(t) = 500\left(e^{-0.4t} - e^{-0.5t}\right)$. **2**

(iii) The concentration of the drug increases to a maximum. **2**

For what value of t does this maximum occur?

End of paper

Replacement questions

with content from the most up-to-date syllabus

Marks

Question 3 (1 mark)

A circle with centre (3, –4) passes through the origin.

Which of these is the equation of the circle? **1**

(A) $x = 3 - 5\cos\theta \quad y = 4 + 5\sin\theta$

(B) $x = 3 - 5\cos\theta \quad y = 4 - 5\sin\theta$

(C) $x = 5\cos\theta - 3 \quad y = 5\sin\theta + 4$

(D) $x = 5\cos\theta + 3 \quad y = 5\sin\theta - 4$

Question 6 (1 mark)

When a standard 6-sided dice is thrown, the probability that it shows a multiple of 3 is $\frac{1}{3}$. If 30 standard dice are thrown, the number of times, X, a multiple of 3 is showing has a binomial distribution.

What is the standard deviation of X, correct to 2 decimal places? **1**

(A) 2.58

(B) 3.26

(C) 3.81

(D) 6.67

Question 11 (1 mark)

(a) A binomial random variable X has a probability of success of p after 4 trials.

Use the table for the binomial distribution to find the value of p. **1**

x	0	1	2	3	4
$P(X = x)$	0.1296	0.3456	0.3456	0.1536	0.0256

REPLACEMENT QUESTIONS

Marks

Question 12 (8 marks)

(a) Solve $\sin 5x = \sin 3x$, where $0 \leq x \leq \frac{\pi}{2}$. **2**

(c) (ii) Hence, explain why $0 < h < \frac{1}{3}$. **1**

(d) A particle moves so that its position vector is given by $\underset{\sim}{r}(t) = \sin 2t\underset{\sim}{i} + 2\cos 2t\underset{\sim}{j}$.

Find its speed, in cms^{-1}, after $\frac{5\pi}{8}$ seconds. **3**

(e) Functions f and g are defined by $f(x) = 2x - 3$ and $g(x) = 5 - 4x$.

Find the inverse of the composite function $f \circ g$, written in terms of x. **2**

Question 13 (8 marks)

(a) Consider the quadrilateral $ABCD$. **3**

The midpoints of AB, BC, CD and DA are M, N, P and Q respectively.

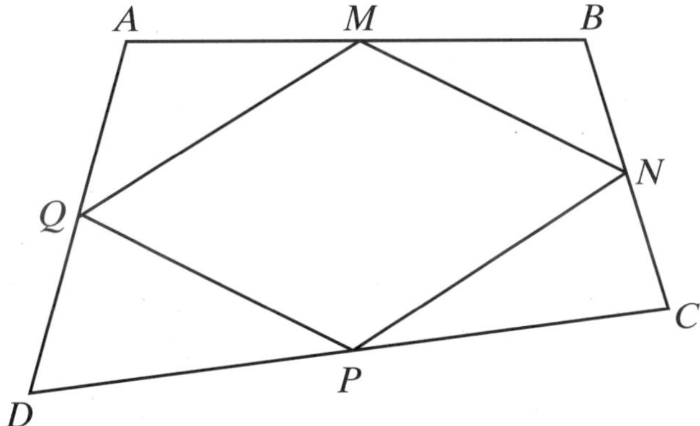

Use vector methods to prove that $MNPQ$ is a parallelogram.

(b) The diagram shows part of the graph of $xy = 4$. A line passes through the points $A(1, 4)$ and $B(2, 2)$ which lie on the curve.

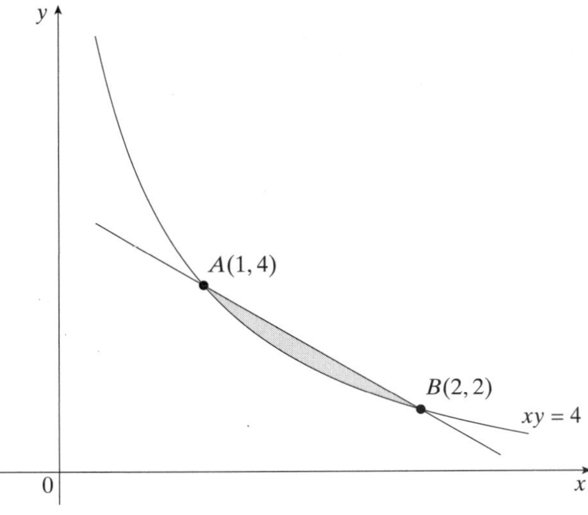

REPLACEMENT QUESTIONS

Marks

Question 13 (continued)

(i) Find the equation of the line AB. **1**

(ii) The shaded area is rotated about the y-axis to form a solid.

Find the volume of the solid formed. **4**

Question 14 (6 marks)

(b)* If $\underset{\sim}{u} = -4\underset{\sim}{i} + 2\underset{\sim}{j}$ and $\underset{\sim}{v} = -3\underset{\sim}{i} + \underset{\sim}{j}$, find the vector projection of $\underset{\sim}{v}$ in the direction of $\underset{\sim}{u}$. **2**

(b)** Solve the differential equation $(x^2 + 5)\dfrac{dy}{dx} - xy = 0$, with the condition $y(2) = 3$ and hence find the value of y when $x = 2\sqrt{5}$. **4**

REPLACEMENT QUESTIONS

2017 Higher School Certificate
Worked answers

Section I (*Total 10 marks*)

1. A	**2.** B	**3.** B	**4.** C	**5.** D
6. D	**7.** A	**8.** C	**9.** C	**10.** B

1. $P(x) = x^3 - 5x^2 + 11x - 10$

$P(2) = 2^3 - 5 \times 2^2 + 11 \times 2 - 10$
$= 0$

$\therefore x - 2$ is a factor of $P(x)$.

Answer A

2. $\log_a 8 = 1.893$

Now $\log_a 8 = \log_a 2^3$
$= 3 \log_a 2$

$\log_a 4 = \log_a 2^2$
$= 2 \log_a 2$

$\therefore \log_a 4 = \frac{2}{3} \log_a 8$
$= 1.26 \quad$ (2 d.p.)

Answer B

3. $\angle TAB = \angle ADB = 65°$
(angle in alternate segment)
$\angle DAB + \angle DCB = 180°$
(opposite angles of cyclic quad.)
$\angle DAB = 180° - 110°$
$= 70°$
$\therefore \angle TAD = 65° + 70°$
$= 135°$

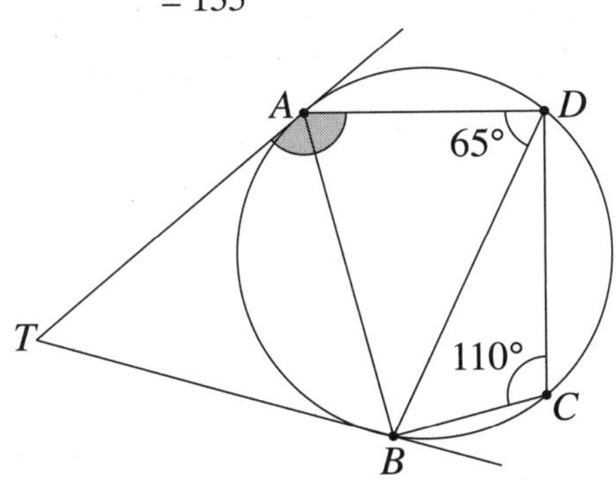

Answer B

4. $\sqrt{5} \sin (x - \alpha) \equiv 2 \sin x - \cos x$

$\sin x \cos \alpha - \cos x \sin \alpha \equiv \frac{2}{\sqrt{5}} \sin x - \frac{1}{\sqrt{5}} \cos x$

$\therefore \cos \alpha = \frac{2}{\sqrt{5}}$ and $\sin \alpha = \frac{1}{\sqrt{5}}$

Now $\tan \alpha = \frac{\sin\alpha}{\cos\alpha}$
$= \frac{1}{\sqrt{5}} \div \frac{2}{\sqrt{5}}$
$= \frac{1}{2}$

Answer C

5. $y = \frac{2x^2}{1-x^2}$
As $x \to \infty$, $y \to -2$

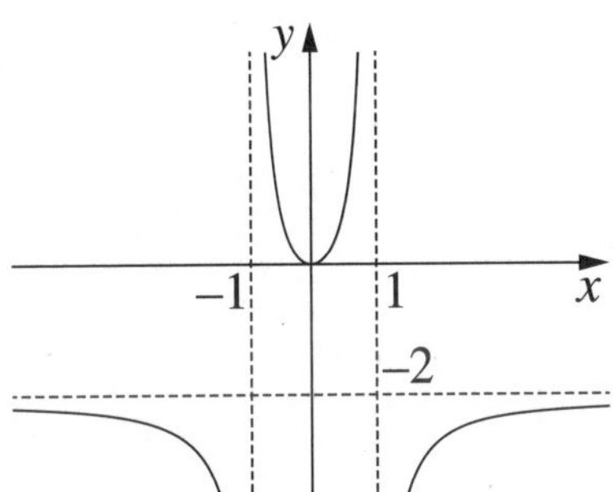

Answer D

6. Equation of normal to $x^2 = 4ay$ at $(2at, at^2)$ is

$x + ty = at^3 + 2at$

At $P\left(\frac{2}{p}, \frac{1}{p^2}\right)$ on $x^2 = 4y$, $a = 1$ and $t = \frac{1}{p}$

$\therefore x + \frac{y}{p} = \frac{1}{p^3} + \frac{2}{p}$

$p^3x + p^2y = 1 + 2p^2$

Answer D

7. $f(x) = \sin^{-1}\left(\frac{3}{x}\right)$

Domain: $-1 \le \frac{3}{x} \le 1$

If $\frac{3}{x} \ge -1$

$\frac{x}{3} \le -1$ or $x > 0$

$x \le -3$ or $x > 0$

If $\frac{3}{x} \le 1$

$\frac{x}{3} \ge 1$ or $x < 0$

$x \ge 3$ or $x < 0$

But $\frac{3}{x} \ge -1$ and $\frac{3}{x} \le 1$ must both hold at the same time.

$\therefore x \le -3$ or $x \ge 3$

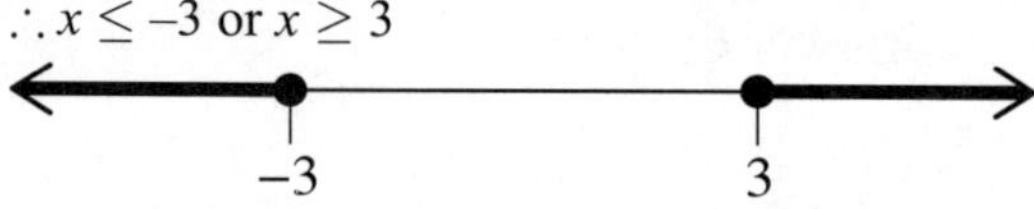

Answer A

8. $\frac{dr}{dt} = 5$

$A = \pi r^2$

$\frac{dA}{dr} = 2\pi r$

$\frac{dA}{dt} = \frac{dA}{dr} \cdot \frac{dr}{dt}$

$= 2 \times \pi \times 15 \times 5$

$= 150\pi$

The area is increasing at 150π cm^2 s^{-1} when the radius is 15 cm.

Answer C

9. Consider $\left(x^3 + \frac{1}{x^4}\right)^7$.

General term $= \binom{7}{k}(x^3)^{7-k}(x^{-4})^k$

$= \binom{7}{k}x^{21-3k}x^{-4k}$

$= \binom{7}{k}x^{21-7k}$

So when $k = 3$ there is a non-zero constant term.

[This is not the case with any of the other options.]

Answer C

10. Combinations of 3 squares from 9

$= {}^9C_3$

$= 84$

Total rows, columns or diagonals = 8

$P(\text{form a line}) = \frac{8}{84}$

$= \frac{2}{21}$

Answer B

Section II

QUESTION 11

(a) $A(-4\,-4)$, $B(1, 6)$ $\quad 2:3$

$P\left(\frac{mx_2 + nx_1}{m+n}, \frac{my_2 + ny_1}{m+n}\right)$

$P\left(\frac{2\times1+3\times-4}{2+3}, \frac{2\times6+3\times-4}{2+3}\right)$

$P(-2, 0)$

$\therefore$ the x-coordinate of P is -2 *(1 mark)*

(b) $y = \tan^{-1}(x^3)$

$$\frac{dy}{dx} = \frac{1}{1+\left(x^3\right)^2} \times 3x^2$$

$$= \frac{3x^2}{1+x^6}$$

(2 marks)

(c) $\dfrac{2x}{x+1} > 1 \quad (x \neq -1)$

Multiply both sides by $(x + 1)^2$

$2x(x + 1) > (x + 1)^2$

$2x^2 + 2x > x^2 + 2x + 1$

$x^2 > 1$

$\therefore x < -1$ or $x > 1$

(3 marks)

(d)

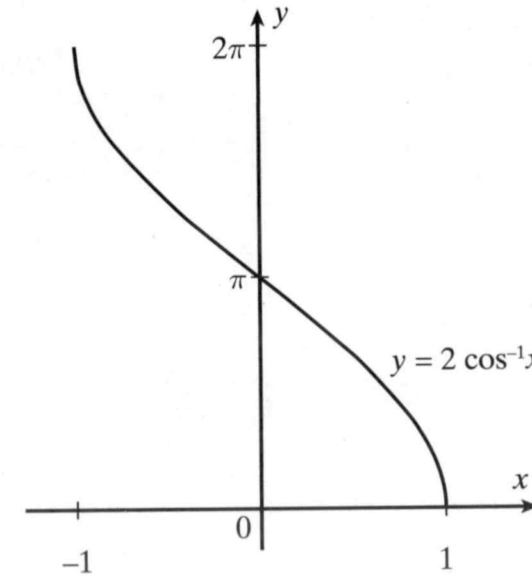

(2 marks)

(e) $\displaystyle\int_0^3 \frac{x}{\sqrt{x+1}}\, dx$

Let $x = u^2 - 1$

$dx = 2u\, du$

$u^2 = x + 1$

$u = \sqrt{x+1} \quad (u > 0)$

When $x = 0, u = 1$

When $x = 3, u = 2$

$$\therefore \int_0^3 \frac{x}{\sqrt{x+1}}\, dx = \int_1^2 \frac{u^2-1}{u} \cdot 2u\, du$$

$$= \int_1^2 \left(2u^2 - 2\right) du$$

$$= \left[\frac{2u^3}{3} - 2u\right]_1^2$$

$$= \frac{2\times 2^3}{3} - 2\times 2 - \left(\frac{2\times 1^3}{3} - 2\times 1\right)$$

$$= 2\frac{2}{3}$$

(3 marks)

(f) Let $u = \sin x$

$du = \cos x \, dx$

$$\int \sin^2 x \cos x \, dx = \int u^2 \, du$$
$$= \frac{u^3}{3} + C$$
$$= \frac{1}{3}\sin^3 x + C$$

(1 mark)

(g) $P(\text{red}) = \frac{1}{5}$, $P(\text{not red}) = \frac{4}{5}$

(i) $P(\text{exactly 3 red in 8}) = \binom{8}{3}\left(\frac{1}{5}\right)^3\left(\frac{4}{5}\right)^5$

(1 mark)

(ii) $P(\text{no red flowers in 8}) = \left(\frac{4}{5}\right)^8$

(1 mark)

(iii) $P(\text{at least 1 red in 8}) = 1 - \left(\frac{4}{5}\right)^8$

(1 mark)

QUESTION 12

(a) Reflex $\angle AOC = 360° - 100°$

(Angles in revolution)

$= 260°$

$\angle ABC = 260° \div 2$

(Angle at centre twice angle at circumference)

$= 130°$

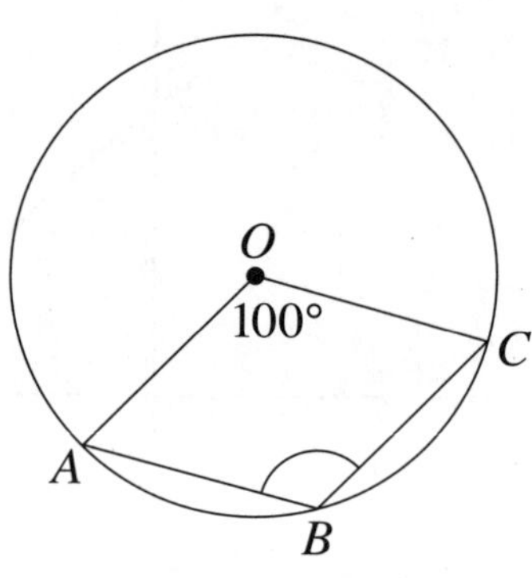

(2 marks)

(b) (i)

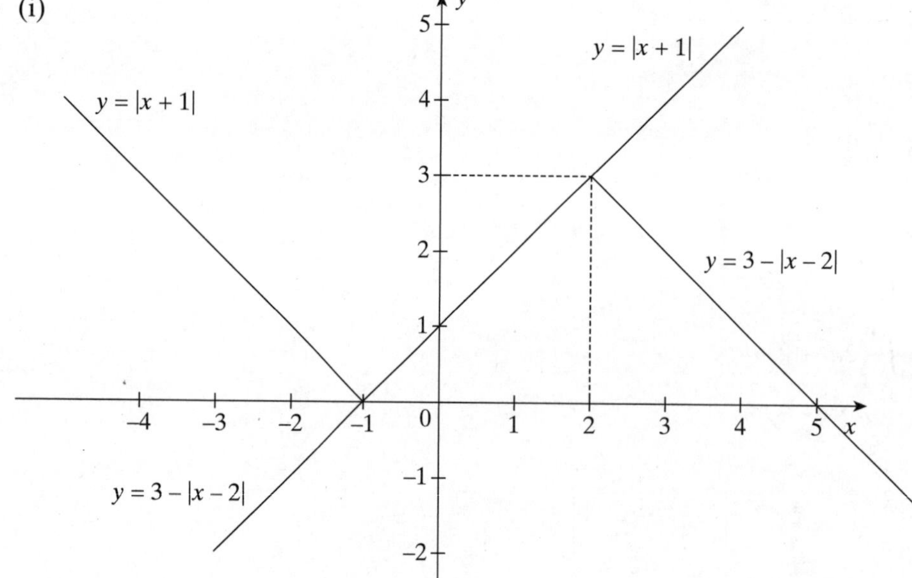

(3 marks)

(ii) $|x+1| + |x-2| = 3$

$$|x+1| = 3 - |x-2|$$

From the graph $-1 \leq x \leq 2$

(1 mark)

(c) (i) $V = \int \pi y^2 \, dx$

Larger volume:

$$V = \int_{-1}^{h} \pi\left(1-x^2\right) dx$$
$$= \pi\left[x - \frac{x^3}{3}\right]_{-1}^{h}$$
$$= \pi\left(h - \frac{h^3}{3} + 1 - \frac{1}{3}\right)$$
$$= \frac{\pi}{3}(3h - h^3 + 2)$$

Smaller volume:

$$V = \int_{h}^{1} \pi\left(1-x^2\right) dx$$
$$= \pi\left[x - \frac{x^3}{3}\right]_{h}^{1}$$
$$= \pi\left(1 - \frac{1}{3} - h + \frac{h^3}{3}\right)$$
$$= \frac{\pi}{3}(2 - 3h + h^3)$$

Now the volumes are in the ratio 2 : 1.

$$\therefore 3h - h^3 + 2 = 2(2 - 3h + h^3)$$
$$3h - h^3 + 2 = 4 - 6h + 2h^3$$
$$0 = 2 - 9h + 3h^3$$
$$\therefore 3h^3 - 9h + 2 = 0$$

(3 marks)

(ii) $f(h) = 3h^2 - 9h + 2$

$f'(h) = 6h - 9$

By Newton's method with $h_1 = 0$

$$h_2 = h_1 - \frac{f(h_1)}{f'(h_1)}$$
$$= 0 - \frac{3\times0^2 - 9\times0 + 2}{6\times0 - 9}$$
$$= \frac{2}{9}$$

A second approximation for h is $\frac{2}{9}$.

(1 mark)

(d)
$$t = 4 - e^{-2x}$$
$$\frac{dt}{dx} = 2e^{-2x}$$
$$\frac{dx}{dt} = \frac{1}{2e^{-2x}}$$
$$\therefore v = \frac{e^{2x}}{2}$$
$$\frac{d^2x}{dt^2} = \frac{d}{dx}\left(\frac{1}{2}v^2\right)$$
$$= \frac{d}{dx}\left(\frac{1}{2}\times\frac{e^{4x}}{4}\right)$$
$$= \frac{1}{2}e^{4x}$$

(3 marks)

(e)
$$\lim_{x\to0}\frac{1-\cos 2x}{x^2} = \lim_{x\to0}\frac{1-\left(\cos^2 x - \sin^2 x\right)}{x^2}$$
$$= \lim_{x\to0}\frac{2\sin^2 x}{x^2}$$
$$= 2\lim_{x\to0}\left(\frac{\sin x}{x}\right)^2$$
$$= 2$$

(2 marks)

QUESTION 13

(a) SHM centred at origin

$$\ddot{x} = -n^2 x$$
$$\therefore \frac{d}{dx}\left(\frac{1}{2}v^2\right) = -n^2 x$$
$$\frac{1}{2}v^2 = \frac{-n^2x^2}{2} + c$$
$$v^2 = -n^2x^2 + C$$

When $x = 2$, $v = 4$

$$4^2 = -n^2 \times 2^2 + C$$
$$16 = -4n^2 + C \qquad \text{(i)}$$

When $x = 5$, $v = 3$

$$3^2 = -n^2 \times 5^2 + C$$
$$9 = -25n^2 + C \qquad \text{(ii)}$$

(i) – (ii)
$$7 = 21n^2$$
$$n^2 = \frac{1}{3}$$
$$n = \frac{1}{\sqrt{3}} \qquad (n > 0)$$

$$\text{Period} = \frac{2\pi}{n} = \frac{2\pi}{\frac{1}{\sqrt{3}}} = 2\sqrt{3}\,\pi$$

(3 marks)

(b) (i) $(1+x)^n = \binom{n}{0} + \binom{n}{1}x + \binom{n}{2}x^2 + \binom{n}{3}x^3 + \ldots + \binom{n}{n}x^n$

$(1-x)^n = \binom{n}{0} - \binom{n}{1}x + \binom{n}{2}x^2 - \binom{n}{3}x^3 + \ldots + (-1)^n\binom{n}{n}x^n$

But if n is even

$(1-x)^n = \binom{n}{0} - \binom{n}{1}x + \binom{n}{2}x^2 - \binom{n}{3}x^3 + \ldots + \binom{n}{n}x^n$

$(1+x)^n + (1-x)^n = 2\left[\binom{n}{0} + \binom{n}{2}x^2 + \binom{n}{4}x^4 + \ldots + \binom{n}{n}x^n\right]$

(2 marks)

(ii) Differentiating both sides:

$$n(1+x)^{n-1} + n(1-x)^{n-1} \times -1 = 2\left[\binom{n}{2}2x + \binom{n}{4}4x^3 + \ldots + \binom{n}{n}nx^{n-1}\right]$$

$$n[(1+x)^{n-1} - (1-x)^{n-1}] = 2\left[2\binom{n}{2}x + 4\binom{n}{4}x^3 + \ldots + n\binom{n}{n}x^{n-1}\right]$$

(1 mark)

(iii) $2\left[2\binom{n}{2}x + 4\binom{n}{4}x^3 + 6\binom{n}{6}x^5 + \ldots + n\binom{n}{n}x^{n-1}\right] = n[(1+x)^{n-1} - (1-x)^{n-1}]$

Let $x = 1$,

$$2\left[2\binom{n}{2}1 + 4\binom{n}{4}1^3 + 6\binom{n}{6}1^5 + \ldots + n\binom{n}{n}1^{n-1}\right] = n[(1+1)^{n-1} - (1-1)^{n-1}]$$

$$2^2\left[\binom{n}{2} + 2\binom{n}{4} + 3\binom{n}{6} + \ldots + \frac{n}{2}\binom{n}{n}\right] = n(2^{n-1})$$

$$\left[\binom{n}{2} + 2\binom{n}{4} + 3\binom{n}{6} + \ldots + \frac{n}{2}\binom{n}{n}\right] = n(2^{n-1}) \div 2^2$$

$$= n2^{n-3}$$

(2 marks)

(c) (i) $y = Vt\sin\theta - \frac{1}{2}gt^2$

When $y = 0$,

$Vt\sin\theta - \frac{1}{2}gt^2 = 0$

$t(V\sin\theta - \frac{1}{2}gt) = 0$

$t = 0$ or $V\sin\theta - \frac{1}{2}gt = 0$

$gt = 2V\sin\theta$

$t = \dfrac{2V\sin\theta}{g}$

Now $x = Vt\cos\theta$

When $t = \dfrac{2V\sin\theta}{g}$,

$x = \dfrac{2V^2\sin\theta\cos\theta}{g}$

$= \dfrac{V^2\sin 2\theta}{g}$

So the horizontal range of the projectile is $\dfrac{V^2\sin 2\theta}{g}$.

(2 marks)

(ii) If $V^2 < 100g$ the horizontal range is less than $100\sin 2\theta$.

But $\sin 2\theta \leq 1$ for all values of θ.

So the horizontal range is less than 100 m.

(1 mark)

(iii) Now horizontal range > 100 m.

So $\dfrac{V^2\sin 2\theta}{g} > 100$

But $V^2 = 200g$

$\therefore \dfrac{200g\sin 2\theta}{g} > 100$

$200\sin 2\theta > 100$

$\sin 2\theta > 0.5$

Now $0 < \theta < \dfrac{\pi}{2}$

$\therefore 0 < 2\theta < \pi$

If $\sin 2\theta = 0.5$

$2\theta = \dfrac{\pi}{6}$ or $\pi - \dfrac{\pi}{6}$

$= \dfrac{\pi}{6}$ or $\dfrac{5\pi}{6}$

$\theta = \dfrac{\pi}{12}$ or $\dfrac{5\pi}{12}$

$\sin 2\theta \geq 0.5$ when $\dfrac{\pi}{12} \leq \theta \leq \dfrac{5\pi}{12}$

(2 marks)

(iv) $y = Vt\sin\theta - \frac{1}{2}gt^2$

$\dot{y} = V\sin\theta - gt$

Maximum height occurs when $\dot{y} = 0$

i.e $0 = V\sin\theta - gt$

$gt = V\sin\theta$

$t = \dfrac{V\sin\theta}{g}$

When $t = \dfrac{V\sin\theta}{g}$,

$y = \dfrac{V^2\sin^2\theta}{g} - \dfrac{1}{2}g\left(\dfrac{V\sin\theta}{g}\right)^2$

$= \dfrac{V^2\sin^2\theta}{2g}$

But $V^2 = 200g$

$\therefore y = 100\sin^2\theta$

For $\dfrac{\pi}{12} \leq \theta \leq \dfrac{5\pi}{12}$, the greatest value of $\sin\theta$ occurs when $\theta = \dfrac{5\pi}{12}$.

When $\theta = \dfrac{5\pi}{12}$,

$y = 100\sin^2\left(\dfrac{5\pi}{12}\right)$

$= 93.301\,2701\ldots$

$= 93.3$ (1 d.p.)

So the greatest possible height that can be reached is 93.3 metres.

(2 marks)

QUESTION 14

(a) To prove $8^{2n+1} + 6^{2n-1}$ is divisible by 7 for all $n \geq 1$

When $n = 1$,

$8^{2 \times 1+1} + 6^{2 \times 1-1} = 8^3 + 6$

$= 518$

$= 7 \times 74.$

It is true for $n = 1$.

Assume it is true for $n = k$.

i.e. assume $8^{2k+1} + 6^{2k-1} = 7m$ for some positive integer m.

When $n = k + 1$,

$8^{2(k+1)+1} + 6^{2(k+1)-1}$

$= 8^{2k+3} + 6^{2k+1}$

$= 8^2 \times 8^{2k+1} + 6^2 \times 6^{2k-1}$

$= 64 \times 8^{2k+1} + 36 \times 6^{2k-1}$

$= 28 \times 8^{2k+1} + 36 \times 8^{2k+1} + 36 \times 6^{2k-1}$

$= 7 \times 4 \times 8^{2k+1} + 36(7m)$

$= 7[4 \times 8^{2k+1} + 36m]$

So if it is true for $n = k$, it is also true for $n = k + 1$.

It is true for $n = 1$, so it is true for $n = 1 + 1 = 2$.

It is true for $n = 2$, so it is true for $n = 2 + 1 = 3$.

And so on.

By the process of induction it is true for all integers $n \geq 1$.

$8^{2n+1} + 6^{2n-1}$ is divisible by 7 for all $n \geq 1$.

(3 marks)

(b) (i) For $x^2 = 4ay$ the tangent at $(2at, at^2)$ is given by:

$y = tx - at^2$

On $x^2 = 4y$ at $P(2p, p^2)$ $a = 1$ and $t = p$

Equation of tangent is $y = px - p^2$.

Q and R are the points of intersection of the tangent with the parabola $x^2 = -4ay$.

At the points of intersection:

$x^2 = -4a(px - p^2)$

$= -4apx + 4ap^2$

$x^2 + 4apx - 4ap^2 = 0$

So the x coordinates of Q and R satisfy $x^2 + 4apx - 4ap^2$.

(2 marks)

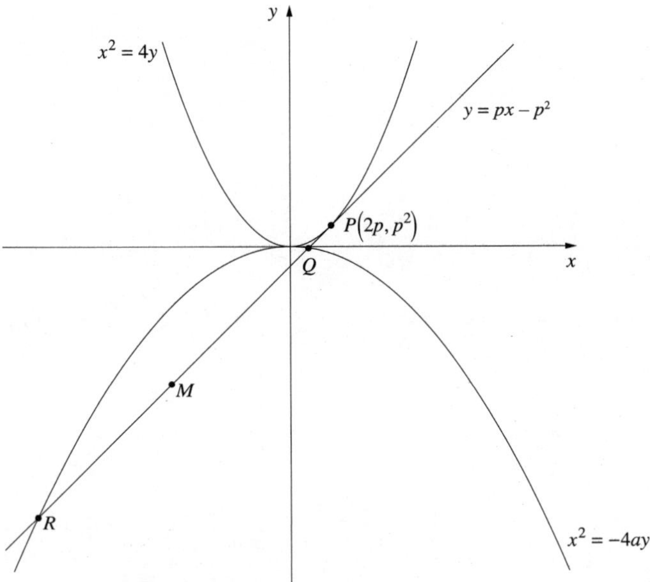

(ii) $x^2 + 4apx - 4ap^2 = 0$

The x coordinates of Q and R are the roots of the equation.

$$\text{Sum of roots} = \frac{-4ap}{1} = -4ap$$

So the x-coordinate of the midpoint, M, is $\frac{-4ap}{2}$ or $-2ap$.

M lies on the tangent $y = px - p^2$.

When $x = -2ap$,

$$\begin{aligned} y &= p(-2ap) - p^2 \\ &= -2ap^2 - p^2 \\ &= -p^2(2a + 1) \end{aligned}$$

$\therefore M$ is the point $(-2ap, -p^2(2a + 1))$.

(2 marks)

(iii) If M lies on $x^2 = -4y$

$$\begin{aligned} (-2ap)^2 &= -4(-p^2(2a + 1)) \\ 4a^2p^2 &= 4p^2(2a + 1) \\ a^2 &= 2a + 1 \end{aligned}$$

$$a^2 - 2a - 1 = 0$$

$$\begin{aligned} a &= \frac{-(-2) \pm \sqrt{(-2)^2 - 4 \times 1 \times -1}}{2 \times 1} \\ &= \frac{2 \pm \sqrt{8}}{2} \\ &= \frac{2(1 \pm \sqrt{2})}{2} \\ &= 1 \pm \sqrt{2} \end{aligned}$$

But $a > 0$

$$\therefore a = 1 + \sqrt{2}$$

(2 marks)

(c) (i) $F'(t) = 50e^{-0.5t} - 0.4F(t)$

$\frac{d}{dt}(\text{F}(t)e^{0.4t})$

$= F(t) \times 0.4e^{0.4t} + e^{0.4t} \times F'(t)$

$= 0.4F(t)e^{0.4t} + e^{0.4t}(50e^{-0.5t} - 0.4F(t))$

$= 0.4F(t)e^{0.4t} + 50e^{-0.1t} - 0.4F(t)e^{0.4t}$

$= 50e^{-0.1t}$

(2 marks)

(ii) $F(t)e^{0.4t} = \int 50e^{-0.1t}\, dt$

$= \dfrac{50e^{-0.1t}}{-0.1} + C$

$= -500e^{-0.1t} + C$

Now $F(0) = 0$

When $t = 0$,

$0 = -500e^{0} + C$

$C = 500$

So $F(t)e^{0.4t} = 500(1 - e^{-0.1t})$

$F(t) = \dfrac{500\left(1-e^{-0.1t}\right)}{e^{0.4t}}$

$= 500(e^{-0.4t} - e^{-0.5t})$

(2 marks)

(iii) $F'(t) = 50e^{-0.5t} - 0.4F(t)$

$= 50e^{-0.5t} - 0.4 \times 500(e^{-0.4t} - e^{-0.5t})$

$= 50e^{-0.5t} - 200e^{-0.4t} + 200e^{-0.5t}$

$= 250e^{-0.5t} - 200e^{-0.4t}$

When $F'(t) = 0$,

$250e^{-0.5t} = 200e^{-0.4t}$

$\dfrac{250}{200} = \dfrac{e^{-0.4t}}{e^{-0.5t}}$

$1.25 = e^{0.1t}$

$\ln 1.25 = 0.1t$

$t = 10 \ln 1.25$

$F''(t) = -125e^{-0.5t} + 80e^{-0.4t}$

$F''(10 \ln 1.25) < 0$ (maximum)

The maximum concentration occurs when $t = 10 \ln 1.25$ hours ($\approx 2\frac{1}{4}$ h).

(2 marks)

Solutions to replacement questions

QUESTION 3

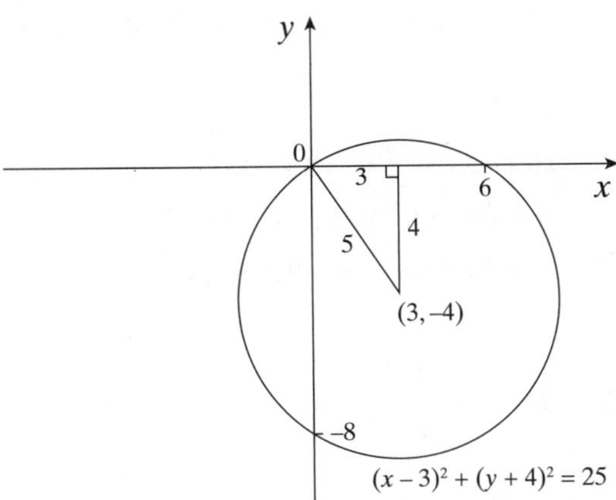

From the diagram, the radius is 5 units.

$\therefore$ the equation of the circle is

$$(x-3)^2 + (y+4)^2 = 25$$

$$\left(\frac{x-3}{5}\right)^2 + \left(\frac{y+4}{5}\right)^2 = 1$$

As $\sin^2\theta + \cos^2\theta = 1$, then consider:

$$\cos\theta = \frac{x-3}{5} \qquad \sin\theta = \frac{y+4}{5}$$

$$5\cos\theta = x-3 \qquad 5\sin\theta = y+4$$

$$x = 5\cos\theta + 3 \qquad y = 5\sin\theta - 4$$

Answer D

(1 mark)

QUESTION 6

$$\begin{aligned} E(X) &= np \\ &= 30 \times \frac{1}{3} \\ &= 10 \end{aligned}$$

$$\begin{aligned} Var(X) &= np(1-p) \\ &= 10 \times \frac{2}{3} \\ &= \frac{20}{3} \end{aligned}$$

$$\begin{aligned} \text{Stand. dev.} &= \sqrt{\frac{20}{3}} \\ &= 2.581\,988\,97\ldots \\ &= 2.58 \text{ (2 dec. pl.)} \end{aligned}$$

Answer A

(1 mark)

QUESTION 11

(a) $\binom{4}{4}p^4(1-p)^0 = 0.0256$

$$p^4 = 0.0256$$

$$p = 0.4$$

(1 mark)

QUESTION 12

(a)

$$\begin{aligned} \sin 5x &= \sin 3x \\ \sin 5x - \sin 3x &= 0 \\ 2\cos 4x \sin x &= 0 \end{aligned}$$

$$\cos 4x = 0 \qquad \sin x = 0$$

$$4x = \frac{\pi}{2}, \frac{3\pi}{2} \qquad x = 0$$

$$x = \frac{\pi}{8}, \frac{3\pi}{8}$$

$$\therefore x = 0, \frac{\pi}{8}, \frac{3\pi}{8}$$

(2 marks)

(c) (ii) Let $P(h) = 3h^3 - 9h + 2$

$$\begin{aligned} P(0) &= 3(0)^3 - 9(0) + 2 \\ &= 2 \end{aligned}$$

$$\begin{aligned} P\left(\frac{1}{3}\right) &= 3\left(\frac{1}{3}\right)^3 - 9\left(\frac{1}{3}\right) + 2 \\ &= \frac{1}{9} - 3 + 2 \\ &= -\frac{8}{9} \end{aligned}$$

As $P(0) > 0$ and $P\left(\frac{1}{3}\right) < 0$ then a root exists in $0 < h < \frac{1}{3}$.

(1 mark)

(d)

$$\begin{aligned} \underset{\sim}{r}(t) &= \sin 2t\underset{\sim}{i} + 2\cos 2t\underset{\sim}{j} \\ \underset{\sim}{v}(t) &= 2\cos 2t\underset{\sim}{i} - 4\sin 2t\underset{\sim}{j} \\ |\underset{\sim}{v}(t)| &= \sqrt{(2\cos 2t)^2 + (-4\sin 2t)^2} \\ &= \sqrt{4\cos^2 2t + 16\sin^2 2t} \end{aligned}$$

$$\begin{aligned} \left|\underset{\sim}{v}\left(\frac{5\pi}{8}\right)\right| &= \sqrt{4\cos^2 2\left(\frac{5\pi}{8}\right) + 16\sin^2 2\left(\frac{5\pi}{8}\right)} \\ &= \sqrt{4\cos^2\frac{5\pi}{4} + 16\sin^2\frac{5\pi}{4}} \\ &= \sqrt{4\left(-\frac{1}{\sqrt{2}}\right)^2 + 16\left(-\frac{1}{\sqrt{2}}\right)^2} \\ &= \sqrt{2+8} \\ &= \sqrt{10} \end{aligned}$$

$\therefore$ a speed of $\sqrt{10}$ cms^{-1}.

(3 marks)

(e) $f \circ g = f(g(x))$

$f \circ g = 2(5 - 4x) - 3$

$= 10 - 8x - 3$

$= 7 - 8x$

Let $y = 7 - 8x$

$(f \circ g)^{-1}$: $x = 7 - 8y$

$8y = 7 - x$

$y = \dfrac{7 - x}{8}$

$\therefore (f \circ g)^{-1} = \dfrac{7 - x}{8}$

(2 marks)

QUESTION 13

(a) Let $\vec{AB} = \underset{\sim}{a}, \vec{BC} = \underset{\sim}{b}, \vec{CD} = \underset{\sim}{c}, \vec{DA} = \underset{\sim}{d}$.

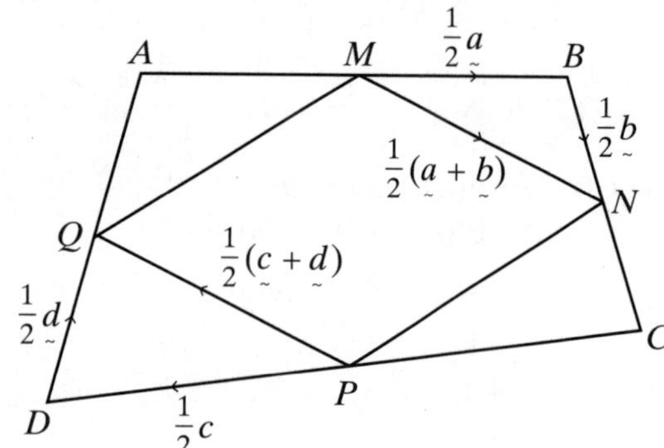

As M, N, P and Q are midpoints, then

$\vec{MB} = \frac{1}{2}\underset{\sim}{a}, \vec{BN} = \frac{1}{2}\underset{\sim}{b}, \vec{PD} = \frac{1}{2}\underset{\sim}{c}, \vec{DQ} = \frac{1}{2}\underset{\sim}{d}$.

Hence, $\vec{MN} = \frac{1}{2}(\underset{\sim}{a} + \underset{\sim}{b})$ and $\vec{PQ} = \frac{1}{2}(\underset{\sim}{c} + \underset{\sim}{d})$.

But $\underset{\sim}{a} + \underset{\sim}{b} + \underset{\sim}{c} + \underset{\sim}{d} = 0$.

Hence, $\underset{\sim}{a} + \underset{\sim}{b} = -(\underset{\sim}{c} + \underset{\sim}{d})$.

$\therefore MN = QP$ and $MN \parallel PQ$.

$\therefore MNPQ$ is a parallelogram.

(3 marks)

(b) (i) $m = \dfrac{4 - 2}{1 - 2}$

$= -2$

Equation of the line is $y = -2x + c$.

Substitute $(1, 4)$:

$4 = -2(1) + c$

$c = 6$

$\therefore y = 6 - 2x$

(1 mark)

(ii) As $y = 6 - 2x$,

$x = \dfrac{6 - y}{2}$

$\therefore x^2 = \dfrac{(6 - y)^2}{4}$

Also, $x = \dfrac{4}{y}$

$\therefore x^2 = \dfrac{16}{y^2}$

$$\text{Volume} = \pi\int_2^4 \frac{(6-y)^2}{4}dy - \pi\int_2^4 \frac{16}{y^2}dy$$
$$= \frac{\pi}{4}\int_2^4 (6-y)^2 dy - 16\pi\int_2^4 y^{-2}dy$$
$$= \frac{\pi}{4}\left[\frac{(6-y)^3}{-3}\right]_2^4 - 16\pi\left[\frac{y^{-1}}{-1}\right]_2^4$$
$$= -\frac{\pi}{12}\left[(6-y)^3\right]_2^4 + 16\pi\left[\frac{1}{y}\right]_2^4$$
$$= -\frac{\pi}{12}\left[(6-4)^3-(6-2)^3\right] + 16\pi\left[\frac{1}{4}-\frac{1}{2}\right]$$
$$= -\frac{\pi}{12}[8-64] + 16\pi\left[-\frac{1}{4}\right]$$
$$= \frac{56\pi}{12} - 4\pi$$
$$= \frac{2\pi}{3}$$

$\therefore$ volume of $\frac{2\pi}{3}$ units^3. (4 marks)

(4 marks)

QUESTION 14

(b)* $\underset{\sim}{v}\bullet\underset{\sim}{u} = (-3)\times(-4) + 1\times 2$

$$= 14$$
$$\left|\underset{\sim}{u}\right|^2 = (-4)^2 + 2^2$$
$$= 20$$
$$\therefore \underset{\sim}{v}_{\underset{\sim}{u}} = \frac{14}{20}(-4\underset{\sim}{i} + 2)\underset{\sim}{j}$$
$$= \frac{7}{5}(-2\underset{\sim}{i} + \underset{\sim}{j})$$

(2 marks)

(b)** $(x^2+5)\,\dfrac{dy}{dx} - xy = 0$

$$(x^2+5)\,\frac{dy}{dx} = xy$$
$$\int \frac{1}{y}dy = \int \frac{x}{x^2+5}dx$$
$$\ln|y| = \frac{1}{2}\ln|x^2+5| + c$$
$$\ln|y| = \ln\sqrt{x^2+5} + c$$
$$c = \ln\left|\frac{y}{\sqrt{x^2+5}}\right|$$

Substitute $y(2) = 3$: $c = \ln\left|\dfrac{3}{\sqrt{2^2+5}}\right|$

$$c = \ln\frac{3}{3}$$
$$c = 0$$
$$\ln|y| = \ln\sqrt{x^2+5}$$
$$y = \sqrt{x^2+5}$$

Substitute $x = 2\sqrt{5}$:

$$y(2\sqrt{5}) = \sqrt{(2\sqrt{5})^2+5}$$
$$= \sqrt{20+5}$$
$$= \sqrt{25}$$
$$= 5$$

(4 marks)

NSW Education Standards Authority

2018 HIGHER SCHOOL CERTIFICATE EXAMINATION

Mathematics Extension 1

General Instructions

- Reading time – 5 minutes
- Working time – 2 hours
- Write using black pen
- Calculators approved by NESA may be used
- A reference sheet is provided at the back of this paper
- In Questions 11–14, show relevant mathematical reasoning and/or calculations

Total marks: 70

Section I – 10 marks

- Attempt Questions 1–10
- Allow about 15 minutes for this section

Section II – 60 marks

- Attempt Questions 11–14
- Allow about 1 hour and 45 minutes for this section

Section I

10 marks
Attempt Questions 1–10
Allow about 15 minutes for this section

Use the multiple-choice answer sheet for Questions 1–10.

---

1 The polynomial $2x^3 + 6x^2 - 7x - 10$ has zeros α, β and γ.

What is the value of $\alpha\beta\gamma(\alpha + \beta + \gamma)$?

A. −60

B. −15

C. 15

D. 60

✗ 2 The acute angle between the lines $y = 3x$ and $y = 5x$ is θ.

What is the value of $\tan\theta$?

A. $\frac{1}{8}$

B. $\frac{1}{7}$

C. $\frac{1}{2}$

D. $\frac{4}{7}$

✗ 3 What is the value of $\lim_{x \to 0} \frac{\sin 3x \cos 3x}{12x}$?

A. $\frac{1}{4}$

B. $\frac{1}{2}$

C. $\frac{3}{4}$

D. 1

4 The diagram shows the graph of $y = a(x + b)(x + c)(x + d)^2$.

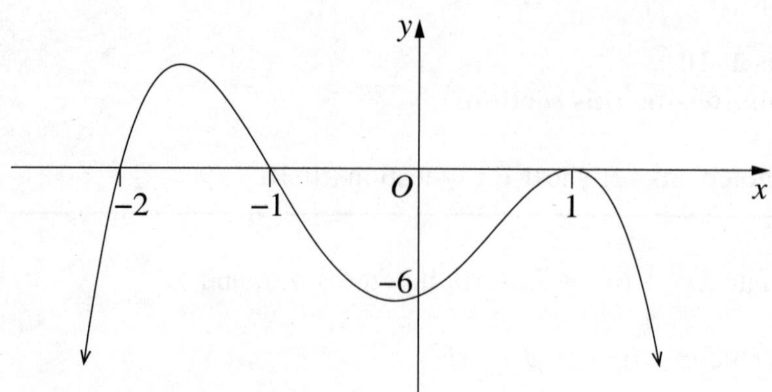

What are possible values of a, b, c and d?

A. $a = -6,\ b = -2,\ c = -1,\ d = 1$

B. $a = -6,\ b = 2,\ c = 1,\ d = -1$

C. $a = -3,\ b = -2,\ c = -1,\ d = 1$

D. $a = -3,\ b = 2,\ c = 1,\ d = -1$

5 The diagram shows the number of penguins, $P(t)$, on an island at time t.

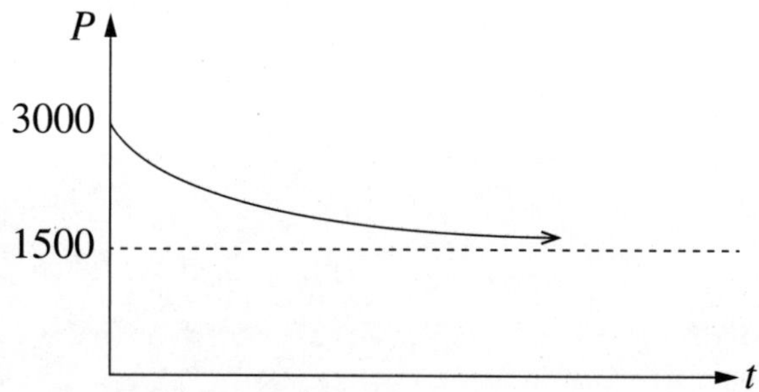

Which equation best represents this graph?

A. $P(t) = 1500 + 1500e^{-kt}$

B. $P(t) = 3000 - 1500e^{-kt}$

C. $P(t) = 3000 + 1500e^{-kt}$

D. $P(t) = 4500 - 1500e^{-kt}$

✗ **6** The diagram shows the graph of $y = f(x)$. The equation $f(x) = 0$ has a solution at $x = w$.

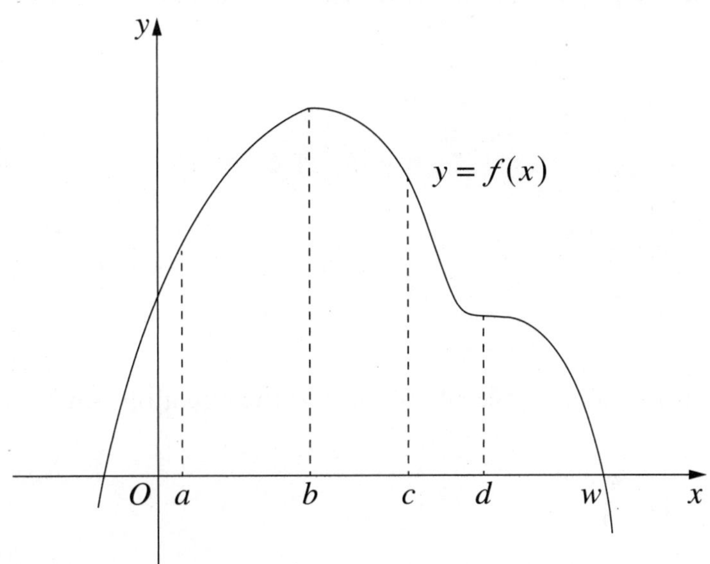

Newton's method can be used to give an approximation close to the solution $x = w$.

Which initial approximation, x_1, will give the second approximation that is closest to the solution $x = w$?

A. $x_1 = a$

B. $x_1 = b$

C. $x_1 = c$

D. $x_1 = d$

✗ **7** The velocity of a particle, in metres per second, is given by $v = x^2 + 2$, where x is its displacement in metres from the origin.

What is the acceleration of the particle at $x = 1$?

A. $2\ \text{m s}^{-2}$

B. $3\ \text{m s}^{-2}$

C. $6\ \text{m s}^{-2}$

D. $12\ \text{m s}^{-2}$

8 Six men and six women are to be seated at a round table.

In how many different ways can they be seated if men and women alternate?

A. 5! 5!

B. 5! 6!

C. 2! 5! 5!

D. 2! 5! 6!

✗ **9** Which of the following is a general solution of the equation $\sin 2x = -\frac{1}{2}$?

A. $x = n\pi + (-1)^n \frac{\pi}{12}$

B. $x = \frac{n\pi}{2} + (-1)^n \frac{\pi}{12}$

C. $x = n\pi + (-1)^{n+1} \frac{\pi}{12}$

D. $x = \frac{n\pi}{2} + (-1)^{n+1} \frac{\pi}{12}$

✗ **10** A particle is moving in simple harmonic motion. The displacement of the particle is x and its velocity, v, is given by the equation $v^2 = n^2\left(2kx - x^2\right)$, where n and k are constants. The particle is initially at $x = k$.

Which function, in terms of time t, could represent the motion of the particle?

A. $x = k\cos(nt)$

B. $x = k\sin(nt) + k$

C. $x = 2k\cos(nt) - k$

D. $x = 2k\sin(nt) + k$

Section II

60 marks
Attempt Questions 11–14
Allow about 1 hour and 45 minutes for this section

Answer each question in the appropriate writing booklet. Extra writing booklets are available.

In Questions 11–14, your responses should include relevant mathematical reasoning and/or calculations.

Question 11 (15 marks) Use the Question 11 Writing Booklet.

(a) Consider the polynomial $P(x) = x^3 - 2x^2 - 5x + 6$.

(i) Show that $x = 1$ is a zero of $P(x)$. **1**

(ii) Find the other zeros. **2**

(b) Solve $\log_2 5 + \log_2 (x - 2) = 3$. **2**

(c) Write $\sqrt{3}\sin x + \cos x$ in the form $R\sin(x + \alpha)$ where $R > 0$ and $0 \le \alpha \le \frac{\pi}{2}$. **2**

(d) Two secants from the point C intersect a circle as shown in the diagram. **2**

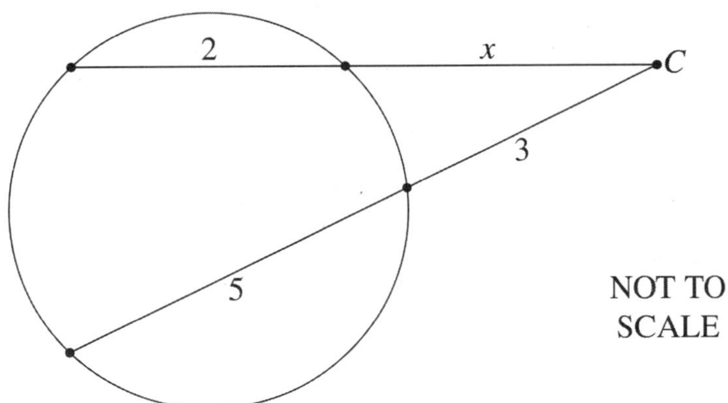

What is the value of x?

Question 11 continues on the following page

Question 11 (continued)

(e) Consider the function $f(x) = \dfrac{1}{4x-1}$.

(i) Find the domain of $f(x)$. **1**

(ii) For what values of x is $f(x) < 1$? **2**

(f) Evaluate $\displaystyle\int_{-3}^{0} \frac{x}{\sqrt{1-x}}\,dx$, using the substitution $u = 1 - x$. **3**

End of Question 11

Question 12 (15 marks) Use the Question 12 Writing Booklet.

(a) Find $\int \cos^2(3x)\,dx$. **2**

(b) A ferris wheel has a radius of 20 metres and is rotating at a rate of 1.5 radians per minute. The top of a carriage is h metres above the horizontal diameter of the ferris wheel. The angle of elevation of the top of the carriage from the centre of the ferris wheel is θ.

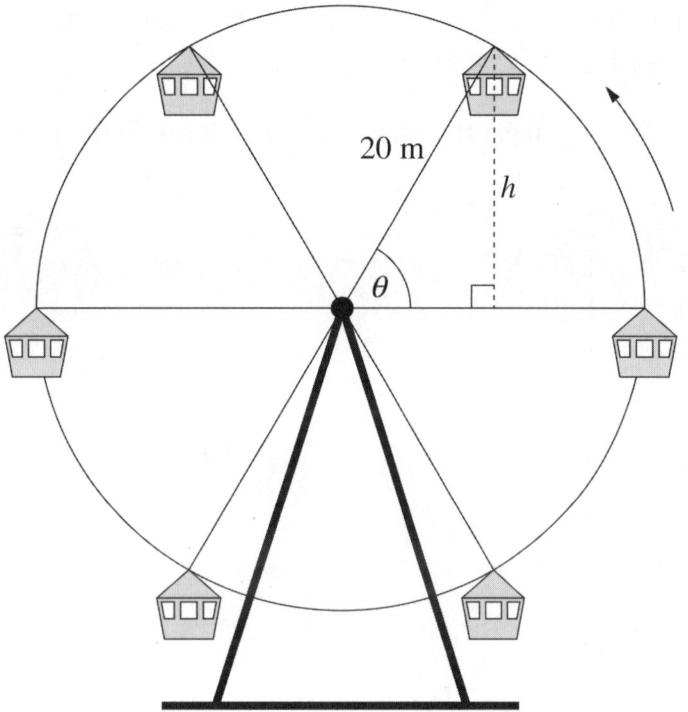

(i) Show that $\frac{dh}{d\theta} = 20\cos\theta$. **1**

(ii) At what speed is the top of the carriage rising when it is 15 metres higher than the horizontal diameter of the ferris wheel? Give your answer correct to one decimal place. **2**

(c) Let $f(x) = \sin^{-1}x + \cos^{-1}x$.

(i) Show that $f'(x) = 0$. **1**

(ii) Hence, or otherwise, prove $\sin^{-1}x + \cos^{-1}x = \frac{\pi}{2}$. **1**

(iii) Hence, sketch $f(x) = \sin^{-1}x + \cos^{-1}x$. **1**

Question 12 continues on the following page

Question 12 (continued)

(d) A group of 12 people sets off on a trek. The probability that a person finishes the trek within 8 hours is 0.75. **2**

Find an expression for the probability that at least 10 people from the group complete the trek within 8 hours.

(e) The points $P(2ap, ap^2)$ and $Q(2aq, aq^2)$ lie on the parabola $x^2 = 4ay$. The focus of the parabola is $S(0, a)$ and the tangents at P and Q intersect at $T(a(p+q), apq)$. (Do NOT prove this.)

The tangents at P and Q meet the x-axis at A and B respectively, as shown.

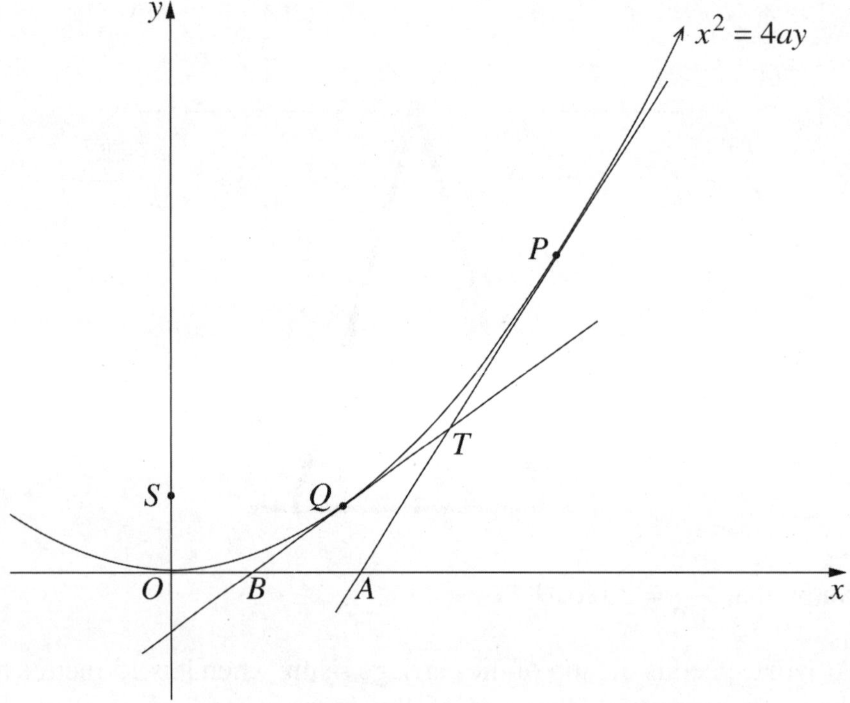

(i) Show that $\angle PAS = 90°$. **2**

(ii) Explain why S, B, A, T are concyclic points. **1**

(iii) Show that the diameter of the circle through S, B, A and T has length **2**

$$a\sqrt{(p^2+1)(q^2+1)}.$$

End of Question 12

Question 13 (15 marks) Use the Question 13 Writing Booklet.

(a) Prove by mathematical induction that, for $n \geq 1$, **3**

$$2 - 6 + 18 - 54 + \cdots + 2(-3)^{n-1} = \frac{1-(-3)^n}{2}.$$

(b) The diagram shows the graph $y = \dfrac{x}{x^2+1}$, for all real x.

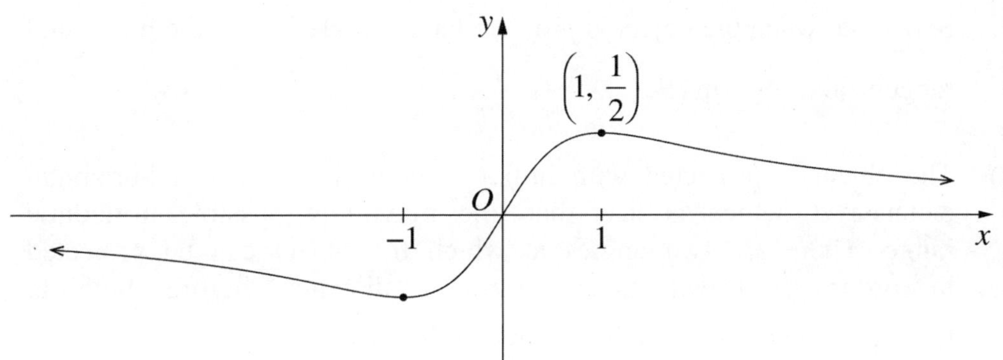

Consider the function $f(x) = \dfrac{x}{x^2+1}$, for $x \geq 1$.

The function $f(x)$ has an inverse. (Do NOT prove this.)

(i) State the domain and range of $f^{-1}(x)$. **2**

(ii) Sketch the graph $y = f^{-1}(x)$. **1**

(iii) Find an expression for $f^{-1}(x)$. **3**

Question 13 continues on the following page

Question 13 (continued)

* (c) An object is projected from the origin with an initial velocity of V at an angle θ to the horizontal. The equations of motion of the object are

$$x(t) = Vt\cos\theta$$
$$y(t) = Vt\sin\theta - \frac{gt^2}{2}. \quad \text{(Do NOT prove these.)}$$

(i) Show that when the object is projected at an angle θ, the horizontal range is $\frac{V^2}{g}\sin 2\theta$. **2**

(ii) Show that when the object is projected at an angle $\frac{\pi}{2} - \theta$, the horizontal range is also $\frac{V^2}{g}\sin 2\theta$. **1**

(iii) The object is projected with initial velocity V to reach a horizontal distance d, which is less than the maximum possible horizontal range. There are two angles at which the object can be projected in order to travel that horizontal distance before landing. Let these angles be α and β, where $\beta = \frac{\pi}{2} - \alpha$. **3**

Let h_α be the maximum height reached by the object when projected at the angle α to the horizontal.

Let h_β be the maximum height reached by the object when projected at the angle β to the horizontal.

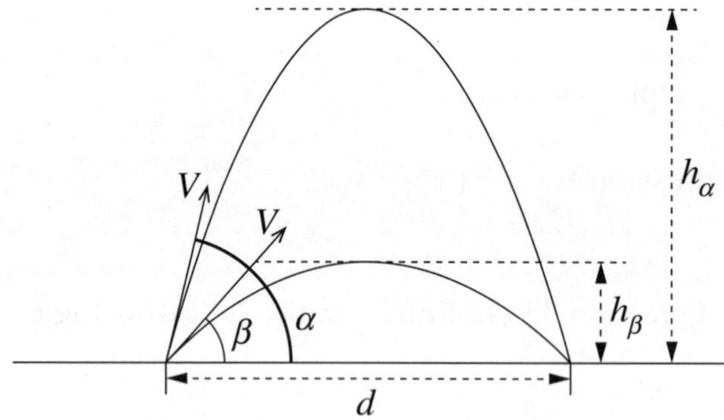

Show that the average of the two heights, $\frac{h_\alpha + h_\beta}{2}$, depends only on V and g.

* In the new Mathematics Extension 1 course, Projectile Motion questions will be expressed in vector form:

$x = Vt\cos\theta$ and $y = Vt\sin\theta - \frac{1}{2}gt^2$ is expressed as a position vector:

$$\underset{\sim}{r}(t) = (Vt\cos\theta)\underset{\sim}{i} + \left(Vt\sin\theta - \frac{1}{2}gt^2\right)\underset{\sim}{j}$$

End of Question 13

Question 14 (15 marks) Use the Question 14 Writing Booklet.

(a) The diagram shows quadrilateral $ABCD$ and the bisectors of the angles at A, B, C and D. The bisectors at A and B intersect at the point P. The bisectors at A and D meet at Q. The bisectors at C and D meet at R. The bisectors at B and C meet at S. **3**

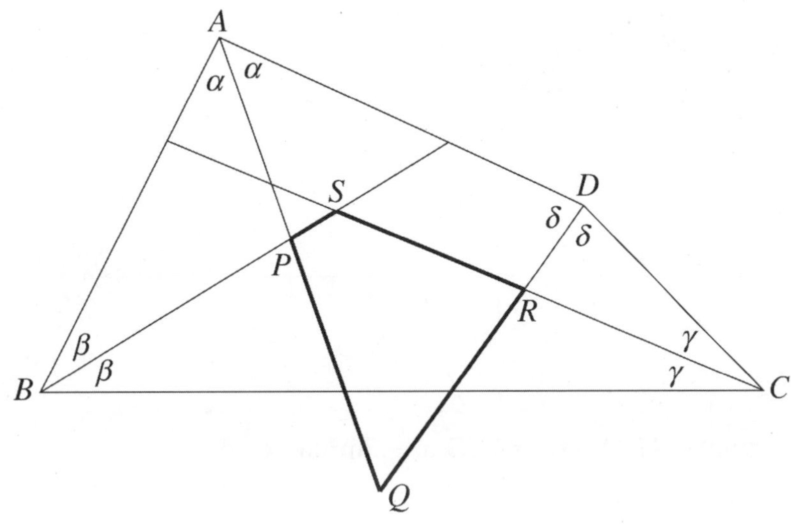

Copy or trace the diagram into your writing booklet.

Show that $PQRS$ is a cyclic quadrilateral.

(b) (i) By considering the expansions of $(1+(1+x))^n$ and $(2+x)^n$, show that **3**

$$\binom{n}{r}\binom{r}{r}+\binom{n}{r+1}\binom{r+1}{r}+\binom{n}{r+2}\binom{r+2}{r}+\cdots+\binom{n}{n}\binom{n}{r}=\binom{n}{r}2^{n-r}.$$

(ii) There are 23 people who have applied to be selected for a committee of 4 people. **2**

The selection process starts with Selector A choosing a group of at least 4 people from the 23 people who applied.

Selector B then chooses the 4 people to be on the committee from the group Selector A has chosen.

In how many ways could this selection process be carried out?

Question 14 continues on the following page

Question 14 (continued)

(c) In triangle ABC, BC is perpendicular to AC. Side BC has length a, side AC has length b and side AB has length c. A quadrant of a circle of radius x, centred at C, is constructed. The arc meets side BC at E. It touches the side AB at D, and meets side AC at F. The interval CD is perpendicular to AB.

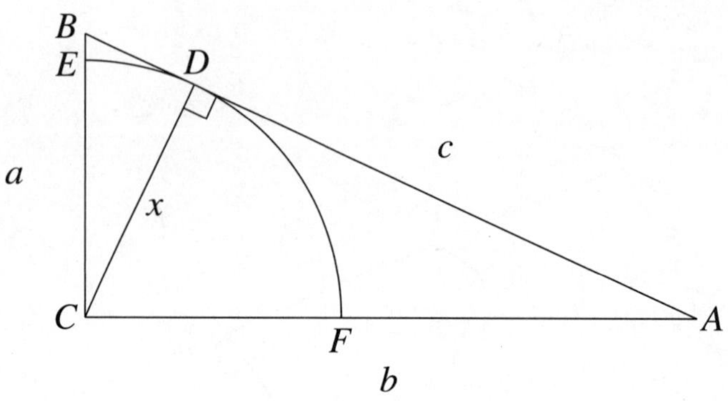

(i) Show that $\triangle ABC$ and $\triangle ACD$ are similar. **1**

(ii) Show that $x = \dfrac{ab}{c}$. **1**

From F, a line perpendicular to AC is drawn to meet AB at G, forming the right-angled triangle GFA. A new quadrant is constructed in triangle GFA touching side AB at H. The process is then repeated indefinitely.

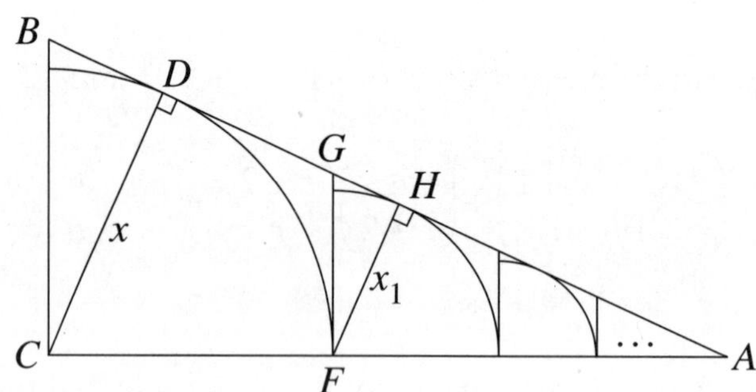

(iii) Show that the limiting sum of the areas of all the quadrants is $\dfrac{\pi ab^2}{4(2c-a)}$. **4**

(iv) Hence, or otherwise, show that $\dfrac{\pi}{2} < \dfrac{2c-a}{b}$. **1**

End of paper

Replacement questions

with content from the most up-to-date syllabus

Marks

Question 2 (1 mark)

Solve $\frac{|x+3|}{2} \le 1$ **1**

A. $-5 \le x \le -1$

B. $-5 \le x \le 1$

C. $-1 \le x \le 5$

D. $1 \le x \le 5$

Question 3 (1 mark)

Consider the parametric equations $x = 4\cos\theta - 3$ and $y = 4\sin\theta + 2$. **1**

Which of these is the corresponding cartesian equation?

A. $x^2 - 6x + y^2 + 4y = 3$

B. $x^2 - 2x + y^2 + 4y = 1$

C. $x^2 + 6x + y^2 - 4y = 3$

D. $x^2 + 6x + y^2 - 4y = 16$

Question 6 (1 mark)

Which of these is equal to $\cos 3\alpha \sin 2\alpha$? **1**

A. $\frac{1}{2}(\sin 5\alpha + \sin\alpha)$

B. $\frac{1}{2}(\sin 5\alpha - \sin\alpha)$

C. $\frac{1}{2}\left(\sin\frac{5\alpha}{2} + \sin\frac{\alpha}{2}\right)$

D. $\frac{1}{2}\left(\sin\frac{5\alpha}{2} - \sin\frac{\alpha}{2}\right)$

Marks

Question 7 (1 mark)

The amount of a drug, A mg, remaining in Emma's bloodstream t hours after taking the drug is given by the differential equation $\frac{dA}{dt} = -0.24A$.

The number of hours needed for the amount A to halve is **1**

A. $2\log_e \frac{6}{25}$

B. $25\log_e \frac{2}{3}$

C. $\frac{6}{25}\log_e 2$

D. $\frac{25}{6}\log_e 2$

REPLACEMENT QUESTIONS

Question 9 (1 mark)

A drawer contains many socks.

There are 6 blue socks, 10 red socks, 5 black socks, 4 yellow socks and 7 white socks.

What is the minimum number of randomly chosen socks to be taken out of the drawer until it is guaranteed that 4 socks are of the same colour? **1**

A. 11

B. 16

C. 20

D. 21

Question 10 (1 mark)

A binomial random variable X has a probability of success of p after 5 trials.

Use the table for the binomial distribution to find the value of p, correct to 2 decimal places. **1**

x	0	1	2	3	4	5
$P(X = x)$	0.00525	0.04877	0.18115	0.33642	0.31239	0.11603

A. $p = 0.56$

B. $p = 0.61$

C. $p = 0.63$

D. $p = 0.65$

Marks

Question 11 (2 marks)

(d) Express the vector $\underset{\sim}{u}$ in the form $x\underset{\sim}{i} + y\underset{\sim}{j}$, if the vector has a magnitude 6 and makes an angle of 120° with the horizontal. **2**

Question 12 (5 marks)

(e)* Evaluate $\int_{\frac{\pi}{2}}^{\frac{3\pi}{4}} \sin x \cos^2 x \, dx$, using the substitution $u = \cos x$. **3**

(e)** An examination consists of 50 multiple choice questions, each question having four possible answers.

A student guesses the answer to every question.

Let X be the number of correct answers.

Find the value of $Var(X)$. **2**

Question 14 (15 marks)

(a) The particles A and B have position vectors given as

$\underset{\sim}{r}_A(t) = (1 + 5t)\underset{\sim}{i} + (t^2 + 3)\underset{\sim}{j}$ and $\underset{\sim}{r}_B(t) = (t^2 + 5)\underset{\sim}{i} + (7t - 9)\underset{\sim}{j}$.

Determine when and where the particles collide. **3**

REPLACEMENT QUESTIONS

Question 14 (continued) **Marks**

(b) Each of the diagrams below shows the graph of $y = f(x)$ which has x-intercepts of $-4, 0, 2$ and 4 passes through the point $(1, 1)$.

On each diagram draw the graphs of the following functions:

(i) $y = |f(x)|$ **1**

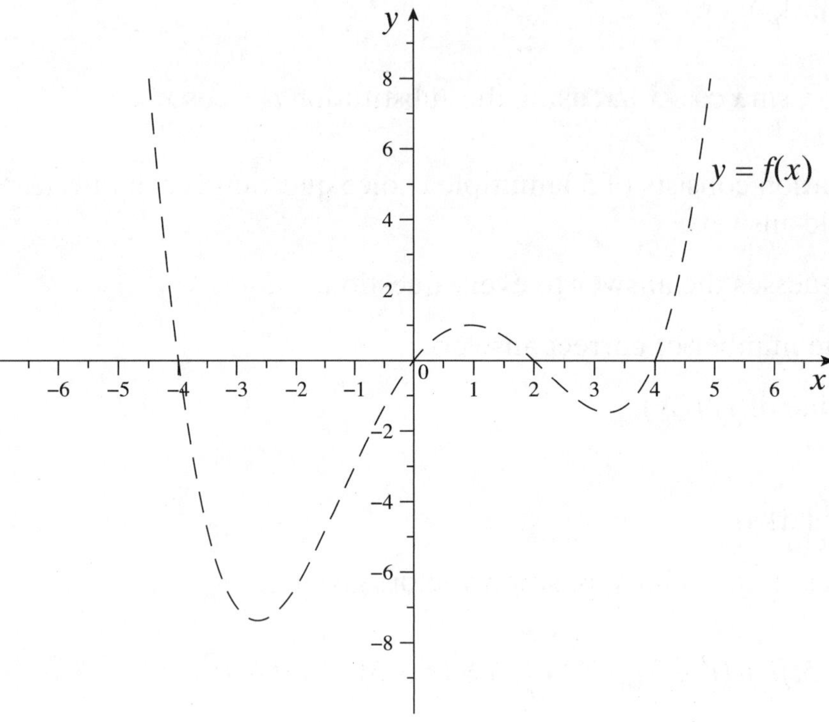

(ii) $y = \sqrt{f(x)}$ **2**

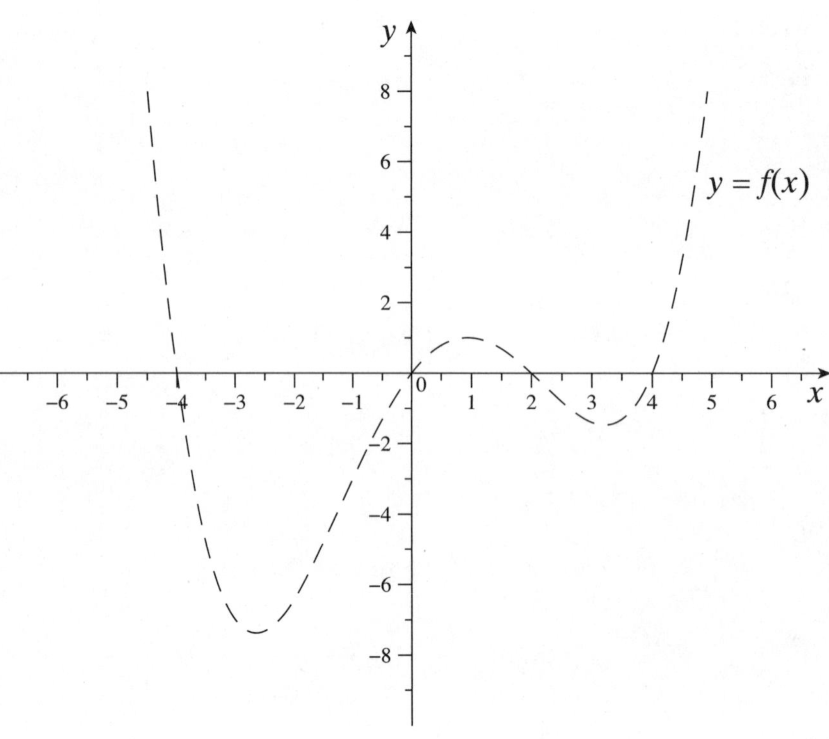

Marks

Question 14 (continued)

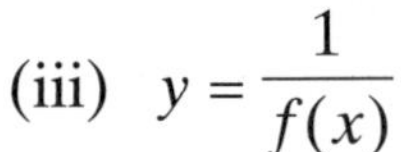

(iii) $y = \dfrac{1}{f(x)}$  **2**

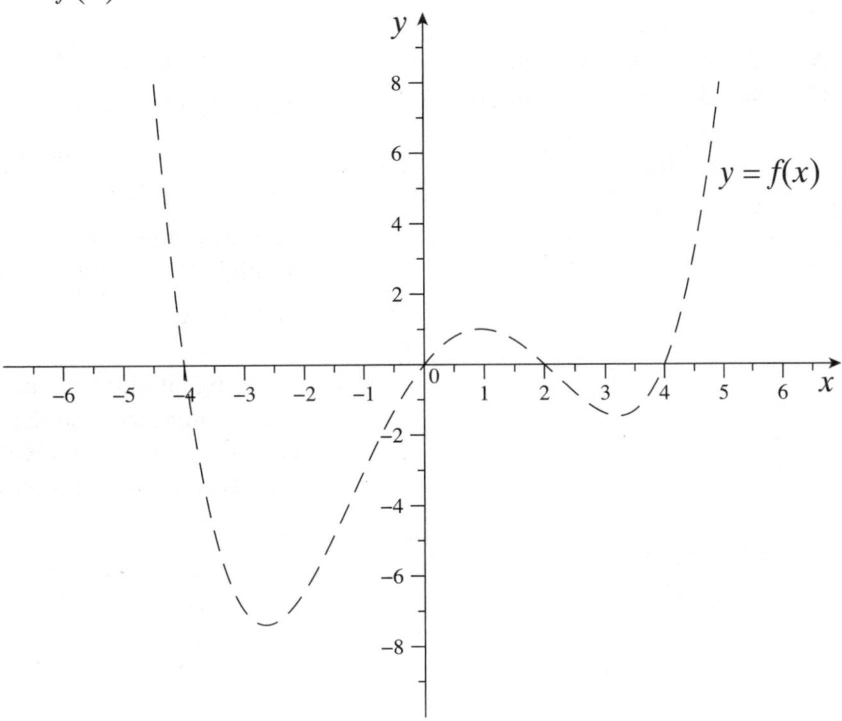

(c)* Find the volume of the solid formed when the curve $y = \sec x$ is rotated around the x-axis between $x = 0$ and $x = \dfrac{\pi}{4}$. **2**

(c)** A population of wild dogs in a national park grows in a way described by the logistical differential equation $\dfrac{dP}{dt} = 0.1P\left(1 - \dfrac{P}{2500}\right)$, where P is the number of wild dogs after t months, and the initial population is $P_0 = 200$. Also, $0 < P < 2500$.

[Use the result $\dfrac{25000}{P(2500-P)} = \left(\dfrac{10}{P} + \dfrac{10}{2500-P}\right)$]

(i) By solving the differential equation, show $P = \dfrac{5000e^{0.1t}}{23 + 2e^{0.1t}}$. **4**

(ii) How many wild dogs are there after 5 years? **1**

2018 Higher School Certificate
Worked answers

Section I (*Total 10 marks*)

1. B	**2.** A	**3.** A	**4.** D	**5.** A
6. C	**7.** C	**8.** B	**9.** D	**10.** B

1. $P(x) = 2x^3 + 6x^2 - 7x - 10$

$$\alpha\beta\gamma = -\frac{d}{a} = -\frac{-10}{2} = 5$$

$$\alpha + \beta + \gamma = -\frac{b}{a} = -\frac{6}{2} = -3$$

So $\alpha\beta\gamma(\alpha + \beta + \gamma) = 5 \times -3 = -15$

Answer B

2. $y = 3x$ has gradient $m_1 = 3$

$y = 5x$ has gradient $m_2 = 5$

$$\text{Now } \tan\theta = \left|\frac{m_1 - m_2}{1 + m_1 m_2}\right| = \left|\frac{3-5}{1+3\times 5}\right| = \frac{1}{8}$$

Answer A

3.
$$\lim_{x\to 0}\frac{\sin 3x\cos 3x}{12x} = \lim_{x\to 0}\frac{\sin 6x}{24x} = \frac{1}{4}\lim_{x\to 0}\frac{\sin 6x}{6x} = \frac{1}{4}$$

Answer A

4. There are single roots at –2 and –1 and a double root at 1.

So possible values of b and c are 2 and 1 and a possible value of d is –1.

So $y = a(x + 2)(x + 1)(x - 1)^2$

When $x = 0$, $y = -6$

$-6 = a(0 + 2)(0 + 1)(0 - 1)^2$

$-6 = 2a$

$a = -3$

Answer D

5. $P(0) = 3000$

Now $1500e^0 = 1500$

So possible equations are A and D.

As $t \to \infty$, $1500e^{-kt} \to 0$

So the equation that best represents the graph is $P(t) = 1500 + 1500e^{-kt}$.

Answer A

6. The tangent at c cuts the x-axis at the closest point to w, so the initial approximation $x_1 = c$, will give the second approximation that is closest to $x = w$.

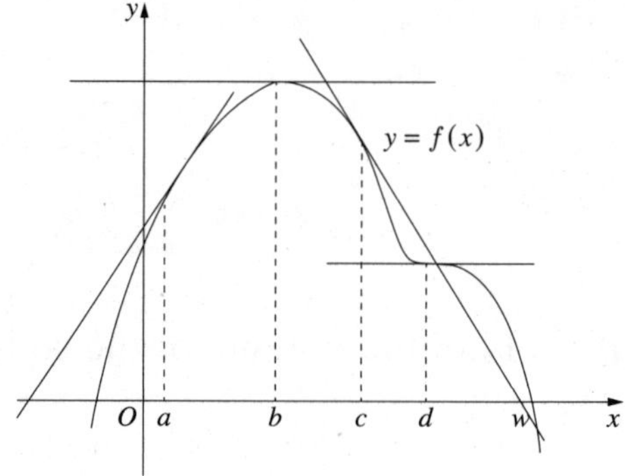

Answer C

7. $v = x^2 + 2$

$$\text{Now } a = v\frac{dv}{dx} = (x^2 + 2) \times 2x = 2x^3 + 4x$$

When $x = 1$,

$$\frac{dv}{dt} = 2 \times 1^3 + 4 \times 1 = 6$$

The acceleration of the particle is 6 m s^{-2}.

Answer C

8. The men can sit in (6 – 1)! or 5! ways.

For each of those the women can sit in 6! ways.

Number of ways = 5! 6!

Answer B

9. $\sin 2x = -\frac{1}{2}$

$$2x = n\pi + (-1)^n \sin^{-1}\left(-\frac{1}{2}\right)$$
$$= n\pi + (-1)^n\left(-\frac{\pi}{6}\right)$$
$$= n\pi + (-1)^{n+1}\frac{\pi}{6}$$
$$x = \frac{n\pi}{2} + (-1)^{n+1}\frac{\pi}{12}$$

Answer D

10. $v^2 = n^2(2kx - x^2)$

$$= n^2(k^2 - k^2 + 2kx - x^2)$$
$$= n^2(k^2 - (x^2 - 2kx + k^2))$$
$$= n^2(k^2 - (x - k)^2)$$

So $x - k = k\cos(nt + \alpha)$

$$x = k\cos(nt + \alpha) + k$$

But $x = k$ when $t = 0$.

So $x = k\sin(nt) + k$ could represent the motion.

Answer B

Section II

QUESTION 11

(a) $P(x) = x^3 - 2x^2 - 5x + 6$

(i) $P(1) = 1^3 - 2 \times 1^2 - 5 \times 1 + 6$
$= 0$
$\therefore x = 1$ is a zero of $P(x)$.

(1 mark)

(ii)
$$\begin{array}{r} x^2 - x - 6 \\ x-1\overline{)x^3 - 2x^2 - 5x + 6} \\ x^3 - x^2 \quad\quad\quad\quad \\ -x^2 - 5x \quad\quad \\ -x^2 + x \quad\quad \\ -6x + 6 \\ -6x + 6 \\ 0 \end{array}$$

So $P(x) = (x - 1)(x^2 - x - 6)$
$= (x - 1)(x + 2)(x - 3)$

$\therefore$ the other zeros are $x = -2$ and $x = 3$.

(2 marks)

(b) $\log_2 5 + \log_2(x - 2) = 3$

$$\log_2(5x - 10) = 3$$
$$\therefore 5x - 10 = 2^3$$
$$5x - 10 = 8$$
$$5x = 18$$
$$x = 3.6$$

(2 marks)

(c) $\sqrt{3}\sin x + \cos x \equiv R\sin(x + \alpha)$

Now $R\sin(x + \alpha)$
$= R\sin x\cos\alpha + R\cos x\sin\alpha$
$\therefore R\cos\alpha = \sqrt{3}$ and $R\sin\alpha = 1$

$$\frac{R\sin\alpha}{R\cos\alpha} = \frac{1}{\sqrt{3}}$$
$$\tan\alpha = \frac{1}{\sqrt{3}}$$
$$\alpha = \frac{\pi}{6} \quad \left(0 \le \alpha \le \frac{\pi}{2}\right)$$
$$R\cos\frac{\pi}{6} = \sqrt{3}$$
$$R \times \frac{\sqrt{3}}{2} = \sqrt{3}$$
$$R = 2$$
$$\therefore \sqrt{3}\sin x + \cos x \equiv 2\sin\left(x + \frac{\pi}{6}\right)$$

(2 marks)

(d) The products of the intercepts of secants from an external point are equal.

$$\therefore x(x + 2) = 3 \times (3 + 5)$$
$$x^2 + 2x = 24$$
$$x^2 + 2x - 24 = 0$$
$$(x - 4)(x + 6) = 0$$
$$x = 4 \text{ or } x = -6$$

But $x > 0$

$\therefore x = 4$

(2 marks)

(e) $f(x) = \dfrac{1}{4x-1}$

(i) Now $4x - 1 \neq 0$

$$4x \neq 1$$

$$x \neq \frac{1}{4}$$

Domain is all real x except $x = \frac{1}{4}$.

(1 mark)

(ii) $f(x) < 1$

If $x < \frac{1}{4}$, the denominator is negative and so $f(x)$ is also negative and hence less than 1.

If $x > \frac{1}{4}$,

$$\frac{1}{4x-1} < 1$$

$$1 < 4x - 1$$

$$2 < 4x$$

$$\frac{1}{2} < x$$

So $f(x) < 1$ when $x < \frac{1}{4}$ and when $x > \frac{1}{2}$.

(2 marks)

(f) Let $u = 1 - x$

$x = 1 - u$

$dx = -du$

When $x = 0$, $u = 1$

When $x = -3$, $u = 4$

$$\therefore \int_{-3}^{0} \frac{x}{\sqrt{1-x}}\,dx = \int_{4}^{1} \frac{1-u}{\sqrt{u}} \cdot -du$$

$$= \int_{1}^{4} \left(u^{-\frac{1}{2}} - u^{\frac{1}{2}}\right) du$$

$$= \left[\frac{u^{\frac{1}{2}}}{\frac{1}{2}} - \frac{u^{\frac{3}{2}}}{\frac{3}{2}}\right]_1^4$$

$$= \left[2\sqrt{u} - \frac{2u\sqrt{u}}{3}\right]_1^4$$

$$= 2\times\sqrt{4} - \frac{2\times 4\times\sqrt{4}}{3} - \left(2\times\sqrt{1} - \frac{2\times 1\times\sqrt{1}}{3}\right)$$

$$= -2\frac{2}{3}$$

(3 marks)

QUESTION 12

(a) $\cos 2\theta = \cos^2\theta - \sin^2\theta$

$= 2\cos^2\theta - 1$

So $\cos^2\theta = \frac{1}{2}(\cos 2\theta + 1)$

$\int \cos^2(3x)\,dx = \int \frac{1}{2}(\cos 6x + 1)dx$

$= \frac{1}{12}\sin 6x + \frac{x}{2} + C$

(2 marks)

(b) (i) $\sin\theta = \frac{h}{20}$

$h = 20\sin\theta$

$\frac{dh}{d\theta} = 20\cos\theta$

20 m

h

θ

(1 mark)

(ii) $\frac{d\theta}{dt} = 1.5$

$\frac{dh}{dt} = \frac{dh}{d\theta}\cdot\frac{d\theta}{dt}$

$= 20\cos\theta \times 1.5$

$= 30\cos\theta$

When $h = 15$,

$\sin\theta = \frac{15}{20}$

$= \frac{3}{4}$

$\cos\theta = \sqrt{1-\left(\frac{3}{4}\right)^2}$ (θ is acute, so $\cos\theta > 0$)

So $\frac{dh}{dt} = 30\sqrt{1-\left(\frac{3}{4}\right)^2}$

$= 19.843\,1348\ldots$

$= 19.8$ (1 d.p.)

The top of the carriage is rising at 19.8 m per minute, correct to one decimal place.

(2 marks)

(c) (i) $f(x) = \sin^{-1}x + \cos^{-1}x$

$f'(x) = \frac{1}{\sqrt{1-x^2}} + \frac{-1}{\sqrt{1-x^2}}$

$= 0$

(1 mark)

(ii) $f'(x) = 0$

$\therefore f(x)$ is constant.

When $x = 0$,

$$f(x) = \sin^{-1}0 + \cos^{-1}0 = 0 + \frac{\pi}{2} = \frac{\pi}{2}$$

$\therefore f(x) = \frac{\pi}{2}$ for all x in the domain.

$\therefore \sin^{-1}x + \cos^{-1}x = \frac{\pi}{2}$ *(1 mark)*

(iii) $\sin^{-1}x$ and $\cos^{-1}x$ have domain $-1 \leq x \leq 1$.

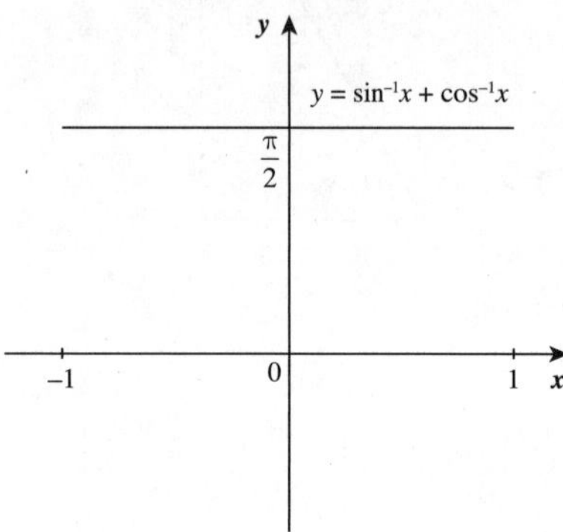

(1 mark)

(d) $P(\text{finishes in 8 h}) = 0.75$

$\therefore P(\text{does not finish in 8 h}) = 0.25$

$P(\text{at least 10 finish in 8 h})$

$$= P(10\text{ finish}) + P(11\text{ finish}) + P(12\text{ finish})$$

$$= \binom{12}{10}(0.75)^{10}(0.25)^2 + \binom{12}{11}(0.75)^{11}(0.25) + \binom{12}{12}(0.75)^{12}$$

$$= 66\left(\frac{3}{4}\right)^{10}\left(\frac{1}{4}\right)^2 + 12\left(\frac{3}{4}\right)^{11}\left(\frac{1}{4}\right) + \left(\frac{3}{4}\right)^{12}$$

$$= \frac{66 \times 3^{10} + 12 \times 3^{11} + 3^{12}}{4^{12}}$$

$$= \frac{3^{11}(22 + 12 + 3)}{4^{12}}$$

$$= \frac{37 \times 3^{11}}{4^{12}}$$

(2 marks)

(e) (i) The tangent at P has equation $y = px - ap^2$.

The gradient of the tangent is $m_1 = p$.

At A, $y = 0$

$$\therefore px - ap^2 = 0$$

$$px = ap^2$$

$$x = ap \quad (p \neq 0)$$

$A(ap, 0)$, $S(0, a)$

$$m = \frac{y_2 - y_1}{x_2 - x_1}$$

$$m_2 = \frac{a-0}{0-ap}$$
$$= -\frac{1}{p}$$

Now $m_1 m_2 = p \times -\frac{1}{p}$
$= -1$

$\therefore AS$ is perpendicular to AP.

$\therefore \angle PAS = 90°$

(2 marks)

(ii) Similarly, BS is perpendicular to BQ.
So $\angle TBS = 90°$

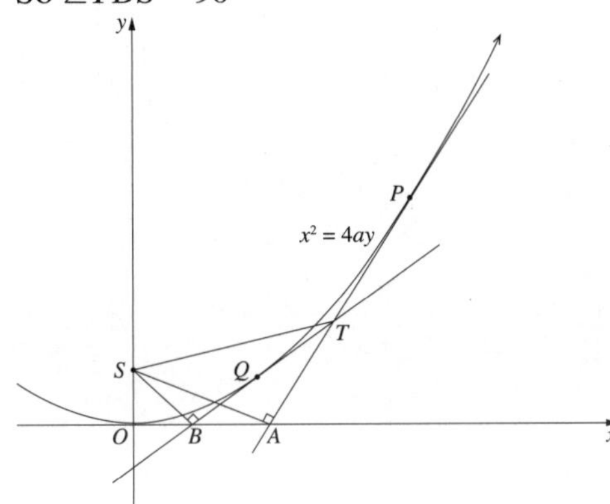

So $\angle TAS = \angle TBS$
Angles in same segment standing on the same arc are equal.
$\therefore S, B, A$ and T are concyclic points.

(1 mark)

(iii) An angle in a semicircle = 90°.
So ST is a diameter of the circle through S, B, A and T.
$S(0, a)$, $T(a(p + q), apq)$

$$d = \sqrt{(x_2 - x_1)^2 + (y_2 - y_1)^2}$$
$$= \sqrt{(a(p+q)-0)^2 + (apq-a)^2}$$
$$= \sqrt{a^2(p+q)^2 + a^2(pq-1)^2}$$
$$= a\sqrt{p^2 + 2pq + q^2 + p^2q^2 - 2pq + 1}$$
$$= a\sqrt{p^2q^2 + p^2 + q^2 + 1}$$
$$= a\sqrt{p^2(q^2+1) + 1(q^2+1)}$$
$$= a\sqrt{(p^2+1)(q^2+1)}$$

(2 marks)

QUESTION 13

(a) To prove $2 - 6 + 18 - 54 + \ldots + 2(-3)^{n-1} = \frac{1-(-3)^n}{2}$.

When $n = 1$,

$\text{LHS} = 2(-3)^{1-1}$
$= 2$

$\text{RHS} = \frac{1-(-3)^1}{2}$
$= 2$

$\therefore$ it is true for $n = 1$.

Assume true for $n = k$

i.e. assume $2 - 6 + 18 - 54 + \ldots + 2(-3)^{k-1} = \frac{1-(-3)^k}{2}$

We want to prove that it is true for $n = k + 1$.

i.e. $2 - 6 + 18 - 54 + \ldots + 2(-3)^k = \frac{1-(-3)^{k+1}}{2}$

$$\begin{aligned}\text{LHS} &= 2 - 6 + 18 - 54 + \ldots + 2(-3)^{k-1} + 2(-3)^k \\ &= \frac{1-(-3)^k}{2} + 2(-3)^k \\ &= \frac{1-(-3)^k + 4(-3)^k}{2} \\ &= \frac{1+3(-3)^k}{2} \\ &= \frac{1-(-3)^{k+1}}{2} \\ &= \text{RHS}\end{aligned}$$

So, if true for $n = k$, it is also true for $n = k + 1$.
It is true for $n = 1$, so it is true for $n = 1 + 1 = 2$.
It is true for $n = 2$, so it is true for $n = 2 + 1 = 3$.
And so on.
By the process of induction it is true for all integers $n \geq 1$.

(3 marks)

(b) (i) $f(x) = \dfrac{x}{x^2+1}, \quad x \geq 1$

Range of $f(x)$ is $0 < y \leq \dfrac{1}{2}$.

So the domain of $f^{-1}(x)$ is $0 < x \leq \dfrac{1}{2}$.

Range of $f^{-1}(x)$ is $y \geq 1$.

(2 marks)

(ii)

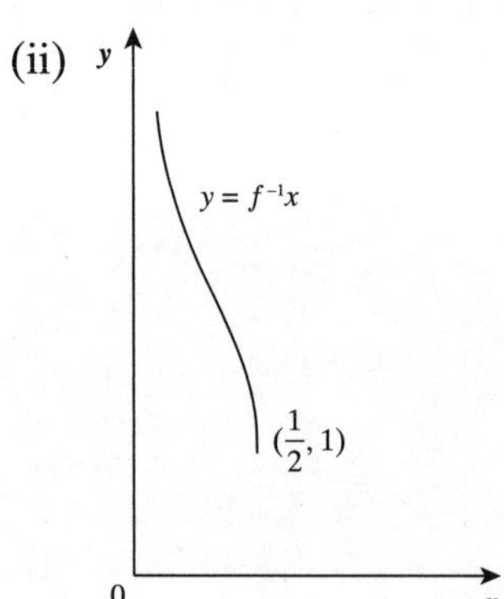

(1 mark)

(iii) $y = \dfrac{x}{x^2+1}$

Inverse function:

$$x = \frac{y}{y^2+1}$$

$$xy^2 + x = y$$

$$xy^2 - y + x = 0$$

$$\begin{aligned} y &= \frac{-b \pm \sqrt{b^2 - 4ac}}{2a} \\ &= \frac{-(-1) \pm \sqrt{(-1)^2 - 4 \times x \times x}}{2x} \\ &= \frac{1 \pm \sqrt{1-4x^2}}{2x}\end{aligned}$$

But, $y \geq 1$

$$\therefore y = \frac{1+\sqrt{1-4x^2}}{2x}$$

(3 marks)

(c) (i) $y = Vt\sin\theta - \dfrac{gt^2}{2}$

When $y = 0$,

$$Vt\sin\theta - \frac{gt^2}{2} = 0$$

$$t\left(V\sin\theta - \frac{gt}{2}\right) = 0$$

$$t = 0 \text{ or } V\sin\theta - \frac{gt}{2} = 0$$

$$gt = 2V\sin\theta$$

$$t = \frac{2V\sin\theta}{g}$$

Now $x = Vt\cos\theta$

When $t = \dfrac{2V\sin\theta}{g}$,

$$x = \frac{2V^2\sin\theta\cos\theta}{g}$$

$$= \frac{V^2\sin 2\theta}{g}$$

So the horizontal range of the projectile is $\dfrac{V^2}{g}\sin 2\theta$.

(2 marks)

(ii) When the object is projected at $\dfrac{\pi}{2} - \theta$:

$$x = \frac{V^2}{g}\sin 2\left(\frac{\pi}{2} - \theta\right)$$

$$= \frac{V^2}{g}\sin(\pi - 2\theta)$$

$$= \frac{V^2}{g}\sin 2\theta$$

So the horizontal range of the projectile is also $\dfrac{V^2}{g}\sin 2\theta$.

(1 mark)

(iii) When projected at angle α:

$$y = Vt\sin\alpha - \frac{gt^2}{2}$$

$$\dot{y} = V\sin\alpha - gt$$

Maximum height occurs when $\dot{y} = 0$

i.e $0 = V\sin\alpha - gt$

$$gt = V\sin\alpha$$

$$t = \frac{V\sin\alpha}{g}$$

When $t = \dfrac{V\sin\alpha}{g}$,

$$h_\alpha = \frac{V^2\sin^2\alpha}{g} - \frac{g\left(\frac{V\sin\alpha}{g}\right)^2}{2}$$

$$= \frac{V^2\sin^2\alpha}{g} - \frac{V^2\sin^2\alpha}{2g}$$

$$= \frac{V^2\sin^2\alpha}{2g}$$

Similarly, when projected at angle β:

$$h_\beta = \frac{V^2\sin^2\beta}{2g}$$

Now $\sin^2\beta = \sin^2\left(\dfrac{\pi}{2} - \alpha\right)$

$$= \cos^2\alpha$$

So $h_\beta = \dfrac{V^2\cos^2\alpha}{2g}$

$$\frac{h_\alpha + h_\beta}{2} = \frac{\frac{V^2\sin^2\alpha}{2g} + \frac{V^2\cos^2\alpha}{2g}}{2}$$

$$= \frac{V^2\left(\sin^2\alpha + \cos^2\alpha\right)}{4g}$$

$$= \frac{V^2}{4g}$$

So the average of the two heights depends only on V and g.

(3 marks)

QUESTION 14

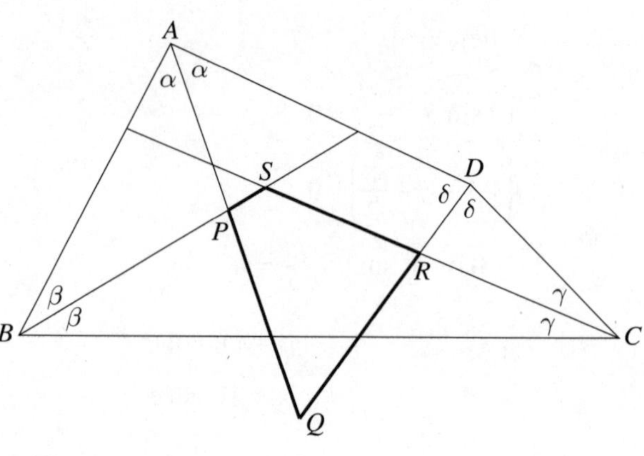

(a) $\angle DAB + \angle ABC + \angle BCD + \angle CDA = 360°$
(angle sum of quadrilateral)

$$\text{So } 2\alpha + 2\beta + 2\gamma + 2\delta = 360°$$
$$\alpha + \beta + \gamma + \delta = 180°$$

$\angle BPQ = \angle PAB + \angle ABP$ (ext. $\angle$ of Δ)
$= \alpha + \beta$
$\therefore \angle SPQ = 180° - (\alpha + \beta)$ (angles in st. line)
$\angle DRS = \angle RCD + \angle CDR$ (ext. $\angle$ of Δ)
$= \gamma + \delta$
$\therefore \angle QRS = 180° - (\gamma + \delta)$ (angles in st. line)

$$\begin{aligned}\angle SPQ + \angle QRS &= 180° - (\alpha + \beta) + 180° - (\gamma + \delta)\\ &= 360° - (\alpha + \beta + \gamma + \delta)\\ &= 360° - 180°\\ &= 180°\end{aligned}$$

$\therefore PQRS$ is a cyclic quadrilateral because the opposite angles are supplementary.

(3 marks)

(b) (i) $(1 + (1 + x))^n$

$$= \binom{n}{0} + \binom{n}{1}(1 + x) + \binom{n}{2}(1 + x)^2 + \binom{n}{3}(1 + x)^3 + \ldots. + \binom{n}{n}(1 + x)^n$$

$$= \binom{n}{0} + \binom{n}{1}\left[\binom{1}{0} + \binom{1}{1}x\right] + \binom{n}{2}\left[\binom{2}{0} + \binom{2}{1}x + \binom{2}{2}x^2\right]$$
$$+ \binom{n}{3}\left[\binom{3}{0} + \binom{3}{1}x + \binom{3}{2}x^2 + \binom{3}{3}x^3\right] + \ldots. + \binom{n}{n}\left[\binom{n}{0} + \binom{n}{1}x + \ldots + \binom{n}{n}x^n\right]$$

So the coefficient of x^r will be $\binom{n}{r}\binom{r}{r} + \binom{n}{r+1}\binom{r+1}{r} + \binom{n}{r+2}\binom{r+2}{r} + \ldots + \binom{n}{n}\binom{n}{r}$

$$(2 + x)^n = \binom{n}{0}2^n + \binom{n}{1}2^{n-1}x + \binom{n}{2}2^{n-2}x^2 + \binom{n}{3}2^{n-3}x^3 + \ldots. + \binom{n}{n}x^n$$

The coefficient of x^r will be $\binom{n}{r}2^{n-r}$.

But $1 + (1 + x) = 2 + x$.

So the coefficient of x^r in both expansions must be equal.

$$\therefore \binom{n}{r}\binom{r}{r} + \binom{n}{r+1}\binom{r+1}{r} + \binom{n}{r+2}\binom{r+2}{r} + \ldots + \binom{n}{n}\binom{n}{r} = \binom{n}{r}2^{n-r}$$

(3 marks)

(ii) Selector A chooses at least 4 from 23.
So the number of ways selector A can choose a group is

$$\binom{23}{4} + \binom{23}{5} + \binom{23}{6} + \binom{23}{7} + \ldots + \binom{23}{23}$$

If selector A chooses a group of 4 then selector B must choose 4 from 4.
If selector A chooses a group of 5 then selector B must choose 4 from 5. So for each of the groups of 5 that selector A has chosen there are $\binom{5}{4}$ possible selections of the group of 4.
If selector A chooses a group of 6 then selector B must choose 4 from 6. So, for each of the groups of 6 that selector A has chosen there are $\binom{6}{4}$ possible selections of the group of 4, and so on.

So the number of ways for the selection process is given by:

$$\binom{23}{4}\binom{4}{4}+\binom{23}{5}\binom{5}{4}+\binom{23}{6}\binom{6}{4}+\binom{23}{7}\binom{7}{4}+\ldots+\binom{23}{23}\binom{23}{4}$$

Using the result from part (i) the selection process could be carried out in $\binom{23}{4}2^{23-4}$ ways.

So the selection process could be carried out in $\binom{23}{4}2^{19}$ or 4 642 570 240 ways.

(2 marks)

(c) (i) In ΔABC, ΔACD

$\angle BAC = \angle CAD$ (common angle)

$\angle ACB = \angle ADC$ (both $90°$)

$\therefore \Delta ABC \;|||\; \Delta ACD$ (equiangular)

(1 mark)

(ii) $\dfrac{CD}{BC} = \dfrac{AC}{AB}$ (corresponding sides similar triangles)

$$\frac{x}{a} = \frac{b}{c}$$

$$x = \frac{ab}{c}$$

(1 mark)

(iii) Total area of all quadrants

$$= \frac{1}{4}\pi x^2 + \frac{1}{4}\pi {x_1}^2 + \frac{1}{4}\pi {x_2}^2 + \ldots$$

$$= \frac{\pi}{4}(x^2 + {x_1}^2 + {x_2}^2 + \ldots)$$

Now $\Delta AHF \;|||\; \Delta ADC$ (equiangular)

So $\dfrac{HF}{DC} = \dfrac{AF}{AC}$

$$\frac{x_1}{x} = \frac{b-x}{b}$$

$$= \frac{b - \frac{ab}{c}}{b}$$

$$= 1 - \frac{a}{c}$$

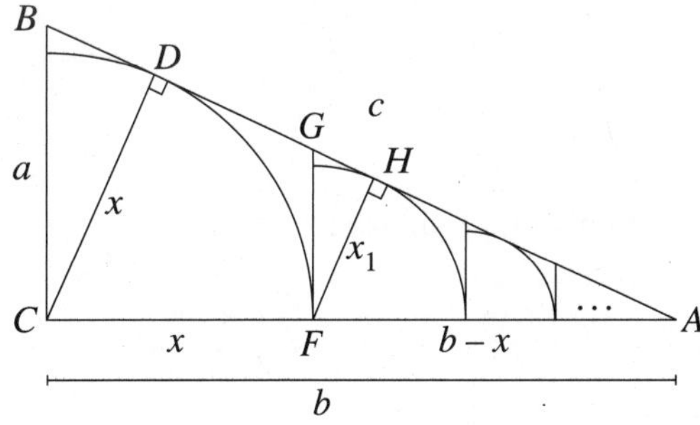

Now as the same process of forming quadrants continues each subsequent triangle will be similar to all the previous ones and the ratios of the radii will be the same.

So $\frac{x_2}{x_1} = 1 - \frac{a}{c}$ and $\frac{x_3}{x_2} = 1 - \frac{a}{c}$ and so on.

$$\left(\frac{x_1}{x}\right)^2 = \left(\frac{x_2}{x_1}\right)^2 = \left(\frac{x_3}{x_2}\right)^2 \ldots$$

So $x^2 + x_1{}^2 + x_2{}^2 + \ldots$ is a geometric series.

$$\begin{aligned} r &= \left(1 - \frac{a}{c}\right)^2 \\ &= 1 - \frac{2a}{c} + \frac{a^2}{c^2} \\ &= 1 - \frac{2ac - a^2}{c^2} \\ &= 1 - \frac{a(2c-a)}{c^2} \end{aligned}$$

As $0 < r < 1, x^2 + x_1{}^2 + x_2{}^2 + \ldots$ is an infinite geometric series.

$$\begin{aligned} \text{Limiting sum} &= \frac{x^2}{1 - \left(1 - \frac{a(2c-a)}{c^2}\right)} \\ &= \frac{\left(\frac{ab}{c}\right)^2}{\frac{a(2c-a)}{c^2}} \\ &= \frac{a^2b^2}{c^2} \times \frac{c^2}{a(2c-a)} \\ &= \frac{ab^2}{2c-a} \end{aligned}$$

$$\begin{aligned} \text{Limiting sum of all areas} &= \frac{\pi}{4} \times \frac{ab^2}{2c-a} \\ &= \frac{\pi ab^2}{4(2c-a)} \end{aligned}$$

(4 marks)

(iv) Limiting sum of all areas must be less than the area of the triangle.

So $\frac{\pi ab^2}{4(2c-a)} < \frac{1}{2}ab$

$\frac{\pi b}{2(2c-a)} < 1 \qquad (a > 0, b > 0)$

$\frac{\pi}{2} < \frac{2c-a}{b} \qquad (c > a \text{ so } 2c - a > 0)$

(1 mark)

Solutions to replacement questions

QUESTION 2

$$\frac{|x+3|}{2} \le 1$$

$$-1 \le \frac{x+3}{2} \le 1$$

$$-2 \le x+3 \le 2$$

$$-5 \le x \le -1$$

Answer B

(1 mark)

QUESTION 3

$$x = 4\cos\theta - 3 \qquad y = 4\sin\theta + 2$$

$$\frac{x+3}{2} = \cos\theta \qquad \frac{y-2}{4} = \sin\theta$$

As $\cos^2\theta + \sin^2\theta = 1$, then

$$\left(\frac{x+3}{4}\right)^2 + \left(\frac{y-2}{4}\right)^2 = 1$$

$$(x+3)^2 + (y-2)^2 = 16$$

$$x^2 + 6x + 9 + y^2 - 4y + 4 = 16$$

$$x^2 + 6x + y^2 - 4y = 16 - 13$$

$$x^2 + 6x + y^2 - 4y = 3$$

Answer C

(1 mark)

QUESTION 6

As $2\cos\frac{A+B}{2}\sin\frac{A-B}{2} = \sin A - \sin B$

$$\cos\frac{A+B}{2}\sin\frac{A-B}{2} = \frac{1}{2}(\sin A - \sin B)$$

Hence, $\frac{A+B}{2} = 3 \quad \therefore A + B = 6 \ldots ①$

Also, $\frac{A-B}{2} = 2 \quad \therefore A - B = 4 \ldots ②$

$① - ②: 2A = 10$

$A = 5$

By substitution, $B = 1$

$\therefore \frac{1}{2}(\sin 5\alpha - \sin\alpha)$

Answer B

(1 mark)

QUESTION 7

As $\frac{dA}{dt} = -0.24A$, then $A = A_0e^{-0.24t}$.

Let $\frac{A}{A_0} = \frac{1}{2}$.

$$\therefore e^{-0.24t} = \frac{1}{2}$$

$$-0.24t = -\log_e 2$$

$$t = \frac{\log_e 2}{0.24}$$

$$= \frac{25}{6}\log_e 2$$

Answer D

(1 mark)

QUESTION 9

As there are 4 socks to be chosen, then $n = 4$.

The number of colours $(k) = 5$

Smallest group $= (n-1)k + 1$

$= 3 \times 5 + 1$

$= 16$

$\therefore$ at least 16 socks need to be selected.

Answer B

(1 mark)

QUESTION 10

$$P(X = 5) = \binom{5}{5}p^5(1-p)^0$$

$$p^5 = 0.11603$$

$$p = 0.65000105\ldots$$

$$= 0.65 \text{ (2 dec. pl.)}$$

Answer D

(1 mark)

QUESTION 11

(d) $x = |\underset{\sim}{u}|\cos\theta$

$= 6\cos 120°$

$= 6 \times -\dfrac{1}{2}$

$= -3$

$y = |\underset{\sim}{u}|\sin\theta$

$= 6\sin 120°$

$= 6 \times \dfrac{\sqrt{3}}{2}$

$= 3\sqrt{3}$

$\therefore \underset{\sim}{u} = x\underset{\sim}{i} + y\underset{\sim}{j}$

$= -3\underset{\sim}{i} + 3\sqrt{3}\,\underset{\sim}{j}$ *(2 marks)*

QUESTION 12

(e)* $u = \cos x \quad \dfrac{du}{dx} = -\sin x$

$dx = \dfrac{du}{-\sin x}$

Also, $x = \dfrac{3\pi}{4}, u = -\dfrac{1}{\sqrt{2}}; x = \dfrac{\pi}{2}, u = 0$

$$\int_{\frac{\pi}{2}}^{\frac{3\pi}{4}} \sin x\cos^2 x\,dx = \int_0^{-\frac{1}{\sqrt{2}}} u^2 \sin x.\frac{du}{-\sin x}$$

$$= \int_{-\frac{1}{\sqrt{2}}}^{0} u^2\,dx$$

$$= \left[\frac{u^3}{3}\right]_{-\frac{1}{\sqrt{2}}}^{0}$$

$$= \frac{0^3}{3} - \frac{(-\frac{1}{\sqrt{2}})^3}{3}$$

$$= \frac{1}{6\sqrt{2}} \times \frac{\sqrt{2}}{\sqrt{2}}$$

$$= \frac{\sqrt{2}}{12}$$

(3 marks)

(e)** The number of correct answers, X, is a binomial variable with $n = 50$ and $p = 0.25$.

$\mu = np$

$= 50(0.25)$

$= 12.5$

$Var(X) = np(1 - p)$

$= 12.5(1 - 0.25)$

$= 9.375$

(5 marks)

QUESTION 14

(a) Equate the components of $\underset{\sim}{i}$: $1 + 5t = t^2 + 5$

$t^2 - 5t + 4 = 0$

$(t-1)(t-4) = 0$

$t = 1, 4$

Equate the components of $\underset{\sim}{j}$: $t^2 + 3 = 7t - 9$

$t^2 - 7t + 12 = 0$

$(t-4)(t-3) = 0$

$t = 4, 3$

$\therefore$ the common solution is $t = 4$.

The particles collide after 4 seconds.

Substituting: $\underset{\sim}{r}_A(4) = (1 + 5(4))\underset{\sim}{i} + (4^2 + 3)\underset{\sim}{j}$

$= 21\underset{\sim}{i} + 19\underset{\sim}{j}$

$\therefore$ the particles collide at $21\underset{\sim}{i} + 19\underset{\sim}{j}$.

(3 marks)

(b) (i)

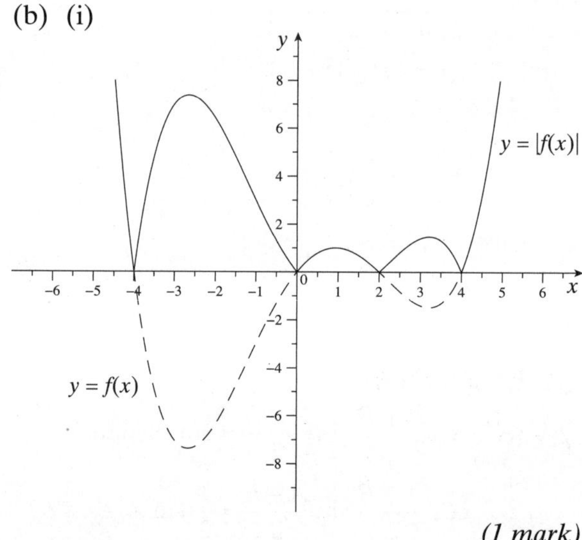

(1 mark)

(ii)

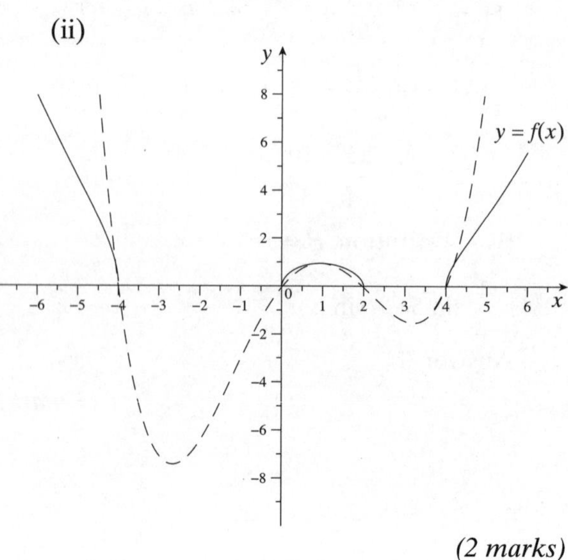

(2 marks)

(iii)

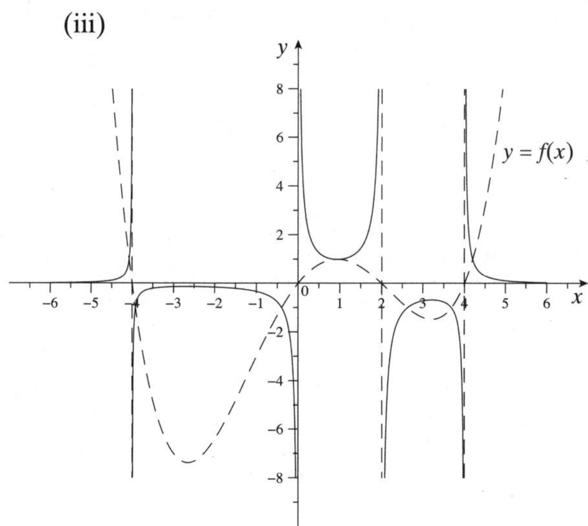

(2 marks)

(c)[*] Volume $= \pi \int_{-\frac{\pi}{4}}^{\frac{\pi}{4}} \sec^2 x \, dx$

$$= 2\pi \int_{0}^{\frac{\pi}{4}} \sec^2 x \, dx$$

$$= 2\pi \left[\tan x\right]_0^{\frac{\pi}{4}}$$

$$= 2\pi \left[\tan \frac{\pi}{4} - \tan 0\right]$$

$$= 2\pi \left[1 - 0\right]$$

$$= 2\pi$$

$\therefore$ the volume is 2π units^3.

(2 marks)

(c)[**] (i) $\dfrac{dt}{dP} = 0.1P\left(1 - \dfrac{P}{2500}\right)$

$$\frac{dP}{dt} = \frac{P}{10}\left(\frac{2500 - P}{2500}\right)$$

$$\frac{dP}{dt} = \frac{P(2500 - P)}{25000}$$

$$\frac{dt}{dP} = \frac{10}{P} + \frac{10}{2500 - P}$$

$$t = 10\int \left(\frac{1}{P} + \frac{10}{2500 - P}\right) dP$$

$$t = 10(\log_e|P| - \log_e|2500 - P|) + c$$

$$t = 10 \log_e \frac{P}{2500 - P} + c \qquad (\text{as } 0 < P < 2500)$$

$$t - c = 10 \log_e \left(\frac{P}{2500 - P}\right)$$

$$\frac{t - c}{10} = \log_e \left(\frac{P}{2500 - P}\right)$$

$$e^{\frac{t-c}{10}} = \frac{P}{2500 - P}$$

Let $A = e^{\frac{-c}{10}}$:

$$Ae^{\frac{t}{10}} = \frac{P}{2500 - P}$$

Substitute $t = 0, P = 200$:

$$0 = 10 \log_e \frac{200}{2500 - 200} + c$$

$$= 10 \log_e \frac{2}{23} + c$$

$$c = -10 \log_e \frac{2}{23}$$

$$c = 10 \log_e \frac{23}{2}$$

$$t = 10 \log_e \frac{P}{2500 - P} + 10 \log_e \frac{23}{2}$$

$$t = 10 \log_e \frac{23P}{5000 - 2P}$$

$$\frac{t}{10} = \log_e \frac{23P}{5000 - 2P}$$

$$e^{0.1t} = \frac{23P}{5000 - 2P}$$

$$23P = 5000e^{0.1t} - 2Pe^{0.1t}$$

$$P(23 + 2e^{0.1t}) = 5000e^{0.1t}$$

$$P = \frac{5000e^{0.1t}}{23 + 2e^{0.1t}}$$

(4 marks)

(ii) Substitute $t = 5$:

$$P = \frac{5000e^{0.1(5)}}{23 + 2e^{0.1(5)}}$$

$= 313.475\,5914\ldots$

$= 313$ (nearest whole)

$\therefore$ there will be 313 wild dogs.

(1 mark)

NSW Education Standards Authority

2019 HIGHER SCHOOL CERTIFICATE EXAMINATION

Mathematics Extension 1

General Instructions

- Reading time – 5 minutes
- Working time – 2 hours
- Write using black pen
- Calculators approved by NESA may be used
- A reference sheet is provided at the back of this paper
- In Questions 11–14, show relevant mathematical reasoning and/or calculations

Total marks: 70

Section I – 10 marks

- Attempt Questions 1–10
- Allow about 15 minutes for this section

Section II – 60 marks

- Attempt Questions 11–14
- Allow about 1 hour and 45 minutes for this section

Section I

10 marks
Attempt Questions 1–10
Allow about 15 minutes for this section

Use the multiple-choice answer sheet for Questions 1–10.

---

1 What is the domain of the function $f(x) = \ln(4 - x)$?

A. $x < 4$

B. $x \leq 4$

C. $x > 4$

D. $x \geq 4$

✗ **2** From the point P outside a circle, a secant and a tangent to the circle are constructed as shown in the diagram.

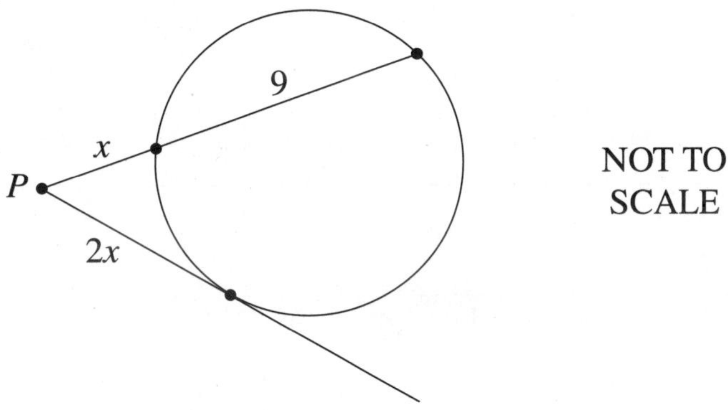

Which equation is satisfied by x?

A. $4x^2 = 9x$

B. $4x^2 = 9 + x$

C. $4x^2 = 9(9 + x)$

D. $4x^2 = x(9 + x)$

3 What is the derivative of $\tan^{-1}\dfrac{x}{2}$?

A. $\dfrac{1}{2\left(4+x^2\right)}$

B. $\dfrac{1}{4+x^2}$

C. $\dfrac{2}{4+x^2}$

D. $\dfrac{4}{4+x^2}$

4 The diagram shows the graph of $y = f(x)$.

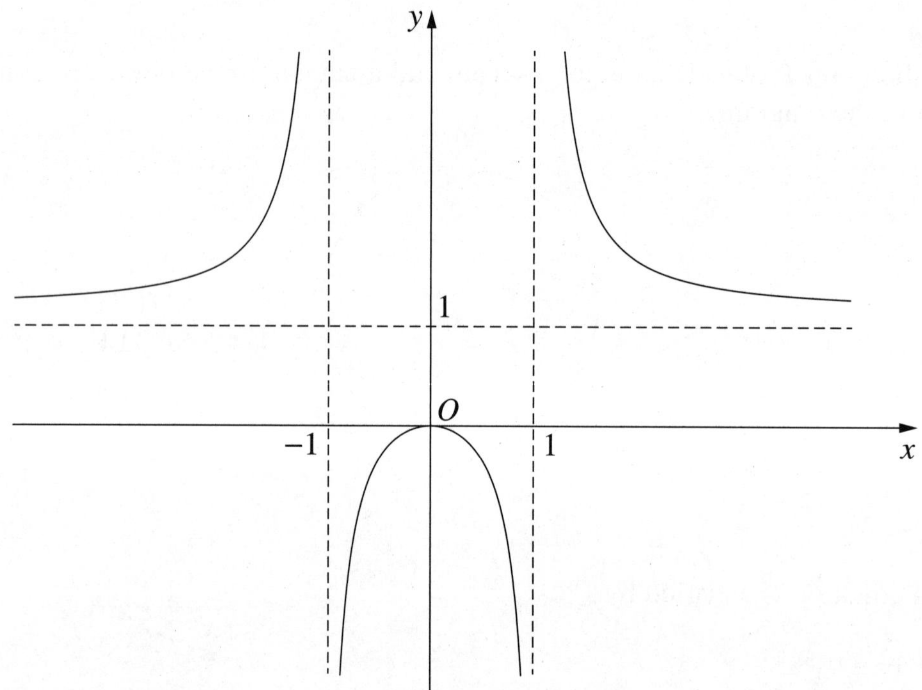

Which equation best describes the graph?

A. $y = \dfrac{x}{x^2 - 1}$

B. $y = \dfrac{x^2}{x^2 - 1}$

C. $y = \dfrac{x}{1 - x^2}$

D. $y = \dfrac{x^2}{1 - x^2}$

5 A particle starts from rest, 2 metres to the right of the origin, and moves along the x-axis in simple harmonic motion with a period of 2 seconds.

Which equation could represent the motion of the particle?

A. $x = 2\cos \pi t$

B. $x = 2\cos 2t$

C. $x = 2 + 2\sin \pi t$

D. $x = 2 + 2\sin 2t$

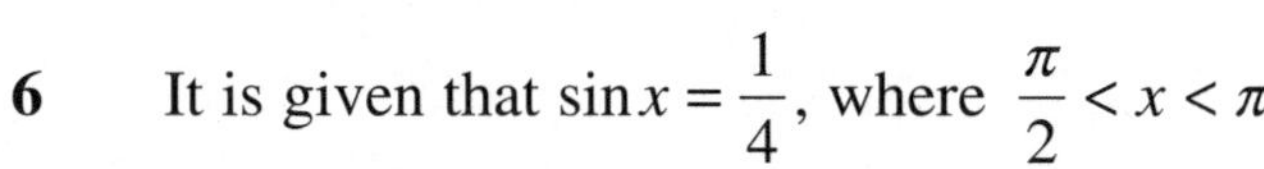

6 It is given that $\sin x = \frac{1}{4}$, where $\frac{\pi}{2} < x < \pi$.

What is the value of $\sin 2x$?

A. $-\frac{7}{8}$

B. $-\frac{\sqrt{15}}{8}$

C. $\frac{\sqrt{15}}{8}$

D. $\frac{7}{8}$

7 Let $P(x) = qx^3 + rx^2 + rx + q$ where q and r are constants, $q \neq 0$. One of the zeros of $P(x)$ is -1.

Given that α is a zero of $P(x)$, $\alpha \neq -1$, which of the following is also a zero?

A. $-\frac{1}{\alpha}$

B. $-\frac{q}{\alpha}$

C. $\frac{1}{\alpha}$

D. $\frac{q}{\alpha}$

8 In how many ways can all the letters of the word PARALLEL be placed in a line with the three Ls together?

A. $\frac{6!}{2!}$

B. $\frac{6!}{2!3!}$

C. $\frac{8!}{2!}$

D. $\frac{8!}{2!3!}$

9 Which graph best represents $y = \cos^{-1}(-\sin x)$, for $-\frac{\pi}{2} \leq x \leq \frac{\pi}{2}$?

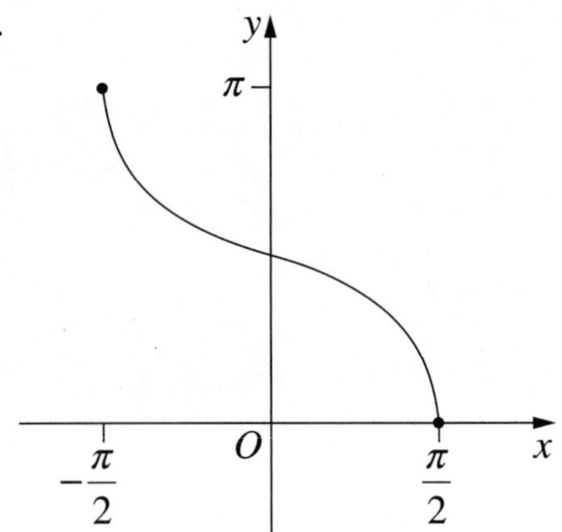

B.

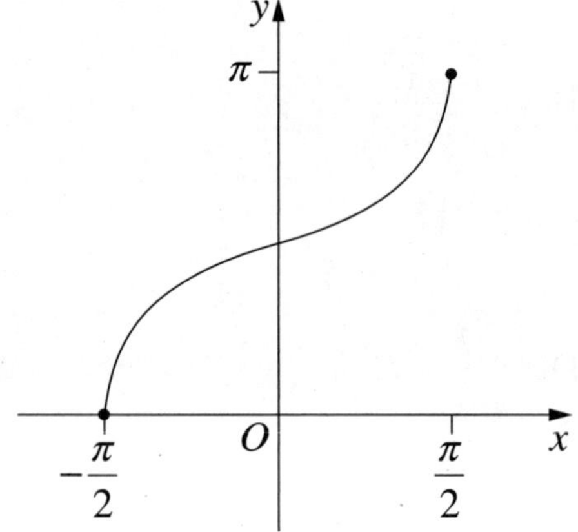

C.

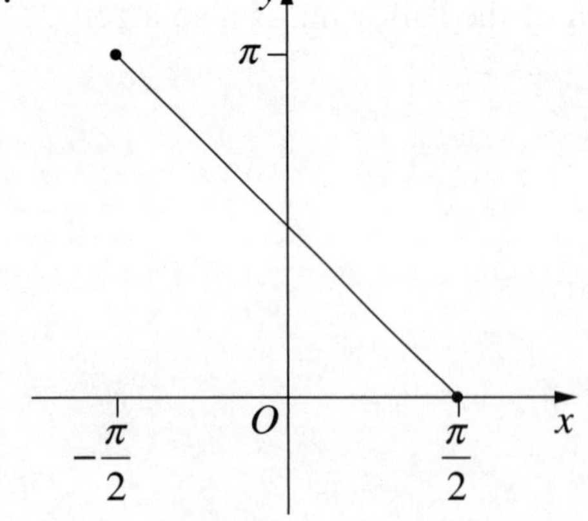

D.

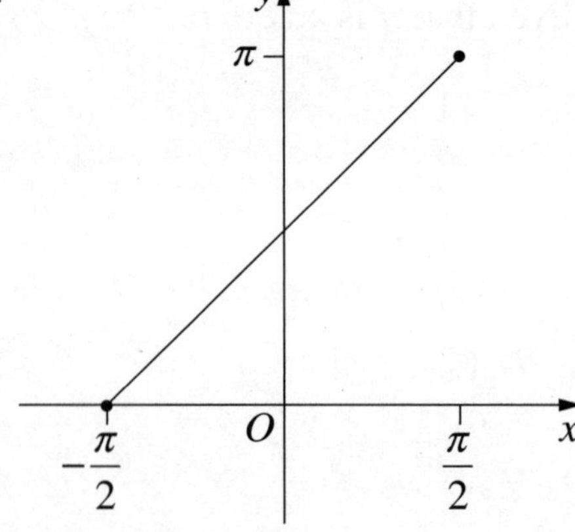

10 The function $f(x) = -\sqrt{1+\sqrt{1+x}}$ has inverse $f^{-1}(x)$.

The graph of $y = f^{-1}(x)$ forms part of the curve $y = x^4 - 2x^2$.

The diagram shows the curve $y = x^4 - 2x^2$.

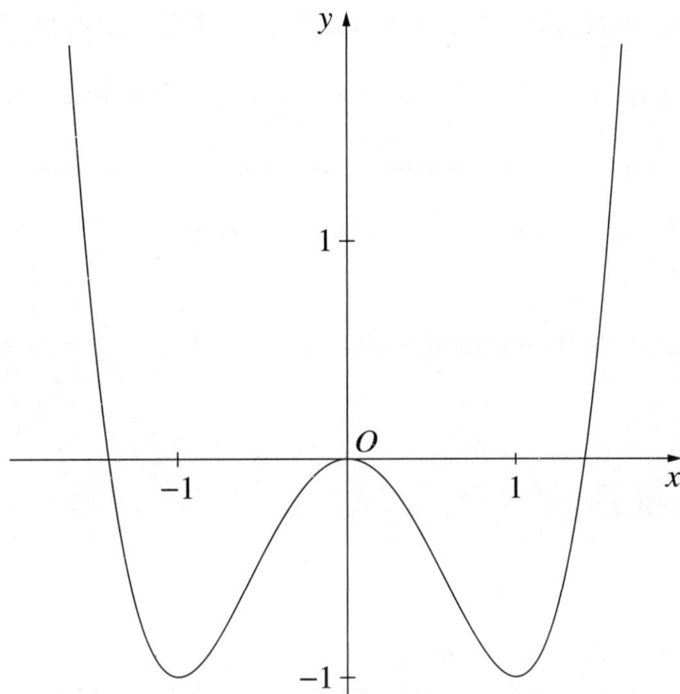

How many points do the graphs of $y = f(x)$ and $y = f^{-1}(x)$ have in common?

A. 1

B. 2

C. 3

D. 4

Section II

60 marks
Attempt Questions 11–14
Allow about 1 hour and 45 minutes for this section

Answer each question in the appropriate writing booklet. Extra writing booklets are available.

In Questions 11–14, your responses should include relevant mathematical reasoning and/or calculations.

Question 11 (15 marks) Use the Question 11 Writing Booklet.

(a) Find the acute angle between the lines $x - 2y + 1 = 0$ and $y = 3x - 4$. **2**

(b) For what values of x is $\dfrac{x}{x+1} < 2$? **3**

(c) The line segment AB is a diameter of the circle, centred at O. The line CD is the tangent to the circle at C and BD is perpendicular to CD, as shown in the diagram. **3**

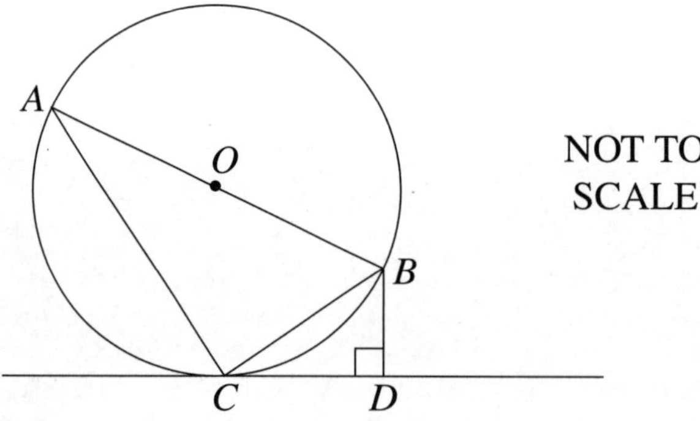

Prove that $\angle ABC = \angle CBD$.

(d) Find the polynomial $Q(x)$ that satisfies $x^3 + 2x^2 - 3x - 7 = (x - 2)Q(x) + 3$. **2**

Question 11 continues on the following page

Question 11 (continued)

(e) Find $\int 2\sin^2 4x\, dx$. **2**

(f) Prize-winning symbols are printed on 5% of ice-cream sticks. The ice-creams are randomly packed into boxes of 8.

(i) What is the probability that a box contains no prize-winning symbols? **1**

(ii) What is the probability that a box contains at least 2 prize-winning symbols? **2**

End of Question 11

Question 12 (15 marks) Use the Question 12 Writing Booklet.

(a) Distance A is inversely proportional to distance B, such that $A = \frac{9}{B}$, where A and B are measured in metres. The two distances vary with respect to time. Distance B is increasing at a rate of 0.2 m s^{-1}. **3**

What is the value of $\frac{dA}{dt}$ when $A = 12$?

(b) A particle is moving along the x-axis in simple harmonic motion. The position of the particle is given by

$$x = \sqrt{2}\cos 3t + \sqrt{6}\sin 3t, \quad \text{for } t \geq 0.$$

(i) Write x in the form $R\cos(3t - \alpha)$, where $R > 0$ and $0 < \alpha < \frac{\pi}{2}$. **2**

(ii) Find the two values for x where the particle comes to rest. **1**

(iii) When is the first time that the speed of the particle is equal to half of its maximum speed? **2**

Question 12 continues on the following page

Question 12 (continued)

(c) The point $P(2ap, ap^2)$ lies on the parabola $x^2 = 4ay$. The focus of the parabola is at $S(0, a)$ and the directrix is the line $y = -a$. The vertical line through P meets the directrix at M. The tangent at P meets the y-axis at R. **3**

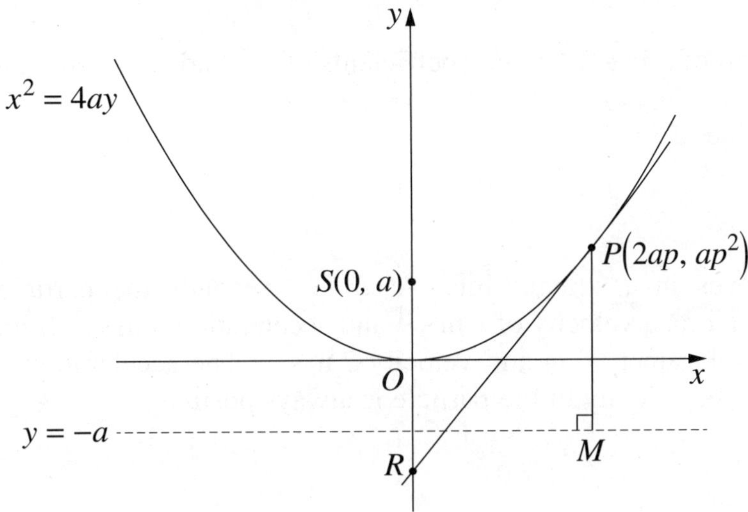

Show that $\angle SPR = \angle SRP$.

(d) A refrigerator has a constant temperature of 3°C. A can of drink with temperature 30°C is placed in the refrigerator.

After being in the refrigerator for 15 minutes, the temperature of the can of drink is 28°C.

The change in the temperature of the can of drink can be modelled by $\frac{dT}{dt} = k(T - 3)$, where T is the temperature of the can of drink, t is the time in minutes after the can is placed in the refrigerator and k is a constant.

(i) Show that $T = 3 + Ae^{kt}$, where A is a constant, satisfies $\frac{dT}{dt} = k(T - 3)$. **1**

(ii) After 60 minutes, at what rate is the temperature of the can of drink changing? **3**

End of Question 12

Question 13 (15 marks) Use the Question 13 Writing Booklet.

(a) Use the substitution $u = \cos^2 x$ to evaluate $\int_0^{\frac{\pi}{4}} \frac{\sin 2x}{4 + \cos^2 x}\, dx$. **3**

(b) In the expansion of $(5x + 2)^{20}$, the coefficients of x^k and x^{k+1} are equal. **3**

What is the value of k?

(c) A particle moves in a straight line. At time t seconds the particle has a displacement of x m, a velocity of v m s^{-1} and acceleration a m s^{-2}. Initially the particle has displacement 0 m and velocity 2 m s^{-1}. The acceleration is given by $a = -2e^{-x}$. The velocity of the particle is always positive.

(i) Show that $v = 2e^{\frac{-x}{2}}$. **2**

(ii) Find an expression for x as a function of t. **2**

Question 13 continues on the following page

Question 13 (continued)

* (d) The point O is on a sloping plane that forms an angle of 45° to the horizontal. A particle is projected from the point O. The particle hits a point A on the sloping plane as shown in the diagram.

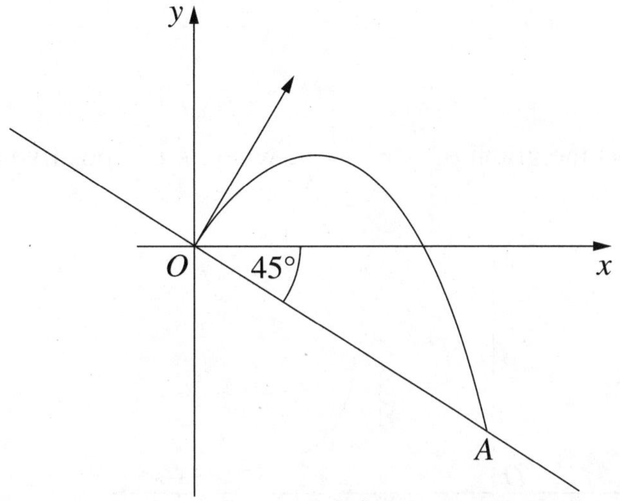

The equation of the line OA is $y = -x$. The equations of motion of the particle are

$$x = 18t$$
$$y = 18\sqrt{3}t - 5t^2,$$

where t is the time in seconds after projection. Do NOT prove these equations.

(i) Find the distance OA between the point of projection and the point where the particle hits the sloping plane. **2**

(ii) What is the size of the acute angle that the path of the particle makes with the sloping plane as the particle hits the point A? **3**

* In the new Mathematics Extension 1 course, Projectile Motion questions will be expressed in vector form:

$x = Vt\cos\theta$ and $y = Vt\sin\theta - \frac{1}{2}gt^2$ is expressed as a position vector:

$$\underset{\sim}{r}(t) = (Vt\cos\theta)\underset{\sim}{i} + \left(Vt\sin\theta - \frac{1}{2}gt^2\right)\underset{\sim}{j}$$

End of Question 13

Question 14 (15 marks) Use the Question 14 Writing Booklet.

(a) Prove by mathematical induction that, for all integers $n \geq 1$, **3**

$$1(1!) + 2(2!) + 3(3!) + \cdots + n(n!) = (n+1)! - 1.$$

(b) The diagram shows the graph of $y = \dfrac{1}{x-k}$, where k is a positive real number.

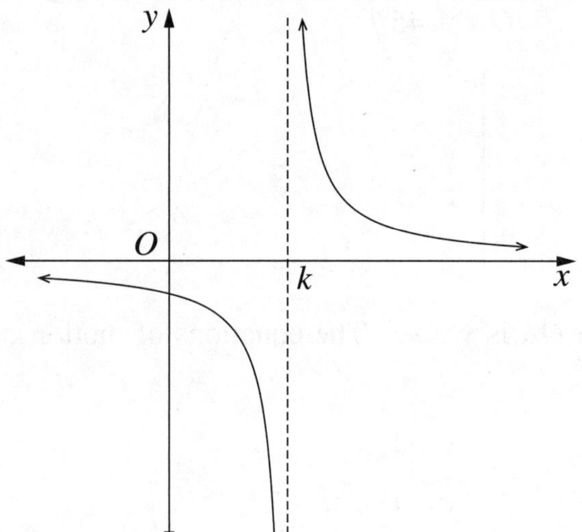

(i) By considering the graphs of $y = x^2$ and $y = \dfrac{1}{x-k}$, explain why the function $f(x) = x^3 - kx^2 - 1$ has exactly one real zero. **2**

Let this zero be α and let $x_1 = k$ be a first approximation to α.

(ii) Show that using one application of Newton's method gives a second approximation, $x_2 = k + \dfrac{1}{k^2}$. **2**

(iii) Show that $x_1 < \alpha < x_2$. **3**

Question 14 continues on the following page

Question 14 (continued)

(c) The diagram shows the two curves $y = \sin x$ and $y = \sin(x - \alpha) + k$, where $0 < \alpha < \pi$ and $k > 0$. The two curves have a common tangent at x_0 where $0 < x_0 < \frac{\pi}{2}$.

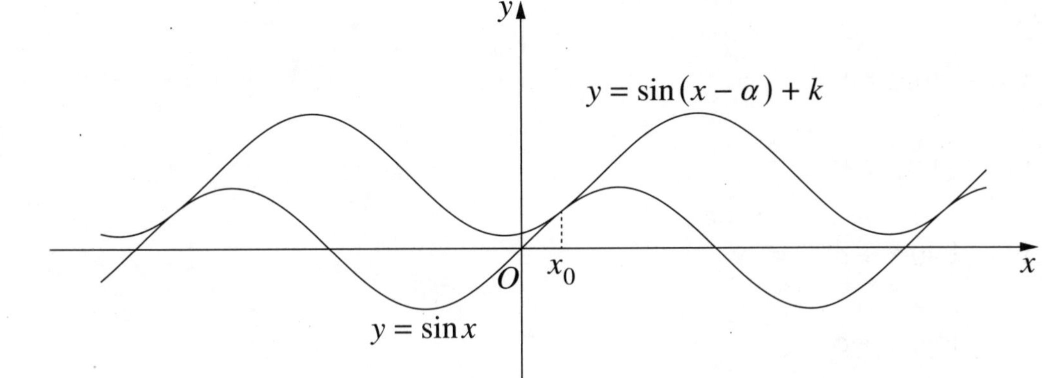

(i) Explain why $\cos x_0 = \cos(x_0 - \alpha)$. **1**

(ii) Show that $\sin x_0 = -\sin(x_0 - \alpha)$. **2**

(iii) Hence, or otherwise, find k in terms of α. **2**

End of paper

Replacement questions

with content from the most up-to-date syllabus

Marks

2 Which of these is the general solution of the differential equation

$2x^2 \frac{dy}{dx} - 3x = 0$? **1**

A. $y = -\frac{3}{2}\log_e x + c$

B. $y = -\frac{2}{3}\log_e x + c$

C. $y = \frac{2}{3}\log_e x + c$

D. $y = \frac{3}{2}\log_e x + c$

5 Let $r(t) = (2 + \sqrt{a-1}\sin t)i + (-1 + \frac{1}{b}\cos t)j$ for $t \geq 0$ be the path of a particle moving in the cartesian plane. **1**

The path of the particle will always be a circle if

A. $b^2\sqrt{a-1} = 1$

B. $b^2(a-1) = 1$

C. $\frac{\sqrt{a-1}}{b} = 1$

D. $\frac{\sqrt{a-1}}{2} - \frac{1}{b} = 1$

REPLACEMENT QUESTIONS

Marks

Question 11 (5 marks)

(a) In a binomial distribution, X is the random variable representing the number of successes in n trial and p is the probability of success.

If the mean is 30 and the variance is 22.5, find the values of n and p. **2**

(c) Evaluate $\int_0^{\frac{\pi}{3}} \cos\frac{3x}{2}\cos\frac{x}{2}\,dx$. **3**

Question 12 (3 marks)

(c) ABC is a triangle. E and D are the midpoints of AC and AB respectively. **3**

Vectors $\vec{AC}$, $\vec{AD}$ and $\vec{CB}$ are shown.

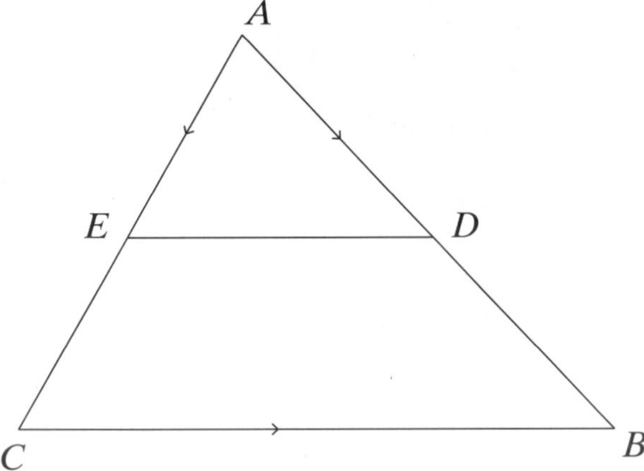

Prove, using vectors, that ED is parallel to CB and half its length.

Question 13 (4 marks)

(c) Find the area under the curve represented by the cartesian equations $x = 2t - 1$ and $y = t^2 + 3$ between the parameters $t = 1$ and $t = 2$. **4**

Question 14 (2 marks)

(b) (ii) Consider the vectors given by $\underset{\sim}{a} = k\underset{\sim}{i} + \underset{\sim}{j}$ and $\underset{\sim}{b} = \underset{\sim}{i} + k\underset{\sim}{j}$ where k is a real number.

If the acute angle between the two vectors is 60°, find the two possible values for k. **2**

REPLACEMENT QUESTIONS

2019 Higher School Certificate
Worked answers

Section I

(*Total 10 marks*)

1 A	**2** D	**3** C	**4** B	**5** A
6 B	**7** C	**8** A	**9** D	**10** A

1 $f(x) = \ln(4 - x)$

Now $4 - x > 0$

$\therefore x < 4$

Answer A

2 The square of a tangent to a circle from an external point equals the product of the intercepts of any secant from the point.

So $(2x)^2 = x(x + 9)$

$4x^2 = x(9 + x)$

Answer D

3 $y = \tan^{-1}\frac{x}{2}$

$$\frac{dy}{dx} = \frac{1}{1+\left(\frac{x}{2}\right)^2} \times \frac{1}{2}$$

$$= \frac{4}{4+x^2} \times \frac{1}{2}$$

$$= \frac{2}{4+x^2}$$

Answer C

4 From the diagram, as $x \to \infty$, $f(x) \to 1^+$

Of the options, this is only true for $y = \frac{x^2}{x^2 - 1}$.

Answer B

5 Period = 2

So $n = \frac{2\pi}{2} = \pi$

The particle starts from rest.

So $\dot{x} = 0$ when $t = 0$.

So $x = 2\cos \pi t$ could represent the motion of the particle.

Answer A

6 $\sin x = \frac{1}{4}$

Now $\sin^2 x + \cos^2 x = 1$

$$\text{So } \cos^2 x = 1 - \left(\frac{1}{4}\right)^2$$

$$= \frac{15}{16}$$

$$\cos x = -\frac{\sqrt{15}}{4} \quad \left(\frac{\pi}{2} < x < \pi\right)$$

$$\sin 2x = 2 \sin x \cos x$$

$$= 2 \times \frac{1}{4} \times -\frac{\sqrt{15}}{4}$$

$$= -\frac{\sqrt{15}}{8}$$

Answer B

7 $P(x) = qx^3 + rx^2 + rx + q$

$$\alpha\beta\gamma = -\frac{d}{a}$$

$$= -\frac{q}{q}$$

$$= -1$$

But one root, γ say, $= -1$

So $\alpha\beta = 1$

$$\beta = \frac{1}{\alpha}$$

Answer C

8 Together the three Ls make one letter unit.

So there are 6 letter units, but two As.

So number of arrangements $= \frac{6!}{2!}$

Answer A

9 $y = \cos^{-1}(-\sin x)$

$$\frac{dy}{dx} = \frac{-1}{\sqrt{1-(-\sin x)^2}} \times -\cos x$$

$$= \frac{\cos x}{\sqrt{1-\sin^2 x}}$$

$$= 1 \quad \left(-\frac{\pi}{2} \le x \le \frac{\pi}{2}\right)$$

So y has a constant gradient of 1.

[Or: Using $\cos^{-1}(-\sin x) = \pi - \cos^{-1}(\sin x)$

it follows that $y = x + \frac{\pi}{2}$]

Answer D

10 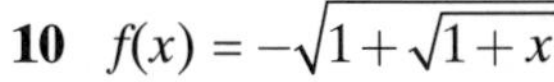

$f(x) = -\sqrt{1+\sqrt{1+x}}$

The range of $y = f(x)$ is $y \leq -1$.

So $y = f^{-1}(x)$ has domain $x \leq -1$.

The graphs have one point in common $(-1, -1)$.

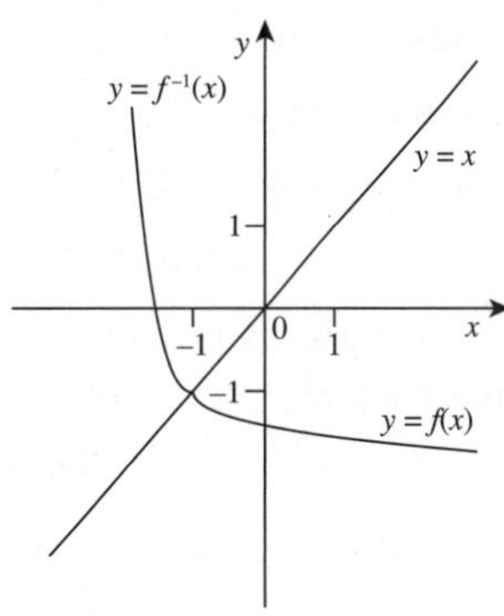

Answer A

Section II

QUESTION 11

(a) $x - 2y + 1 = 0$ has gradient $m_1 = \dfrac{1}{2}$

$y = 3x - 4$ has gradient $m_2 = 3$

Now $\tan\theta = \left|\dfrac{m_1 - m_2}{1 + m_1 m_2}\right|$

$$= \left|\dfrac{\frac{1}{2} - 3}{1 + \frac{1}{2}\times 3}\right|$$

$$= 1$$

$$\theta = 45^\circ$$

The acute angle between the lines is 45°.

(2 marks)

(b) $\dfrac{x}{x+1} < 2$

If $x + 1 > 0$

i.e. $x > -1$,

then $x < 2(x + 1)$

$x < 2x + 2$

$-x < 2$

$x > -2$

But if $x > -1$, it is greater than -2.

If $x < -1$,

then $x > 2x + 2$

$-x > 2$

$x < -2$

So $x < -2$ or $x > -1$.

(3 marks)

(c) $\angle ACB = 90^\circ$ (angle in semicircle)

NOT TO SCALE

$\angle BCD = \angle BAC$

(angle between tangent and chord equals angle in alt. segment)

So $\angle ABC = \angle CBD$

(complements of equal angles; angle sum of triangle is 180°)

(3 marks)

(d) $x^3 + 2x^2 - 3x - 7 = (x - 2)\,Q(x) + 3$

$$\begin{array}{r}
x^2 + 4x + 5 \\
x - 2\,\overline{)\,x^3 + 2x^2 - 3x - 7} \\
\underline{x^3 - 2x^2} \\
4x^2 - 3x \\
\underline{4x^2 - 8x} \\
5x - 7 \\
\underline{5x - 10} \\
3
\end{array}$$

$\therefore Q(x) = x^2 + 4x + 5$

(2 marks)

(e) $\cos 8x = \cos^2 4x - \sin^2 4x$

$= 1 - 2\sin^2 4x$

$\therefore 2\sin^2 4x = 1 - \cos 8x$

$\int 2\sin^2 4x\,dx = \int (1 - \cos 8x)\,dx$

$= x - \dfrac{1}{8}\sin 8x + C$ *(2 marks)*

(f) (i) $P(\text{prize on stick}) = 0.05$

$\therefore P(\text{no prize on stick}) = 0.95$

$P(\text{no prize on 8 sticks}) = 0.95^8$

$= 0.663\,420\,431\ldots$

So the probability that a box has no prize-winning sticks is about 66%.

(1 mark)

(ii) $P(\text{1 prize stick in box}) = 8 \times 0.95^7 \times 0.05$

P(at least 2 prize-winning sticks in a box)

$= 1 - P(\text{0 prize or 1 prize})$

$= 1 - (0.95^8 + 0.4(0.95)^7)$

$= 0.057\,244\,6502\ldots$

$= 5.7\%$ (1 d.p.)

The probability that a box holds at least 2 prize-winning symbols is about 5.7%.

(2 marks)

QUESTION 12

(a)

$$A = \frac{9}{B}$$

$$= 9B^{-1}$$

$$\frac{dA}{dB} = -9B^{-2}$$

$$= -\frac{9}{B^2}$$

$$\text{Now } \frac{dA}{dt} = \frac{dA}{dB} \times \frac{dB}{dt}$$

$$= -\frac{9}{B^2} \times 0.2$$

$$= -\frac{1.8}{B^2}$$

When $A = 12$,

$$12 = \frac{9}{B}$$

$$12B = 9$$

$$B = 0.75$$

$$\text{So } \frac{dA}{dt} = -\frac{1.8}{(0.75)^2}$$

$$= -3.2$$

(3 marks)

(b) (i) $x = \sqrt{2}\cos 3t + \sqrt{6}\sin 3t$

Now $R\cos(3t - \alpha) = R\cos 3t\cos\alpha + R\sin 3t\sin\alpha$

$\therefore R\cos\alpha = \sqrt{2}$ and $R\sin\alpha = \sqrt{6}$

$$\frac{R\sin\alpha}{R\cos\alpha} = \frac{\sqrt{6}}{\sqrt{2}}$$

$$\tan\alpha = \sqrt{3}$$

$$\alpha = \frac{\pi}{3} \quad \left(0 < \alpha < \frac{\pi}{2}\right)$$

$$R\cos\frac{\pi}{3} = \sqrt{2}$$

$$R \times \frac{1}{2} = \sqrt{2}$$

$$R = 2\sqrt{2}$$

$$\therefore x = 2\sqrt{2}\cos\left(3t - \frac{\pi}{3}\right)$$

(2 marks)

(ii) $x = 2\sqrt{2}\cos\left(3t - \frac{\pi}{3}\right)$

The particle comes to rest at the endpoints of its domain.

i.e. when $\cos\left(3t - \frac{\pi}{3}\right) = \pm 1$

The particle is at rest at $x = -2\sqrt{2}$ and $x = 2\sqrt{2}$.

(1 mark)

(iii) $x = 2\sqrt{2}\cos\left(3t - \frac{\pi}{3}\right)$

$v = -6\sqrt{2}\sin\left(3t - \frac{\pi}{3}\right)$

The maximum speed is $6\sqrt{2}$.

The speed of the particle will be half the maximum speed when $\sin\left(3t - \frac{\pi}{3}\right) = \pm\frac{1}{2}$.

This first occurs when:

$$3t - \frac{\pi}{3} = -\frac{\pi}{6} \quad (t \geq 0)$$

$$3t = \frac{\pi}{6}$$

$$t = \frac{\pi}{18}$$

(2 marks)

(c) The equation of the tangent is $y = px - ap^2$.

At $R, x = 0$

$\therefore y = -ap^2$

$SR = a - (-ap^2)$

$= a + ap^2$

$S(0, a) \quad P(2ap, ap^2)$

$[SP]^2 = (2ap - 0)^2 + (ap^2 - a)^2$

$= 4a^2p^2 + a^2p^4 - 2a^2p^2 + a^2$

$= a^2p^4 + 2a^2p^2 + a^2$

$= (ap^2 + a)^2$

So $SP = ap^2 + a$

$\therefore SR = SP$

$\therefore \angle SPR = \angle SRP$ (base angles isos. triangle)

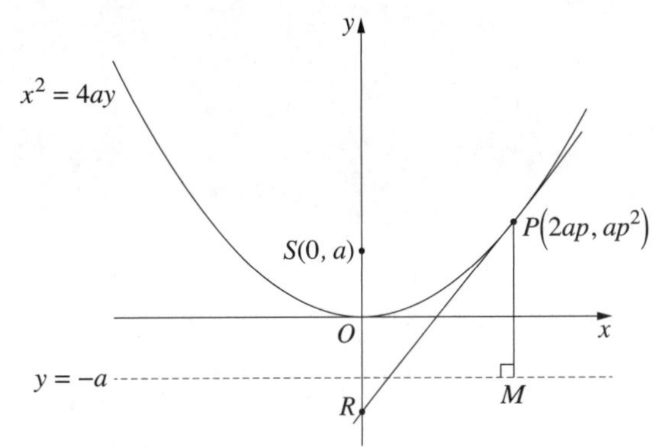

(3 marks)

(d) (i) $T = 3 + Ae^{kt}$

$\frac{dT}{dt} = Ae^{kt} \times k$

But $Ae^{kt} = T - 3$

So $\frac{dT}{dt} = (T - 3) \times k$

$= k(T - 3)$

(1 mark)

(ii) $T = 3 + Ae^{kt}$

When $t = 0, T = 30$

$30 = 3 + Ae^0$

$3 + A = 30$

$A = 27$

So $T = 3 + 27e^{kt}$

When $t = 15, T = 28$

$28 = 3 + 27e^{k \times 15}$

$27e^{15k} = 25$

$e^{15k} = \frac{25}{27}$

$15k = \ln\frac{25}{27}$

$k = \frac{1}{15}\ln\frac{25}{27}$

When $t = 60$,

$T = 3 + 27e^{\left(\frac{1}{15}\ln\frac{25}{27}\right)\times 60}$

$= 22.845\,8060\ldots$

$\frac{dT}{dt} = \frac{1}{15}\ln\frac{25}{27} \times (22.845\,8060\ldots - 3)$

$= -0.101\,823\,592\ldots$

So after 60 minutes the temperature of the can is decreasing at about 0.1° C per minute.

(3 marks)

QUESTION 13

(a) Let $u = \cos^2 x$

$$du = -2\cos x \sin x\, dx$$
$$= -\sin 2x\, dx$$

When $x = 0, u = 1$

When $x = \frac{\pi}{4}, u = \frac{1}{2}$

$$\text{So } \int_0^{\frac{\pi}{4}} \frac{\sin 2x}{4+\cos^2 x}dx = \int_1^{\frac{1}{2}} \frac{-1}{4+u}du$$
$$= \int_{\frac{1}{2}}^{1} \frac{1}{4+u}du$$
$$= \left[\ln(4+u)\right]_{\frac{1}{2}}^{1}$$
$$= \ln(4+1) - \ln\left(4+\frac{1}{2}\right)$$
$$= \ln\frac{5}{4.5}$$
$$= \ln\frac{10}{9}$$

(3 marks)

(b) $(5x+2)^{20} = \sum_{k=0}^{20}\binom{20}{k}(5x)^k 2^{20-k}$

The coefficients of x^k and x^{k+1} are equal.

$$\text{So } \binom{20}{k}(5)^k 2^{20-k} = \binom{20}{k+1}(5)^{k+1}2^{20-(k+1)}$$
$$\frac{20!}{k!(20-k)!} \times 5^k \times 2^{20-k} = \frac{20!}{(k+1)!(20-(k+1))!} \times 5^{k+1} \times 2^{19-k}$$
$$\frac{2}{k!(20-k)!} = \frac{5}{(k+1)!(19-k)!}$$
$$2(k+1) = 5(20-k)$$
$$2k+2 = 100-5k$$
$$7k = 98$$
$$k = 14$$

(3 marks)

(c) (i)

$$a = -2e^{-x}$$
$$\text{So } \frac{d}{dx}\left(\frac{1}{2}v^2\right) = -2e^{-x}$$
$$\frac{1}{2}v^2 = 2e^{-x} + C$$

When $t = 0, x = 0$ and $v = 2$

$$\frac{1}{2} \times 2^2 = 2e^0 + C$$
$$2 = 2 + C$$
$$C = 0$$
$$\frac{1}{2}v^2 = 2e^{-x}$$
$$v^2 = 4e^{-x}$$
$$v = \sqrt{4e^{-x}} \quad (v > 0)$$
$$= 2\left(e^{-x}\right)^{\frac{1}{2}}$$
$$= 2e^{-\frac{x}{2}}$$

(2 marks)

(ii)

$$\frac{dx}{dt} = 2e^{-\frac{x}{2}}$$
$$\frac{dt}{dx} = \frac{1}{2e^{-\frac{x}{2}}}$$
$$= \frac{1}{2}e^{\frac{x}{2}}$$
$$t = \frac{\frac{1}{2}e^{\frac{x}{2}}}{\frac{1}{2}} + K$$
$$t = e^{\frac{x}{2}} + K$$

(cont. over page)

When $t = 0, x = 0$

$$0 = e^0 + K$$
$$K = -1$$
$$t = e^{\frac{x}{2}} - 1$$
$$t + 1 = e^{\frac{x}{2}}$$
$$\ln (t+1) = \frac{x}{2}$$

So $x = 2 \ln (t + 1)$

(2 marks)

(d) (i) On OA, $y = -x$

$$\therefore 18\sqrt{3}t - 5t^2 = -18t$$
$$5t^2 - (18 + 18\sqrt{3})t = 0$$
$$t(5t - (18 + 18\sqrt{3})) = 0$$
$$t = 0 \quad \text{or} \quad 5t = 18 + 18\sqrt{3}$$
$$t = \frac{18(1+\sqrt{3})}{5}$$

Now $t = 0$ when the particle is at O.

So at A, $t = \dfrac{18(1+\sqrt{3})}{5}$

$$x = 18t$$
$$= \frac{18^2(1+\sqrt{3})}{5}$$
$$y = -x$$

So $y = -\dfrac{18^2(1+\sqrt{3})}{5}$

$$x^2 + y^2 = 2\left(\frac{18^2(1+\sqrt{3})}{5}\right)^2$$

So $OA = \dfrac{18^2\sqrt{2}(1+\sqrt{3})}{5}$

$$= 250.367974\ldots$$

The distance OA is about 250 units.

(2 marks)

(ii) $x = 18t$

$$\dot{x} = 18$$
$$y = 18\sqrt{3}t - 5t^2$$
$$\dot{y} = 18\sqrt{3} - 10t$$

When $t = \dfrac{18(1+\sqrt{3})}{5}$,

$$\dot{y} = 18\sqrt{3} - 36(1 + \sqrt{3})$$
$$= 18\sqrt{3} - 36 - 36\sqrt{3}$$
$$= -36 - 18\sqrt{3}$$
$$= -18(2 + \sqrt{3})$$

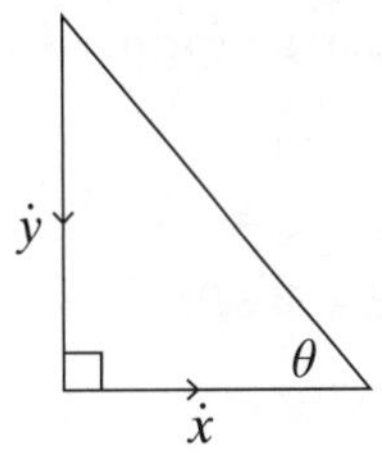

$$\tan \theta = \left|\frac{\dot{y}}{\dot{x}}\right|$$
$$= \left|\frac{-18(2+\sqrt{3})}{18}\right|$$
$$= 2 + \sqrt{3}$$
$$\theta = 75^\circ$$

So the required angle $= 75^\circ - 45^\circ$

$$= 30^\circ$$

(3 marks)

QUESTION 14

(a) $1(1!) + 2(2!) + 3(3!) + \ldots + n(n!) = (n + 1)! - 1$

If $n = 1$,

$$\text{LHS} = 1(1!)$$
$$= 1$$
$$\text{RHS} = (1 + 1)! - 1$$
$$= 2! - 1$$
$$= 1$$

The statement is true for $n = 1$.

Assume it is true for $n = k$.

i.e. assume $1(1!) + 2(2!) + 3(3!) + \ldots + k(k!) = (k + 1)! - 1$

We need to show that it is also true for $n = k + 1$.

We wish to prove that:

$1(1!) + 2(2!) + 3(3!) + \ldots + k(k!) + (k + 1)(k + 1)! = (k + 2)! - 1$

Now $1(1!) + 2(2!) + 3(3!) + \ldots + k(k!) + (k + 1)(k + 1)!$

$= (k + 1)! - 1 + (k + 1)(k + 1)!$

$= (k + 1)! - 1 + k(k + 1)! + (k + 1)!$

$= k(k + 1)! + 2(k + 1)! - 1$

$= (k + 1)!(k + 2) - 1$

$= (k + 2)! - 1$

So if the statement is true for $n = k$ it is also true for $n = k + 1$.

It is true for $n = 1$.

So it is true for $n = 1 + 1 = 2$.

So it is true for $n = 2 + 1 = 3$.

And so on.

By the process of induction it is true for all integers n, $n \geq 1$.

(3 marks)

(b) (i) $y = x^2$ is a parabola, concave up and with vertex the origin, so it will only intersect $y = \dfrac{1}{x-k}$ in one place.

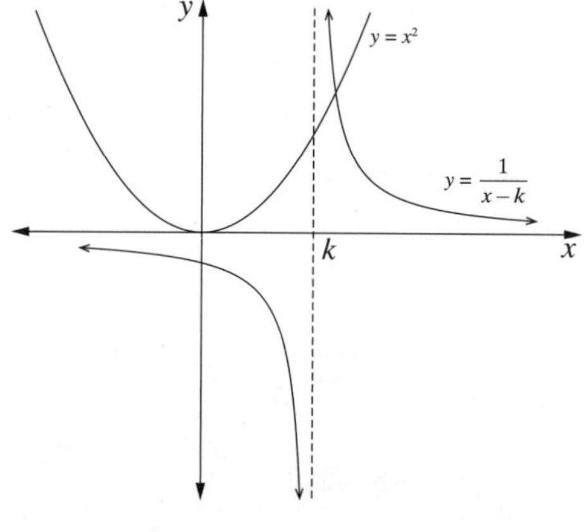

So there is only one real solution of $x^2 = \dfrac{1}{x-k}$.

Now zeroes of $f(x) = x^3 - kx^2 - 1$ occur when $f(x) = 0$

i.e. $x^3 - kx^2 - 1 = 0$

$x^3 - kx^2 = 1$

$x^2(x - k) = 1$

$$x^2 = \frac{1}{x-k} \quad (x \neq k)$$

$f(k) \neq 0$

So $f(x)$ has exactly one real zero.

(2 marks)

(ii) $f(x) = x^3 - kx^2 - 1$

$f'(x) = 3x^2 - 2kx$

$x_1 = k$

$$x_2 = x_1 - \frac{f(x_1)}{f'(x_1)}$$

$$= k - \frac{k^3 - k^3 - 1}{3k^2 - 2k^2}$$

$$= k - \frac{-1}{k^2}$$

$$= k + \frac{1}{k^2}$$

(2 marks)

(iii) $f(x) = x^3 - kx^2 - 1$

$f(x_1) = f(k)$

$= k^3 - k(k^2) - 1$

$= -1$

$f(x_2) = f\left(k + \frac{1}{k^2}\right)$

$= \left(k + \frac{1}{k^2}\right)^3 - k\left(k + \frac{1}{k^2}\right)^2 - 1$

$= k^3 + 3k^2\left(\frac{1}{k^2}\right) + 3k\left(\frac{1}{k^2}\right)^2 + \left(\frac{1}{k^2}\right)^3 - k\left(k^2 + 2k\left(\frac{1}{k^2}\right) + \left(\frac{1}{k^2}\right)^2\right) - 1$

$= k^3 + 3 + \frac{3}{k^3} + \frac{1}{k^6} - k^3 - 2 - \frac{1}{k^3} - 1$

$= \frac{2}{k^3} + \frac{1}{k^6}$

Now $k > 0$

$\therefore f(x_2) > 0$

As the function has exactly one zero, α, and $f(x_1) < 0$ but $f(x_2) > 0$, then $x_1 < \alpha < x_2$.

(3 marks)

(c) (i) $y = \sin x$

$\frac{dy}{dx} = \cos x$

At $x = x_0$, $m = \cos x_0$.

$y = \sin(x - \alpha) + k$

$\frac{dy}{dx} = \cos(x - \alpha)$

At $x = x_0$, $m = \cos(x_0 - \alpha)$.

The gradients must be equal as there is a common tangent.

$\therefore \cos x_0 = \cos(x_0 - \alpha)$

(1 mark)

(ii) $\cos(-\theta) = \cos\theta$ and $\sin(-\theta) = -\sin\theta$ for all values of θ.

Now $\cos x_0 = \cos(x_0 - \alpha)$

As $\alpha \neq 0$, $x_0 \neq x_0 - \alpha$

$\therefore x_0 = -(x_0 - \alpha)$

So $\sin x_0 = \sin(-(x_0 - \alpha))$

$= -\sin(x_0 - \alpha)$

(2 marks)

(iii) $x_0 = -(x_0 - \alpha)$

$2x_0 = \alpha$

$x_0 = \frac{\alpha}{2}$

The curves intersect when $x = x_0$.

$\therefore \sin(x_0 - \alpha) + k = \sin x_0$

$= -\sin(x_0 - \alpha)$

$k = -2\sin(x_0 - \alpha)$

$= -2\sin\left(\frac{\alpha}{2} - \alpha\right)$

$= -2\sin\left(-\frac{\alpha}{2}\right)$

So $k = 2\sin\left(\frac{\alpha}{2}\right)$

(2 marks)

Solutions to replacement questions

QUESTION 2

$$2x^2\frac{dy}{dx} - 3x = 0$$

$$2x^2\frac{dy}{dx} = 3x$$

$$\frac{dy}{dx} = \frac{3}{2x}$$

$$y = \frac{3}{2}\int \frac{1}{x}\,dx$$

$$y = \frac{3}{2}\log_e x + c$$

Answer D

(1 mark)

QUESTION 5

Path is a circle when $\sqrt{a-1} = \frac{1}{b}$

$$a - 1 = \frac{1}{b^2}$$

$$b^2(a-1) = 1$$

Answer B

(1 mark)

QUESTION 11

(a)

$$\mu = np$$

$$30 = np$$

$$\sigma^2 = np(1-p)$$

$$22.5 = 30(1-p)$$

$$1 - p = \frac{22.5}{30}$$

$$1 - p = 0.75$$

$$p = 0.25$$

Hence, $30 = n(0.25)$

$$n = 120$$

$\therefore n = 120, p = 0.25$

(2 marks)

(c) $\int_0^{\frac{\pi}{3}} \cos\frac{3x}{2}\cos\frac{x}{2}\,dx$

$$= \frac{1}{2}\int_0^{\frac{\pi}{3}}(\cos 2x + \cos x)\,dx$$

$$= \frac{1}{2}\left[\frac{1}{2}\sin 2x + \sin x\right]_0^{\frac{\pi}{3}}$$

$$= \frac{1}{2}\left[\frac{1}{2}\sin 2\left(\frac{\pi}{3}\right) + \sin\frac{\pi}{3} - 0\right]$$

$$= \frac{1}{2}\left[\frac{1}{2}\left(\frac{\sqrt{3}}{2}\right) + \frac{\sqrt{3}}{2} - 0\right]$$

$$= \frac{1}{2}\left[\frac{\sqrt{3}}{4} + \frac{\sqrt{3}}{2}\right]$$

$$= \frac{1}{2}\left[\frac{3\sqrt{3}}{4}\right]$$

$$= \frac{3\sqrt{3}}{8}$$

(3 marks)

QUESTION 12

(c)

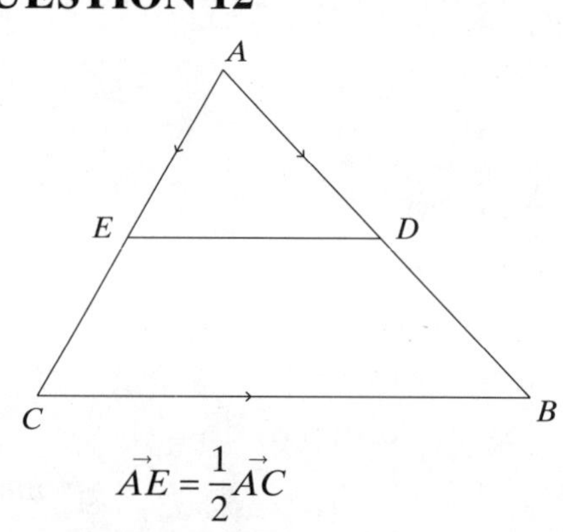

$$\vec{AE} = \frac{1}{2}\vec{AC}$$

Also, $\vec{AD} = \frac{1}{2}\vec{AB}$

Now, $\vec{CB} = \vec{AB} - \vec{AC}$

Hence, $\vec{ED} = \frac{1}{2}\vec{AB} - \frac{1}{2}\vec{AC}$

$$= \frac{1}{2}(\vec{AB} - \vec{AC})$$

$$= \frac{1}{2}\vec{CB}$$

$\therefore \vec{ED} = \frac{1}{2}\vec{CB}$

Hence, $CB \parallel ED$.

Also, $|\vec{ED}| = |\frac{1}{2}\vec{CB}| = \frac{1}{2}|\vec{CB}|$ and so CB is half the length of ED.

$\therefore ED$ is parallel to CB and half its length.

(3 marks)

QUESTION 13

(c) $x = 2t - 1$

$\therefore \quad t = \frac{x+1}{2}$

Substitute into y:

$$y = \left(\frac{x+1}{2}\right)^2 + 3$$

$$= \frac{x^2+2x+1}{4} + 3$$

$$= \frac{x^2+2x+1+12}{4}$$

$$= \frac{x^2+2x+13}{4}$$

As $t = 1$, then $x = 1; t = 2, x = 3$.

$$\text{Area} = \int_1^3 \frac{x^2+2x+13}{4} dx$$

$$= \frac{1}{4}\int_1^3 \left(x^2+2x+13\right) dx$$

$$= \frac{1}{4}\left[\frac{x^3}{3} + x^2 + 13x\right]_1^3$$

$$= \frac{1}{4}\left[\frac{3^3}{3} + 3^2 + 13(3) - \left(\frac{1^3}{3} + 1^2 + 13(1)\right)\right]$$

$$= \frac{1}{4}\left[9 + 9 + 39 - (\frac{1}{3} + 1 + 13)\right]$$

$$= \frac{32}{3}$$

$\therefore$ area of $\frac{32}{3}$ units^2. *(4 marks)*

QUESTION 14

(b) (ii) $\underset{\sim}{a} \cdot \underset{\sim}{b} = k \times 1 + 1 \times k$

$$= 2$$

$$|\underset{\sim}{a}||\underset{\sim}{b}|\cos\theta = \sqrt{k^2+1^2}.\sqrt{1^2+k^2}\ \sqrt{1^2+k^2}.\cos 60^\circ$$

$$= \frac{1}{2}(k^2 + 1)$$

Using $\underset{\sim}{a} \cdot \underset{\sim}{b} = |\underset{\sim}{a}||\underset{\sim}{b}|\cos\theta$,

$$2k = \frac{1}{2}(k^2 + 1)$$

$$k^2 + 1 = 4k$$

$$k^2 - 4k + 1 = 0$$

$$k = \frac{4 \pm \sqrt{16 - 4(1)(1)}}{2(1)}$$

$$= \frac{4 \pm \sqrt{12}}{2}$$

$$= 2 \pm \sqrt{3}$$

(2 marks)

NSW Education Standards Authority

2020 **HIGHER SCHOOL CERTIFICATE EXAMINATION**

Mathematics Extension 1

General Instructions

- Reading time – 10 minutes
- Working time – 2 hours
- Write using black pen
- Calculators approved by NESA may be used
- A reference sheet is provided at the back of this paper
- For questions in Section II, show relevant mathematical reasoning and/or calculations

Total marks: 70

Section I – 10 marks

- Attempt Questions 1–10
- Allow about 15 minutes for this section

Section II – 60 marks

- Attempt Questions 11–14
- Allow about 1 hour and 45 minutes for this section

Section I

10 marks
Attempt Questions 1–10
Allow about 15 minutes for this section

Use the multiple-choice answer sheet for Questions 1–10.

---

1 Which diagram best represents the solution set of $x^2 - 2x - 3 \geq 0$?

A.

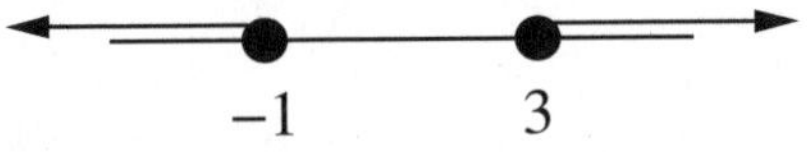

B.

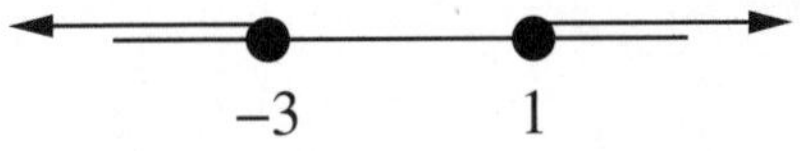

C.

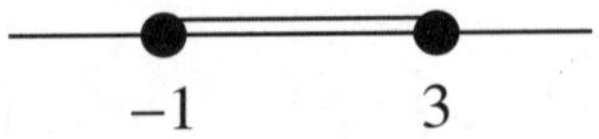

D.

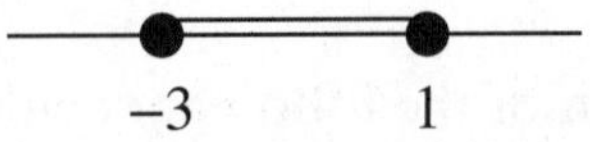

2 Given $f(x) = 1 + \sqrt{x}$, what are the domain and range of $f^{-1}(x)$?

A. $x \geq 0, \quad y \geq 0$

B. $x \geq 0, \quad y \geq 1$

C. $x \geq 1, \quad y \geq 0$

D. $x \geq 1, \quad y \geq 1$

3 Which of the following is an anti-derivative of $\dfrac{1}{4x^2+1}$?

A. $2\tan^{-1}\left(\dfrac{x}{2}\right) + c$

B. $\dfrac{1}{2}\tan^{-1}\left(\dfrac{x}{2}\right) + c$

C. $2\tan^{-1}(2x) + c$

D. $\dfrac{1}{2}\tan^{-1}(2x) + c$

4 Maria starts at the origin and walks along all of the vector $2\underset{\sim}{i} + 3\underset{\sim}{j}$, then walks along all of the vector $3\underset{\sim}{i} - 2\underset{\sim}{j}$ and finally along all of the vector $4\underset{\sim}{i} - 3\underset{\sim}{j}$.

How far from the origin is she?

A. $\sqrt{77}$

B. $\sqrt{85}$

C. $2\sqrt{13} + \sqrt{5}$

D. $\sqrt{5} + \sqrt{7} + \sqrt{13}$

5 A monic polynomial $p(x)$ of degree 4 has one repeated zero of multiplicity 2 and is divisible by $x^2 + x + 1$.

Which of the following could be the graph of $p(x)$?

A.

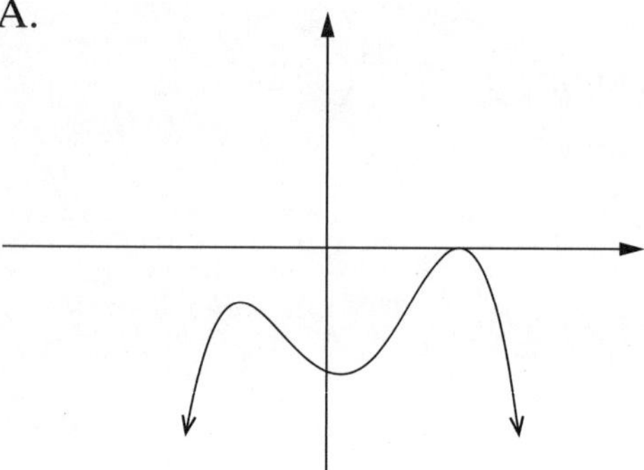

B.

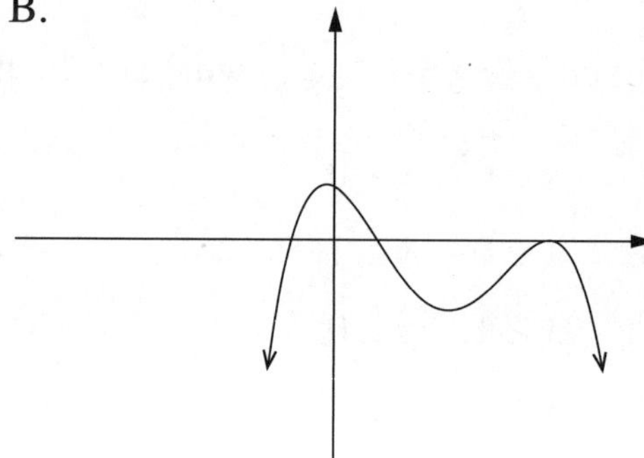

C.

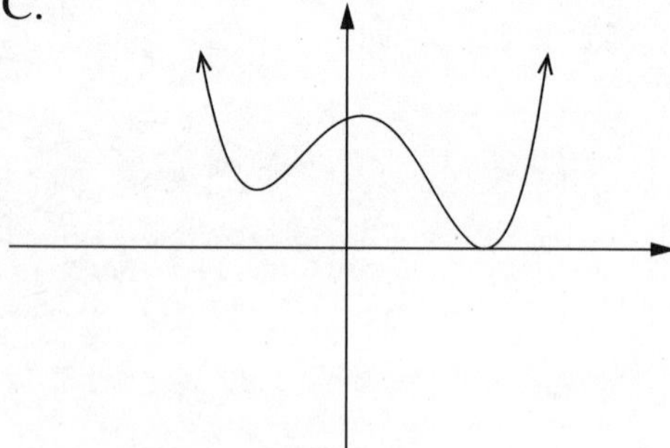

D.

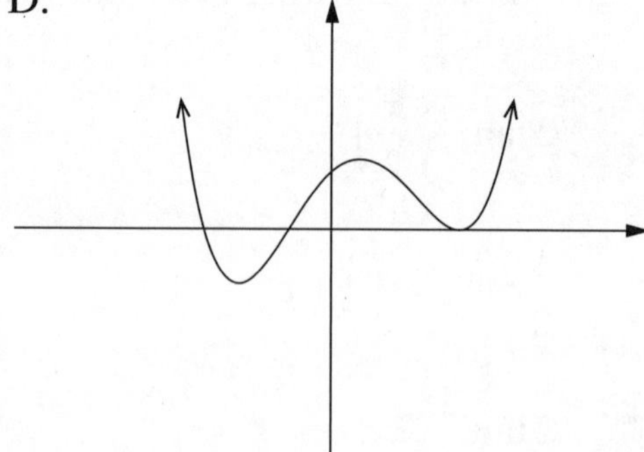

6 The vectors $\underset{\sim}{a}$ and $\underset{\sim}{b}$ are shown.

Which diagram below shows the vector $\underset{\sim}{v} = \underset{\sim}{a} - \underset{\sim}{b}$?

A. 

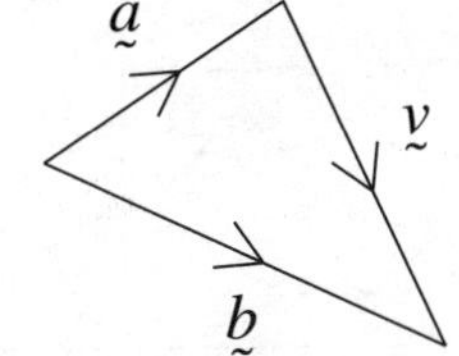

B.

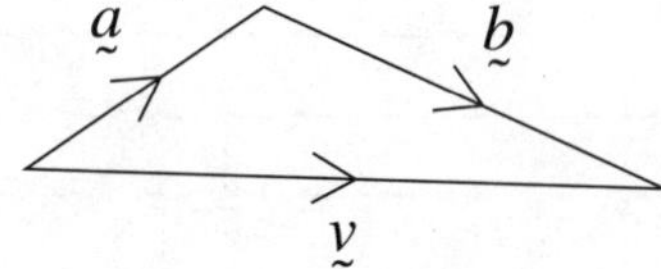

C.

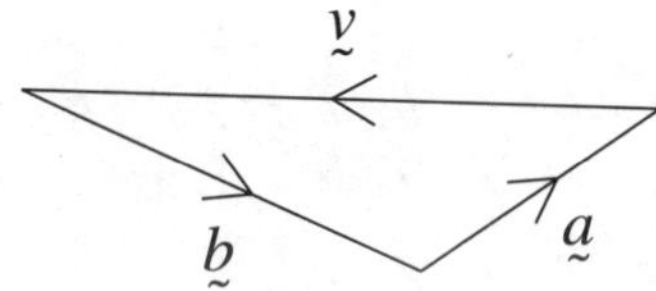

D.

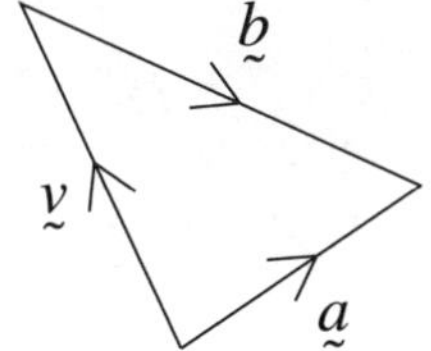

7 Which of the following best represents the direction field for the differential equation $\dfrac{dy}{dx} = -\dfrac{x}{4y}$?

A.

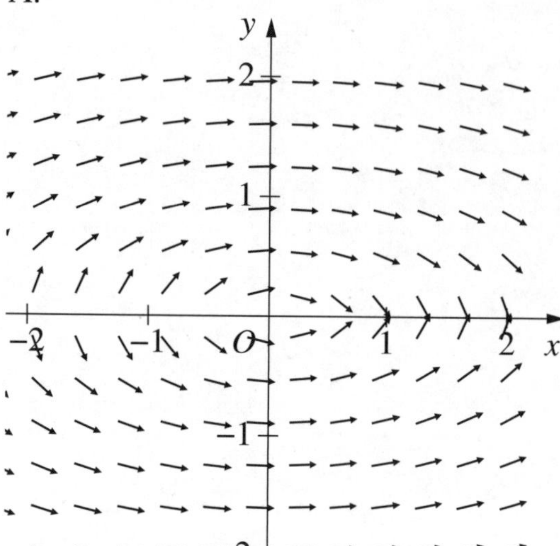

B.

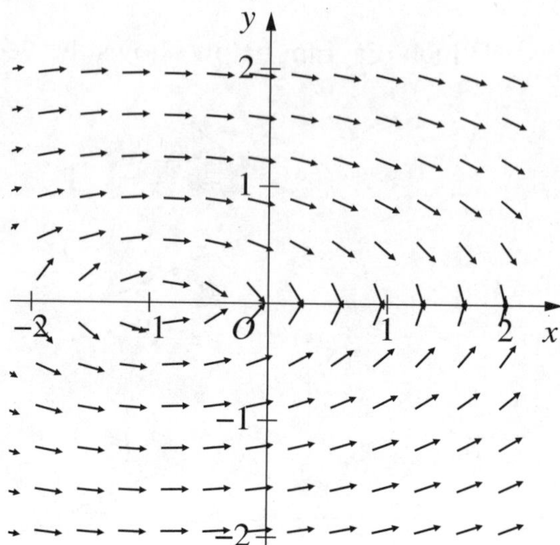

C.

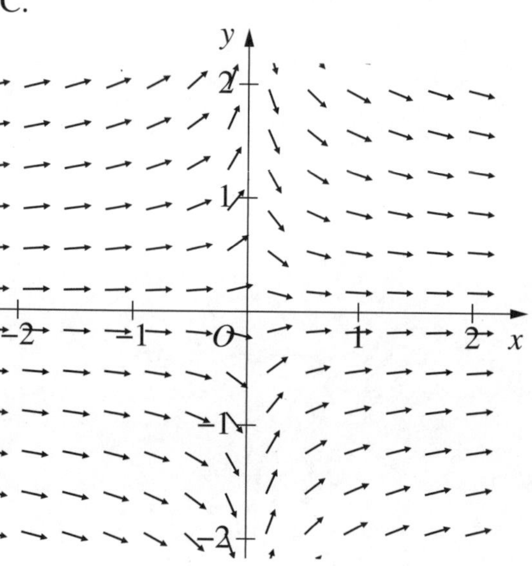

D.

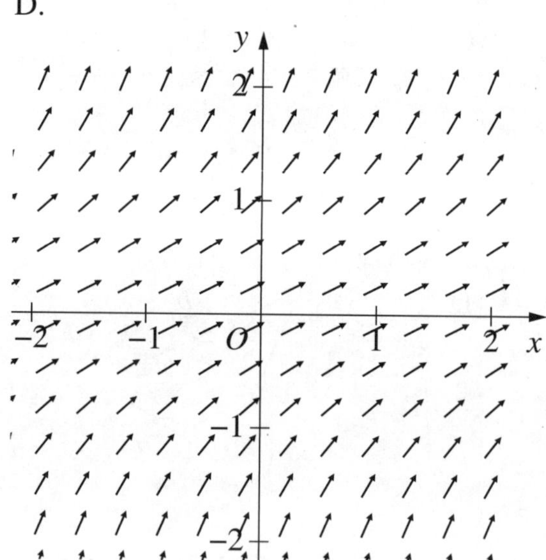

8 Out of 10 contestants, six are to be selected for the final round of a competition. Four of those six will be placed 1st, 2nd, 3rd and 4th.

In how many ways can this process be carried out?

A. $\dfrac{10!}{6!4!}$

B. $\dfrac{10!}{6!}$

C. $\dfrac{10!}{4!2!}$

D. $\dfrac{10!}{4!4!}$

9 The projection of the vector $\begin{pmatrix} 6 \\ 7 \end{pmatrix}$ onto the line $y = 2x$ is $\begin{pmatrix} 4 \\ 8 \end{pmatrix}$.

The point (6, 7) is reflected in the line $y = 2x$ to a point A.

What is the position vector of the point A?

A. $\begin{pmatrix} 6 \\ 12 \end{pmatrix}$

B. $\begin{pmatrix} 2 \\ 9 \end{pmatrix}$

C. $\begin{pmatrix} -6 \\ 7 \end{pmatrix}$

D. $\begin{pmatrix} -2 \\ 1 \end{pmatrix}$

10 The quantities P, Q and R are connected by the related rates,

$$\frac{dR}{dt} = -k^2$$

$$\frac{dP}{dt} = -l^2 \times \frac{dR}{dt}$$

$$\frac{dP}{dt} = m^2 \times \frac{dQ}{dt}$$

where k, l and m are non-zero constants.

Which of the following statements is true?

A. P is increasing and Q is increasing

B. P is increasing and Q is decreasing

C. P is decreasing and Q is increasing

D. P is decreasing and Q is decreasing

Section II

60 marks
Attempt Questions 11–14
Allow about 1 hour and 45 minutes for this section

Answer each question in the appropriate writing booklet. Extra writing booklets are available.

For questions in Section II, your responses should include relevant mathematical reasoning and/or calculations.

Question 11 (15 marks) Use the Question 11 Writing Booklet

(a) Let $P(x) = x^3 + 3x^2 - 13x + 6$.

(i) Show that $P(2) = 0$. **1**

(ii) Hence, factor the polynomial $P(x)$ as $A(x)B(x)$, where $B(x)$ is a quadratic polynomial. **2**

(b) For what value(s) of a are the vectors $\begin{pmatrix} a \\ -1 \end{pmatrix}$ and $\begin{pmatrix} 2a-3 \\ 2 \end{pmatrix}$ perpendicular? **3**

(c) The diagram shows the graph of $y = f(x)$. **3**

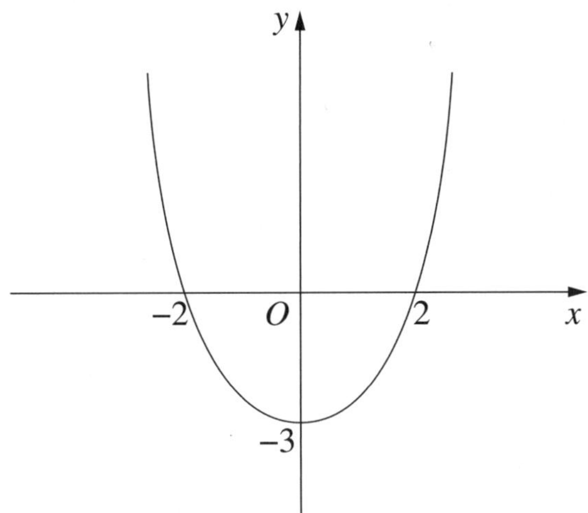

Sketch the graph of $y = \dfrac{1}{f(x)}$.

Question 11 continues on the following page

Question 11 (continued)

(d) By expressing $\sqrt{3}\sin x + 3\cos x$ in the form $A\sin(x + \alpha)$, solve $\sqrt{3}\sin x + 3\cos x = \sqrt{3}$, for $0 \le x \le 2\pi$. **4**

(e) Solve $\frac{dy}{dx} = e^{2y}$, finding x as a function of y. **2**

End of Question 11

Please turn over

Question 12 (14 marks) Use the Question 12 Writing Booklet

(a) Use the principle of mathematical induction to show that for all integers $n \geq 1$, **3**

$$1 \times 2 + 2 \times 5 + 3 \times 8 + \cdots + n(3n-1) = n^2(n+1).$$

(b) When a particular biased coin is tossed, the probability of obtaining a head is $\frac{3}{5}$.

This coin is tossed 100 times.

Let X be the random variable representing the number of heads obtained. This random variable will have a binomial distribution.

(i) Find the expected value, $E(X)$. **1**

(ii) By finding the variance, $\text{Var}(X)$, show that the standard deviation of X is approximately 5. **1**

(iii) By using a normal approximation, find the approximate probability that X is between 55 and 65. **1**

(c) To complete a course, a student must choose and pass exactly three topics. **2**

There are eight topics from which to choose.

Last year 400 students completed the course.

Explain, using the pigeonhole principle, why at least eight students passed exactly the same three topics.

(d) Find $\int_0^{\frac{\pi}{2}} \cos 5x \sin 3x \, dx$. **3**

(e) Find the curve which satisfies the differential equation $\frac{dy}{dx} = -\frac{x}{y}$ and passes through the point $(1, 0)$. **3**

Question 13 (16 marks) Use the Question 13 Writing Booklet

(a) (i) Find $\dfrac{d}{d\theta}\left(\sin^3\theta\right)$. **1**

(ii) Use the substitution $x = \tan\theta$ to evaluate $\displaystyle\int_0^1 \frac{x^2}{\left(1+x^2\right)^{\frac{5}{2}}}\,dx$. **4**

(b) The region R is bounded by the y-axis, the graph of $y = \cos(2x)$ and the graph of $y = \sin x$, as shown in the diagram. **4**

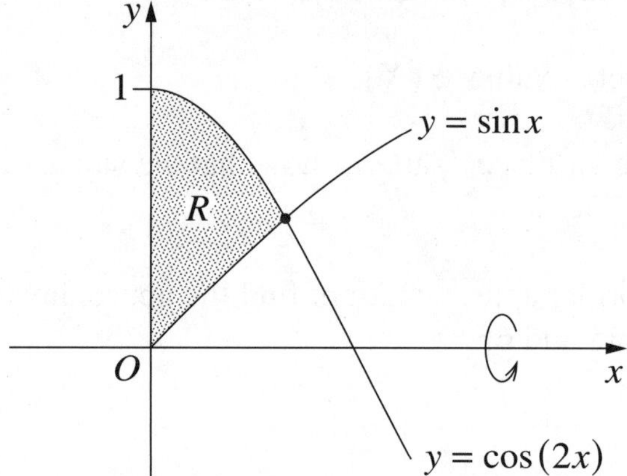

Find the volume of the solid of revolution formed when the region R is rotated about the x-axis.

Question 13 continues on the following page

Question 13 (continued)

(c) Suppose $f(x) = \tan\left(\cos^{-1}(x)\right)$ and $g(x) = \dfrac{\sqrt{1 - x^2}}{x}$.

The graph of $y = g(x)$ is given.

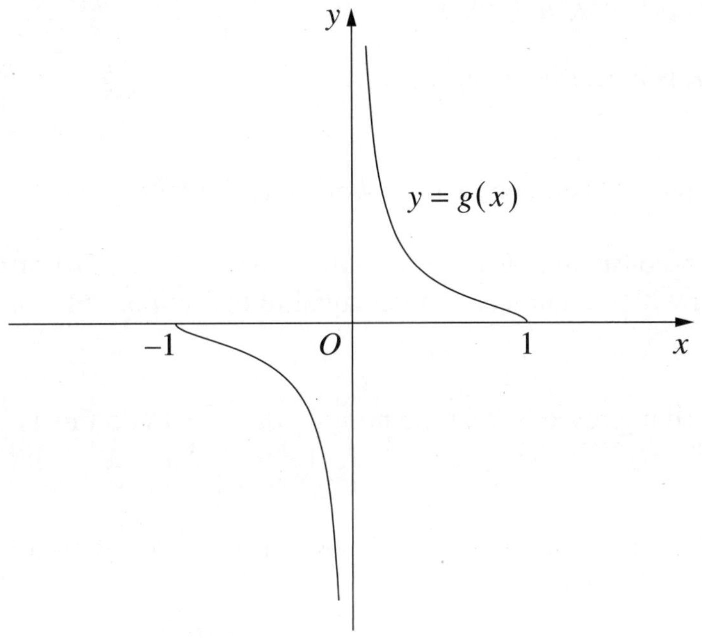

(i) Show that $f'(x) = g'(x)$. **4**

(ii) Using part (i), or otherwise, show that $f(x) = g(x)$. **3**

End of Question 13

Question 14 (15 marks) Use the Question 14 Writing Booklet

(a) (i) Use the identity $(1+x)^{2n} = (1+x)^n (1+x)^n$ **2**

to show that

$$\binom{2n}{n} = \binom{n}{0}^2 + \binom{n}{1}^2 + \cdots + \binom{n}{n}^2,$$

where n is a positive integer.

(ii) A club has $2n$ members, with n women and n men. **2**

A group consisting of an even number $(0, 2, 4, \ldots, 2n)$ of members is chosen, with the number of men equal to the number of women.

Show, giving reasons, that the number of ways to do this is $\binom{2n}{n}$.

(iii) From the group chosen in part (ii), one of the men and one of the women are selected as leaders. **2**

Show, giving reasons, that the number of ways to choose the even number of people and then the leaders is

$$1^2\binom{n}{1}^2 + 2^2\binom{n}{2}^2 + \cdots + n^2\binom{n}{n}^2.$$

(iv) The process is now reversed so that the leaders, one man and one woman, are chosen first. The rest of the group is then selected, still made up of an equal number of women and men. **2**

By considering this reversed process and using part (ii), find a simple expression for the sum in part (iii).

Question 14 continues on the following page

Question 14 (continued)

(b) (i) Show that $\sin^3\theta - \frac{3}{4}\sin\theta + \frac{\sin(3\theta)}{4} = 0.$ **2**

(ii) By letting $x = 4\sin\theta$ in the cubic equation $x^3 - 12x + 8 = 0.$ **2**

Show that $\sin(3\theta) = \frac{1}{2}.$

(iii) Prove that $\sin^2\frac{\pi}{18} + \sin^2\frac{5\pi}{18} + \sin^2\frac{25\pi}{18} = \frac{3}{2}.$ **3**

End of paper

2020 Higher School Certificate
Worked answers

Section I (*Total 10 marks*)

1	A	**2**	C	**3**	D	**4**	B	**5**	C
6	D	**7**	A	**8**	C	**9**	B	**10**	A

1 $x^2 - 2x - 3 \geq 0$

$(x-3)(x+1) \geq 0$

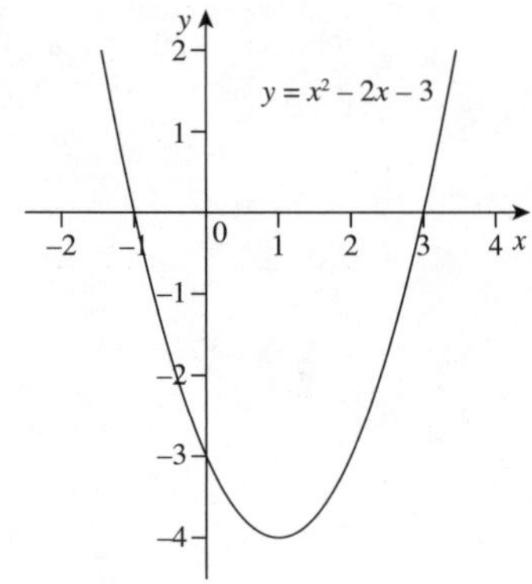

From the graph

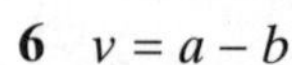

$x \leq -1$ or $x \geq 3$

Answer A

2 $f(x) = 1 + \sqrt{x}$

Domain: $x \geq 0$

Range: $y \geq 1$

So $f^{-1}(x)$ has domain $x \geq 1$ and range $y \geq 0$.

Answer C

3 $\int \frac{1}{4x^2+1}dx = \int \frac{1}{1+(2x)^2}dx$

$$= \frac{1}{2}\int \frac{2}{1+(2x)^2}dx$$

$$= \frac{1}{2}\tan^{-1}(2x) + c$$

Answer D

4 Maria's position is

$(2+3+4)\underset{\sim}{i} + (3-2-3)\underset{\sim}{j}$ or $9\underset{\sim}{i} - 2\underset{\sim}{j}$.

Distance from origin $= \sqrt{9^2+(-2)^2}$

$= \sqrt{85}$

Answer B

5 $x^2 + x + 1$ is not divisible by 2.

So $p(x) = (x-2)^2(x^2+x+1)$.

As $x^2 + x + 1$ has no real roots, $p(x)$ has no other zeroes.

As $x \to \pm\infty, y \to \infty$

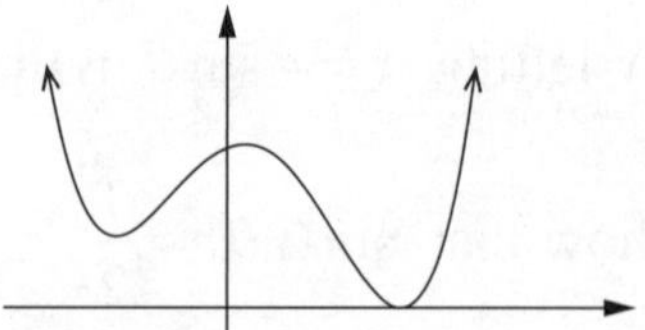

Answer C

6 $\underset{\sim}{v} = \underset{\sim}{a} - \underset{\sim}{b}$

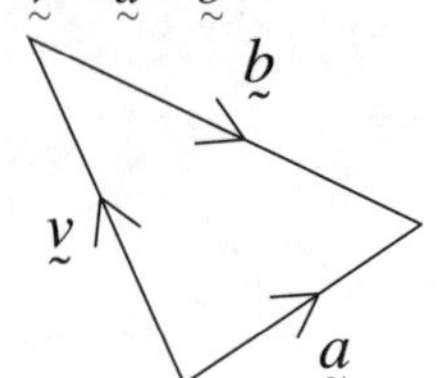

Answer D

7 $\frac{dy}{dx} = -\frac{x}{4y}$

$\frac{dy}{dx}$ is zero when $x = 0$ and undefined when $y = 0$.

At $(-0.5, 0.25)$, $\frac{dy}{dx} = 0.5$.

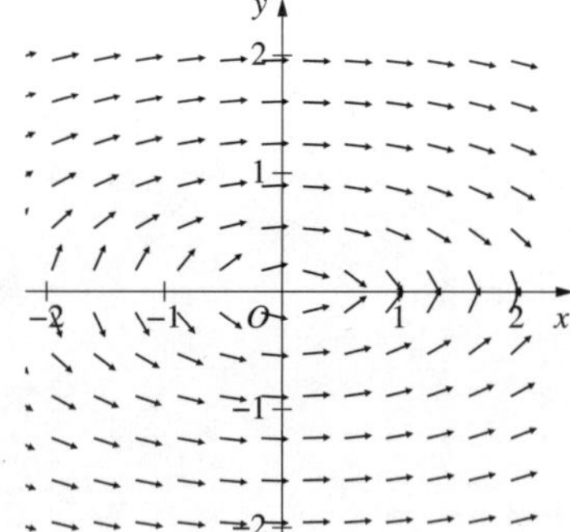

Answer A

8 The six contestants can be selected in $^{10}C_6$ ways.

The four positions can be organised in $^6P_4$ ways.

Number of ways $= {}^{10}C_6 \times {}^6P_4$

$$= \frac{10!}{6!(10-6)!} \times \frac{6!}{(6-4)!}$$

$$= \frac{10!}{6!4!} \times \frac{6!}{2!}$$

$$= \frac{10!}{4!2!}$$

Answer C

9 A is at $(2, 9)$.

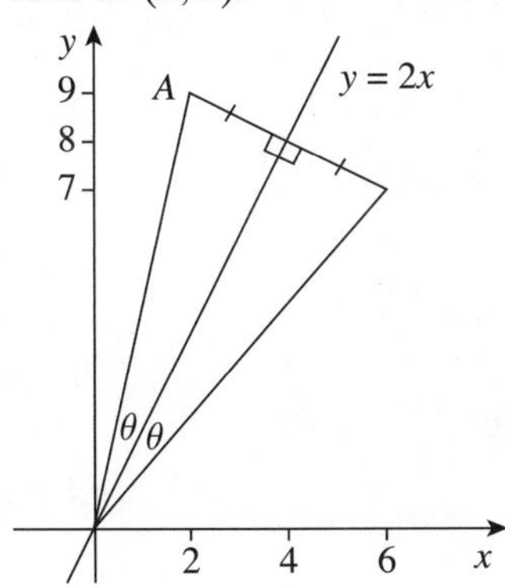

Answer B

10 $\frac{dR}{dt} = -k^2$

So $\frac{dR}{dt} < 0$

$\frac{dP}{dt} = -l^2 \times \frac{dR}{dt}$

So $\frac{dP}{dt} > 0$

$\therefore P$ is increasing.

$\frac{dP}{dt} = m^2 \times \frac{dQ}{dt}$

So $\frac{dQ}{dt} > 0$

$\therefore Q$ is increasing.

Answer A

Section II

QUESTION 11

(a) $P(x) = x^3 + 3x^2 - 13x + 6$

(i) $P(2) = 2^3 + 3 \times 2^2 - 13 \times 2 + 6$
$= 0$

(1 mark)

(ii) As $P(2) = 0$, $(x - 2)$ is a factor of $P(x)$.

$$\begin{array}{r} x^2 + 5x - 3 \\ x - 2 \overline{) x^3 + 3x^2 - 13x + 6} \\ \underline{x^3 - 2x^2} \\ 5x^2 - 13x \\ \underline{5x^2 - 10x} \\ -3x + 6 \\ \underline{-3x + 6} \\ 0 \end{array}$$

$\therefore P(x) = (x - 2)(x^2 + 5x - 3)$

(2 marks)

(b) $\begin{pmatrix} a \\ -1 \end{pmatrix}, \begin{pmatrix} 2a - 3 \\ 2 \end{pmatrix}$

If perpendicular,

$a(2a - 3) + (-1) \times 2 = 0$

$2a^2 - 3a - 2 = 0$

$(2a + 1)(a - 2) = 0$

$2a + 1 = 0$ or $a - 2 = 0$

$a = -\frac{1}{2}$ or $a = 2$

(3 marks)

(c) $f(2) = 0$ and $f(-2) = 0$ so $y = \frac{1}{f(x)}$ will be undefined when $x = \pm 2$.

$\frac{1}{f(x)} \neq 0$ for any value of x.

So there are asymptotes at $x = \pm 2$ and $y = 0$.

$f(x) > 0$, and so $\frac{1}{f(x)} > 0$, when $x < -2$ or $x > 2$.

$f(x) < 0$, and so $\frac{1}{f(x)} < 0$, when $-2 < x < 2$.

$y = f(x)$ has a stationary point at $x = 0$, so $y = \frac{1}{f(x)}$ will also have a stationary point at $x = 0$.

$f(0) = -3$ so $\frac{1}{f(0)} = -\frac{1}{3}$.

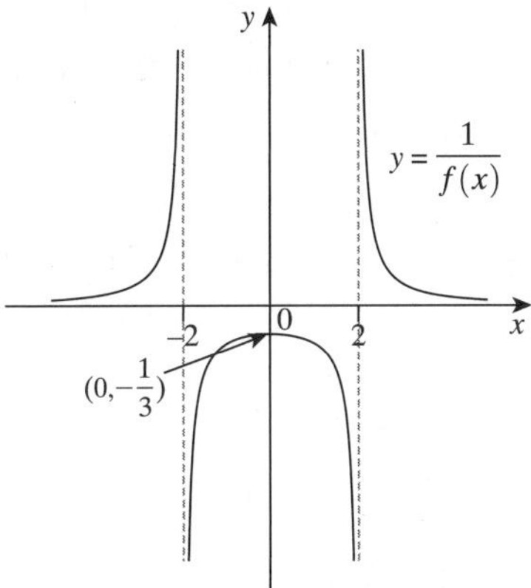

(3 marks)

(d) $A \sin(x+\alpha) = A \sin x \cos\alpha + A \cos x \sin\alpha$

Now $\sqrt{3}\sin x + 3\cos x \equiv A\sin(x+\alpha)$

$\therefore A\cos\alpha = \sqrt{3}$ and $A\sin\alpha = 3$

$$\frac{A\sin\alpha}{A\cos\alpha} = \frac{3}{\sqrt{3}}$$

$$\tan\alpha = \sqrt{3}$$

$$\alpha = \frac{\pi}{3} \quad \left(0 < \alpha < \frac{\pi}{2}\right)$$

$$A\cos\frac{\pi}{3} = \sqrt{3}$$

$$A \times \frac{1}{2} = \sqrt{3}$$

$$A = 2\sqrt{3}$$

So $\sqrt{3}\sin x + 3\cos x \equiv 2\sqrt{3}\sin\left(x+\frac{\pi}{3}\right)$.

Now $0 \le x \le 2\pi$

$$\frac{\pi}{3} \le x + \frac{\pi}{3} \le \frac{7\pi}{3}$$

If $\sqrt{3}\sin x + 3\cos x = \sqrt{3}$

$$2\sqrt{3}\sin\left(x+\frac{\pi}{3}\right) = \sqrt{3}$$

$$\sin\left(x+\frac{\pi}{3}\right) = \frac{1}{2}$$

So $x + \frac{\pi}{3} = \frac{5\pi}{6}$ or $\frac{13\pi}{6}$

$$x = \frac{\pi}{2} \text{ or } \frac{11\pi}{6}$$

(4 marks)

(e) $\frac{dy}{dx} = e^{2y}$

$$\frac{dx}{dy} = \frac{1}{e^{2y}}$$

$$= e^{-2y}$$

$$x = \int e^{-2y}\,dy$$

$$= \frac{e^{-2y}}{-2} + c$$

$$= \frac{-1}{2e^{2y}} + c$$

(2 marks)

QUESTION 12

(a) $1 \times 2 + 2 \times 5 + 3 \times 8 + \ldots + n(3n-1) = n^2(n+1)$

If $n = 1$,

$$\text{LHS} = 1 \times 2 = 2$$

$$\text{RHS} = 1^2(1+1) = 2$$

The statement is true for $n = 1$.

Assume it is true for $n = k$,

i.e. $1 \times 2 + 2 \times 5 + 3 \times 8 + \ldots + k(3k-1) = k^2(k+1)$

We need to show that it is also true for $n = k + 1$.

We wish to prove that:

$1 \times 2 + 2 \times 5 + 3 \times 8 + \ldots + k(3k-1) + (k+1)(3(k+1)-1) = (k+1)^2(k+1+1)$

i.e. $1 \times 2 + 2 \times 5 + 3 \times 8 + \ldots + k(3k-1) + (k+1)(3k+2) = (k+1)^2(k+2)$

$$\begin{aligned}\text{LHS} &= k^2(k+1) + (k+1)(3k+2)\\ &= (k+1)(k^2+3k+2)\\ &= (k+1)(k+2)(k+1)\\ &= (k+1)^2(k+2)\\ &= \text{RHS}\end{aligned}$$

So, if the statement is true for $n = k$ it is also true for $n = k + 1$.

It is true for $n = 1$.

So it is true for $n = 1 + 1 = 2$.

So it is true for $n = 2 + 1 = 3$.

And so on.

By the process of induction it is true for all integers $n, n \ge 1$. *(3 marks)*

(b) (i) $P(\text{head}) = \frac{3}{5}$

So $p = \frac{3}{5}$

The coin is tossed 100 times so $n = 100$.

Binomial distribution.

$$\begin{aligned} E(X) &= np \\ &= 100 \times \frac{3}{5} \\ &= 60 \end{aligned}$$

(1 mark)

(ii) $$\begin{aligned} \text{Var}(X) &= np(1-p) \\ &= 60 \times \left(1 - \frac{3}{5}\right) \\ &= 24 \end{aligned}$$

$$\begin{aligned} \text{Standard deviation} &= \sqrt{24} \\ &= 4.898\,979\,48\ldots \end{aligned}$$

So the standard deviation is approximately 5.

(1 mark)

(iii) A score of 55 is approximately one standard deviation below the mean and a score of 65 is approximately one standard deviation above the mean.

So the approximate probability that X is between 55 and 65 is 68%.

(1 mark)

(c) $$\begin{aligned} \text{Number of choices} &= {}^8C_3 \\ &= 56 \end{aligned}$$

Now $400 \div 56 = 7.1428\ldots$

So, if exactly seven students passed every possible choice that doesn't account for all 400 students. By the pigeonhole principle, at least eight students must have passed exactly the same three topics.

(2 marks)

(d) $$\begin{aligned} &\int_0^{\frac{\pi}{2}} \cos 5x \sin 3x \, dx \\ &= \int_0^{\frac{\pi}{2}} \frac{1}{2}(\sin 8x - \sin 2x)\, dx \\ &= \left[\frac{1}{2}\left(\frac{-\cos 8x}{8} - \frac{-\cos 2x}{2}\right)\right]_0^{\frac{\pi}{2}} \\ &= \left[\frac{-\cos 8x}{16} + \frac{\cos 2x}{4}\right]_0^{\frac{\pi}{2}} \\ &= \frac{-\cos 4\pi}{16} + \frac{\cos \pi}{4} - \left(\frac{-\cos 0}{16} + \frac{\cos 0}{4}\right) \\ &= \frac{-1}{16} + \frac{-1}{4} + \frac{1}{16} - \frac{1}{4} \\ &= -\frac{1}{2} \end{aligned}$$

(3 marks)

(e) $$\begin{aligned} \frac{dy}{dx} &= -\frac{x}{y} \\ y\,dy &= -x\,dx \\ \frac{y^2}{2} &= -\frac{x^2}{2} + c \\ y^2 &= -x^2 + C \quad (\text{where } C = 2c) \end{aligned}$$

When $x = 1, y = 0$

$$\begin{aligned} 0 &= -1^2 + C \\ C &= 1 \\ y^2 &= -x^2 + 1 \\ \therefore x^2 + y^2 &= 1 \end{aligned}$$

(3 marks)

QUESTION 13

(a) (i) $\frac{d}{d\theta}(\sin^3 \theta) = 3\sin^2\theta \cos\theta$

(1 mark)

(ii) $x = \tan\theta$

$dx = \sec^2\theta\, d\theta$

When $x = 0, \theta = 0$

When $x = 1, \theta = \frac{\pi}{4}$

$$\int_0^1 \frac{x^2}{(1+x^2)^{\frac{5}{2}}}\,dx = \int_0^{\frac{\pi}{4}} \frac{\tan^2\theta}{(1+\tan^2\theta)^{\frac{5}{2}}}\sec^2\theta\, d\theta$$
$$= \int_0^{\frac{\pi}{4}} \frac{\tan^2\theta}{(\sec^2\theta)^{\frac{5}{2}}}\sec^2\theta\, d\theta$$
$$= \int_0^{\frac{\pi}{4}} \frac{\tan^2\theta}{\sec^3\theta}\, d\theta$$
$$= \int_0^{\frac{\pi}{4}} \left(\frac{\sin^2\theta}{\cos^2\theta}\times\cos^3\theta\right) d\theta$$
$$= \int_0^{\frac{\pi}{4}} \sin^2\theta\cos\theta\, d\theta$$
$$= \frac{1}{3}\left[\sin^3\theta\right]_0^{\frac{\pi}{4}} \quad \text{(from (i))}$$
$$= \frac{1}{3}\left(\sin^3\frac{\pi}{4} - \sin^3 0\right)$$
$$= \frac{1}{3}\left(\left(\frac{1}{\sqrt{2}}\right)^3 - 0^3\right)$$
$$= \frac{1}{6\sqrt{2}}$$

(4 marks)

(b) $\cos 2x = \cos^2 x - \sin^2 x$

$= 1 - 2\sin^2 x$

At the point of intersection:

$$\sin x = 1 - 2\sin^2 x$$

$2\sin^2 x + \sin x - 1 = 0$

$(2\sin x - 1)(\sin x + 1) = 0$

$\sin x = \frac{1}{2}$ or $\sin x = -1$

From the diagram $0 < x < \frac{\pi}{2}$ at the point of intersection.

So $\sin x = \frac{1}{2}$

$\therefore x = \frac{\pi}{6}$

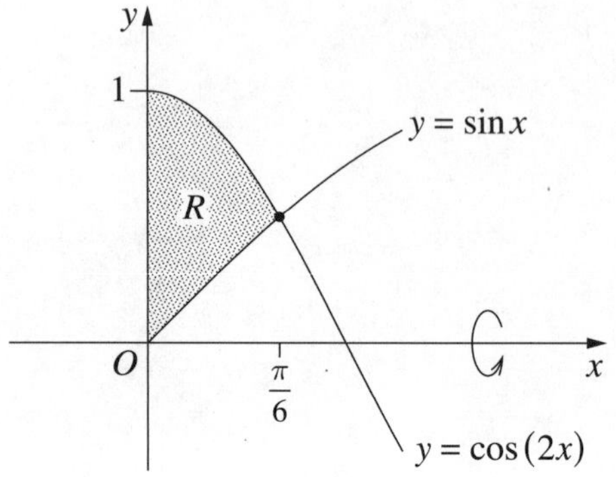

y
1
R
O
$\frac{\pi}{6}$
x
$y = \sin x$
$y = \cos(2x)$

$$V = \int_0^{\frac{\pi}{6}} \pi(\cos^2(2x) - \sin^2 x)\,dx$$
$$= \pi\int_0^{\frac{\pi}{6}} \left(\frac{1}{2}(1+\cos(4x)) - \frac{1}{2}(1-\cos(2x))\right)dx$$
$$= \frac{\pi}{2}\int_0^{\frac{\pi}{6}} (\cos(4x)+\cos(2x))\,dx$$
$$= \frac{\pi}{2}\left[\frac{\sin(4x)}{4} + \frac{\sin(2x)}{2}\right]_0^{\frac{\pi}{6}}$$
$$= \frac{\pi}{2}\left(\frac{\sin\frac{2\pi}{3}}{4} + \frac{\sin\frac{\pi}{3}}{2}\right)$$
$$= \frac{\pi}{2}\left(\frac{\sqrt{3}}{8} + \frac{\sqrt{3}}{4}\right)$$
$$= \frac{3\sqrt{3}\pi}{16}\text{ units}^3$$

(4 marks)

(c) (i) $f(x) = \tan(\cos^{-1}(x))$

$$f'(x) = \sec^2(\cos^{-1}(x)) \times \frac{-1}{\sqrt{1-x^2}}$$
$$= \frac{1}{(\cos(\cos^{-1}x))^2} \times \frac{-1}{\sqrt{1-x^2}}$$
$$= \frac{-1}{x^2\sqrt{1-x^2}}$$

Now $g(x) = \frac{\sqrt{1-x^2}}{x}$

$$g'(x) = \frac{x\times\frac{1}{2}(1-x^2)^{-\frac{1}{2}}\times -2x - \sqrt{1-x^2}\times 1}{x^2}$$
$$= \frac{\frac{-x^2}{\sqrt{1-x^2}} - \sqrt{1-x^2}}{x^2}$$
$$= \frac{\frac{-x^2}{\sqrt{1-x^2}} - \frac{1-x^2}{\sqrt{1-x^2}}}{x^2}$$
$$= \frac{\frac{-1}{\sqrt{1-x^2}}}{x^2}$$
$$= \frac{-1}{x^2\sqrt{1-x^2}}$$

$\therefore f'(x) = g'(x)$

(4 marks)

(ii) $\cos^{-1}x$ has domain $[-1, 1]$.

$\cos^{-1}0 = \frac{\pi}{2}$ and $\tan\frac{\pi}{2}$ is undefined.

So both $f(x)$ and $g(x)$ have domain $[-1, 0) \cup (0, 1]$.

From (i) the curves have the same gradient function.

So they only, at most, differ by a constant.

$\therefore f(x) = g(x) + c$

From the graph $g(1) = 0$.

$$f(x) = \tan(\cos^{-1}(x))$$
$$f(1) = \tan(\cos^{-1}1) = \tan 0 = 0$$

So $c = 0$

$\therefore f(x) = g(x)$

OR: Let $\theta = \cos^{-1}(x)$

So $\cos\theta = x$

$$f(x) = \tan(\cos^{-1}(x)) = \tan\theta = \frac{\sin\theta}{\cos\theta} \quad (\cos\theta \neq 0)$$

Now $0 \leq \theta \leq \pi$.

So $\sin\theta \geq 0$.

Now $\sin^2\theta = 1 - \cos^2\theta$.

So $\sin\theta = \sqrt{1-\cos^2\theta} = \sqrt{1-x^2}$

$$\therefore f(x) = \frac{\sqrt{1-x^2}}{x} = g(x)$$

(3 marks)

QUESTION 14

(a) (i) $(1+x)^{2n} = (1+x)^n(1+x)^n$

Now $(1+x)^n = \binom{n}{0} + \binom{n}{1}x + \binom{n}{2}x^2 + \ldots + \binom{n}{n}x^n$

$$(1+x)^{2n} = \binom{2n}{0} + \binom{2n}{1}x + \binom{2n}{2}x^2 + \ldots + \binom{2n}{n}x^n + \ldots + \binom{2n}{2n}x^{2n}$$

The coefficient of x^n is $\binom{2n}{n}$.

$(1+x)^n(1+x)^n$

$$= \left(\binom{n}{0} + \binom{n}{1}x + \binom{n}{2}x^2 + \ldots + \binom{n}{n}x^n\right)\left(\binom{n}{0} + \binom{n}{1}x + \binom{n}{2}x^2 + \ldots + \binom{n}{n}x^n\right)$$

Term involving x^n is $\binom{n}{0}\binom{n}{n}x^n + \binom{n}{1}x\binom{n}{n-1}x^{n-1} + \binom{n}{2}x^2\binom{n}{n-2}x^{n-2} + \ldots$

So the coefficient of x^n is $\binom{n}{0}\binom{n}{n} + \binom{n}{1}\binom{n}{n-1} + \binom{n}{2}\binom{n}{n-2} + \ldots + \binom{n}{n}\binom{n}{n-n}$.

But $\binom{n}{n-k} = \binom{n}{k}$.

So the coefficient of x^n is $\binom{n}{0}^2 + \binom{n}{1}^2 + \ldots + \binom{n}{n}^2$.

Equating coefficients $\binom{2n}{n} = \binom{n}{0}^2 + \binom{n}{1}^2 + \ldots + \binom{n}{n}^2$.

(2 marks)

(ii) n women and n men

If k women are chosen, the number of ways the women can be chosen is $\binom{n}{k}$.

As there are equal numbers of men and women to be chosen, the number of ways that the men can be chosen will be equal to the number of ways that the women can be chosen.

So for each way that the women can be chosen there are $\binom{n}{k}$ ways that the men can be chosen.

So the number of ways the people can be chosen is $\binom{n}{k}^2$.

The total number of ways is $\binom{n}{0}^2 + \binom{n}{1}^2 + \ldots + \binom{n}{n}^2$.

So, from part (i), the number of ways is $\binom{2n}{n}$.

(2 marks)

(iii) If k women are chosen, the number of ways that one leader can be chosen is $\binom{k}{1}$ or k.

For each way that a female leader can be chosen, there are k ways that a male leader can be chosen.

So the number of ways to choose k women and k men and then the leaders is $k^2\binom{n}{k}^2$.

The total number of ways to choose the people and then the leaders

$= 0^2\binom{n}{0}^2 + 1^2\binom{n}{1}^2 + 2^2\binom{n}{2}^2 + \ldots + n^2\binom{n}{n}^2$

$= 1^2\binom{n}{1}^2 + 2^2\binom{n}{2}^2 + \ldots + n^2\binom{n}{n}^2$

(2 marks)

(iv) If the leader is chosen first, then there is one leader chosen from n women so there are n ways that woman can be chosen.

For each of those there are n ways that the male leader can be chosen.

So the number of ways the two leaders can be chosen is n^2.

Once the leader has been chosen there are $n - 1$ women, and men, left to choose from.

From part (ii) the number of ways to choose those is $\binom{2(n-1)}{n-1}$.

Total ways to choose the leaders first and then the rest of the group $= n^2\binom{2n-2}{n-1}$.

The total number of ways must be the same whether the leaders are chosen first or last.

So $1^2\binom{n}{1}^2 + 2^2\binom{n}{2}^2 + \ldots + n^2\binom{n}{n}^2 = n^2\binom{2n-2}{n-1}$.

(2 marks)

(b) (i) $\sin(3\theta) = \sin((2\theta) + \theta)$

$= \sin(2\theta)\cos\theta + \cos(2\theta)\sin\theta$

$= (2\sin\theta\cos\theta)\cos\theta + (\cos^2\theta - \sin^2\theta)\sin\theta$

$= 2\sin\theta\cos^2\theta + \cos^2\theta\sin\theta - \sin^3\theta$

$= 3\sin\theta\cos^2\theta - \sin^3\theta$

$= 3\sin\theta(1 - \sin^2\theta) - \sin^3\theta$

$= 3\sin\theta - 4\sin^3\theta$

Now $\sin^3\theta - \frac{3}{4}\sin\theta + \frac{\sin(3\theta)}{4}$

$= \sin^3\theta - \frac{3}{4}\sin\theta + \frac{3\sin\theta - 4\sin^3\theta}{4}$

$= \sin^3\theta - \frac{3}{4}\sin\theta + \frac{3}{4}\sin\theta - \sin^3\theta$

$= 0$

(2 marks)

(ii) $x^3 - 12x + 8 = 0$

Let $x = 4\sin\theta$.

$(4\sin\theta)^3 - 12(4\sin\theta) + 8 = 0$

$64\sin^3\theta - 48\sin\theta + 8 = 0$

Dividing both sides by 64:

$\sin^3\theta - \frac{3}{4}\sin\theta + \frac{1}{8} = 0$

But $\sin^3\theta - \frac{3}{4}\sin\theta + \frac{\sin(3\theta)}{4} = 0$ from part (i)

So $\dfrac{\sin(3\theta)}{4} = \dfrac{1}{8}$

$\sin(3\theta) = \dfrac{1}{2}$

(2 marks)

(iii) If $\sin(3\theta) = \dfrac{1}{2}$

$$3\theta = \frac{\pi}{6}, \frac{5\pi}{6}, \frac{13\pi}{6}, \frac{17\pi}{6}, \frac{25\pi}{6}, \ldots \quad (\theta > 0)$$

$$\theta = \frac{\pi}{18}, \frac{5\pi}{18}, \frac{13\pi}{18}, \frac{17\pi}{18}, \frac{25\pi}{18}, \ldots \quad (\theta > 0)$$

Now $4\sin\theta$ is a root of the equation $x^3 - 12x + 8 = 0$.

$4\sin\dfrac{\pi}{18}$, $4\sin\dfrac{5\pi}{18}$ and $4\sin\dfrac{25\pi}{18}$ are distinct roots.

Now $\alpha + \beta + \gamma = -\dfrac{b}{a}$

So $4\sin\dfrac{\pi}{18} + 4\sin\dfrac{5\pi}{18} + 4\sin\dfrac{25\pi}{18} = 0$

$$\sin\frac{\pi}{18} + \sin\frac{5\pi}{18} + \sin\frac{25\pi}{18} = 0$$

$\alpha\beta + \alpha\gamma + \beta\gamma = \dfrac{c}{a}$

$$\left(4\sin\frac{\pi}{18}\right)\left(4\sin\frac{5\pi}{18}\right) + \left(4\sin\frac{\pi}{18}\right)\left(4\sin\frac{25\pi}{18}\right) + \left(4\sin\frac{5\pi}{18}\right)\left(4\sin\frac{25\pi}{18}\right) = -12$$

$$16\left[\left(\sin\frac{\pi}{18}\right)\left(\sin\frac{5\pi}{18}\right) + \left(\sin\frac{\pi}{18}\right)\left(\sin\frac{25\pi}{18}\right) + \left(\sin\frac{5\pi}{18}\right)\left(\sin\frac{25\pi}{18}\right)\right] = -12$$

$$\left(\sin\frac{\pi}{18}\right)\left(\sin\frac{5\pi}{18}\right) + \left(\sin\frac{\pi}{18}\right)\left(\sin\frac{25\pi}{18}\right) + \left(\sin\frac{5\pi}{18}\right)\left(\sin\frac{25\pi}{18}\right) = -\frac{3}{4}$$

$$\sin^2\frac{\pi}{18} + \sin^2\frac{5\pi}{18} + \sin^2\frac{25\pi}{18}$$

$$= \left(\sin\frac{\pi}{18} + \sin\frac{5\pi}{18} + \sin\frac{25\pi}{18}\right)^2 - 2\left(\left(\sin\frac{\pi}{18}\right)\left(\sin\frac{5\pi}{18}\right) + \left(\sin\frac{\pi}{18}\right)\left(\sin\frac{25\pi}{18}\right) + \left(\sin\frac{5\pi}{18}\right)\left(\sin\frac{25\pi}{18}\right)\right)$$

$$= 0^2 - 2 \times -\frac{3}{4}$$

$$= \frac{3}{2}$$

(3 marks)

NSW Education Standards Authority

2021 HIGHER SCHOOL CERTIFICATE EXAMINATION

Mathematics Extension 1

General Instructions

- Reading time – 10 minutes
- Working time – 2 hours
- Write using black pen
- Calculators approved by NESA may be used
- A reference sheet is provided at the back of this paper
- For questions in Section II, show relevant mathematical reasoning and/or calculations
- Write your Centre Number and Student Number on the Question 12 Writing Booklet attached

Total marks: 70

Section I – 10 marks

- Attempt Questions 1–10
- Allow about 15 minutes for this section

Section II – 60 marks

- Attempt Questions 11–14
- Allow about 1 hour and 45 minutes for this section

Section I

10 marks
Attempt Questions 1–10
Allow about 15 minutes for this section

Use the multiple-choice answer sheet for Questions 1–10.

---

1 Given that $\overrightarrow{OP} = \begin{pmatrix} -3 \\ 1 \end{pmatrix}$ and $\overrightarrow{OQ} = \begin{pmatrix} 2 \\ 5 \end{pmatrix}$, what is $\overrightarrow{PQ}$?

A. $\begin{pmatrix} 1 \\ -6 \end{pmatrix}$

B. $\begin{pmatrix} -1 \\ 6 \end{pmatrix}$

C. $\begin{pmatrix} 5 \\ 4 \end{pmatrix}$

D. $\begin{pmatrix} -5 \\ -4 \end{pmatrix}$

2 Which of the following integrals is equivalent to $\int \sin^2 3x \, dx$?

A. $\int \frac{1+\cos 6x}{2} dx$

B. $\int \frac{1-\cos 6x}{2} dx$

C. $\int \frac{1+\sin 6x}{2} dx$

D. $\int \frac{1-\sin 6x}{2} dx$

3 What is the remainder when $P(x) = -x^3 - 2x^2 - 3x + 8$ is divided by $x + 2$?

A. -14

B. -2

C. 2

D. 14

4 Consider the differential equation $\frac{dy}{dx} = \frac{x}{y}$.

Which of the following equations best represents this relationship between x and y?

A. $y^2 = x^2 + c$

B. $y^2 = \frac{x^2}{2} + c$

C. $y = x\ln|y| + c$

D. $y = \frac{x^2}{2}\ln|y| + c$

5 For the two vectors $\overrightarrow{OA}$ and $\overrightarrow{OB}$ it is known that

$$\overrightarrow{OA} \cdot \overrightarrow{OB} < 0.$$

Which of the following statements MUST be true?

A. Either, $\overrightarrow{OA}$ is negative and $\overrightarrow{OB}$ is positive, or, $\overrightarrow{OA}$ is positive and $\overrightarrow{OB}$ is negative.

B. The angle between $\overrightarrow{OA}$ and $\overrightarrow{OB}$ is obtuse.

C. The product $\left|\overrightarrow{OA}\right|\left|\overrightarrow{OB}\right|$ is negative.

D. The points O, A and B are collinear.

6 The random variable X represents the number of successes in 10 independent Bernoulli trials. The probability of success is $p = 0.9$ in each trial.

Let $r = P(X \geq 1)$.

Which of the following describes the value of r?

A. $r > 0.9$

B. $r = 0.9$

C. $0.1 < r < 0.9$

D. $r \leq 0.1$

7 Which curve best represents the graph of the function $f(x) = -a\sin x + b\cos x$ given that the constants a and b are both positive?

A.

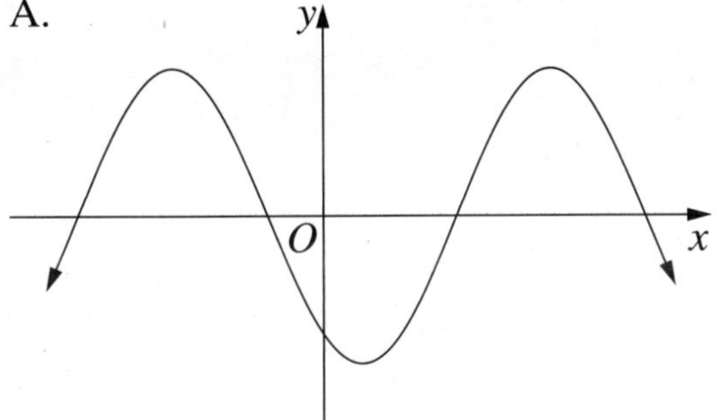

B.

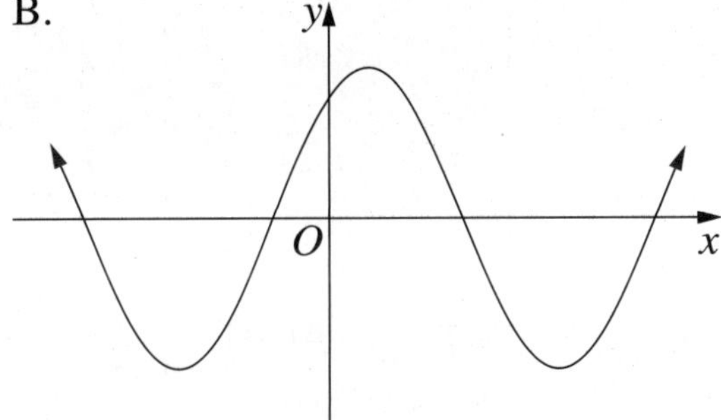

C.

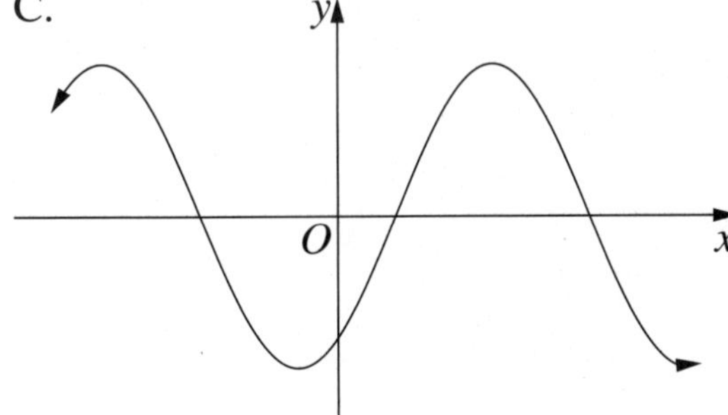

D.

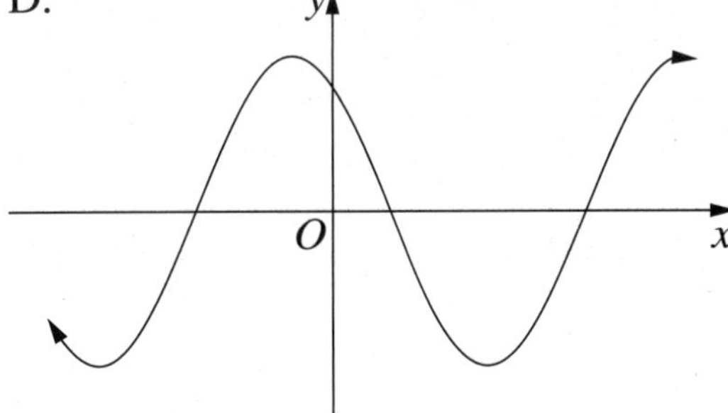

8 The diagram shows a semicircle.

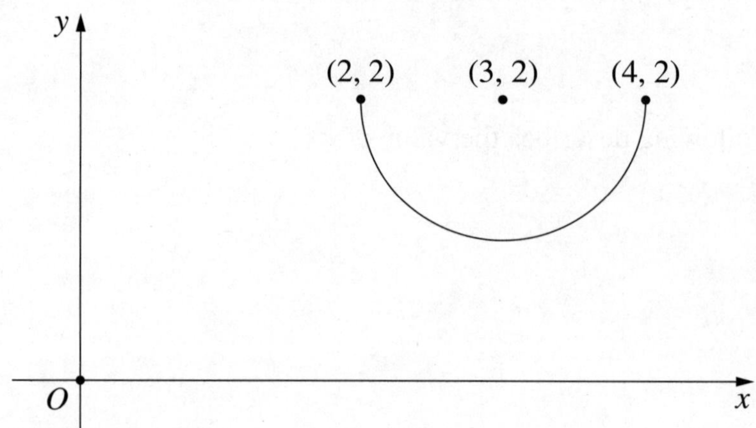

Which pair of parametric equations represents the semicircle shown?

A. $\begin{cases} x = 3 + \sin t \\ y = 2 + \cos t \end{cases}$ for $-\dfrac{\pi}{2} \le t \le \dfrac{\pi}{2}$

B. $\begin{cases} x = 3 + \cos t \\ y = 2 + \sin t \end{cases}$ for $-\dfrac{\pi}{2} \le t \le \dfrac{\pi}{2}$

C. $\begin{cases} x = 3 - \sin t \\ y = 2 - \cos t \end{cases}$ for $-\dfrac{\pi}{2} \le t \le \dfrac{\pi}{2}$

D. $\begin{cases} x = 3 - \cos t \\ y = 2 - \sin t \end{cases}$ for $-\dfrac{\pi}{2} \le t \le \dfrac{\pi}{2}$

9 Which graph represents the function $y = \sin^{-1}(\sin x)$?

A.

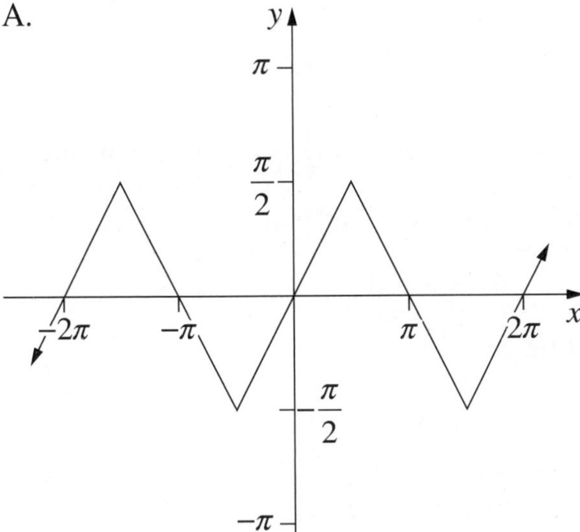

B.

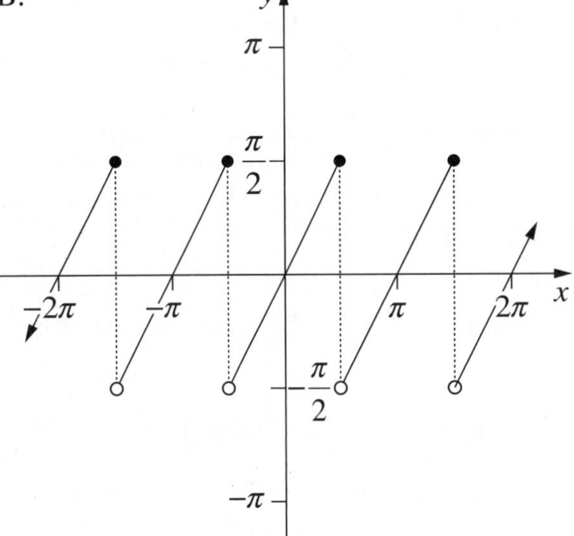

C.

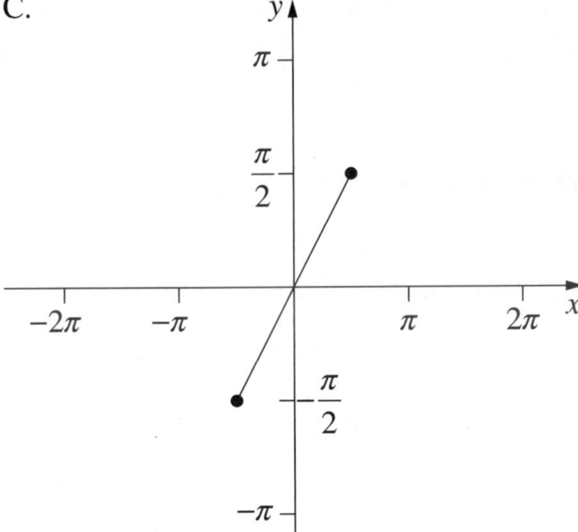

D.

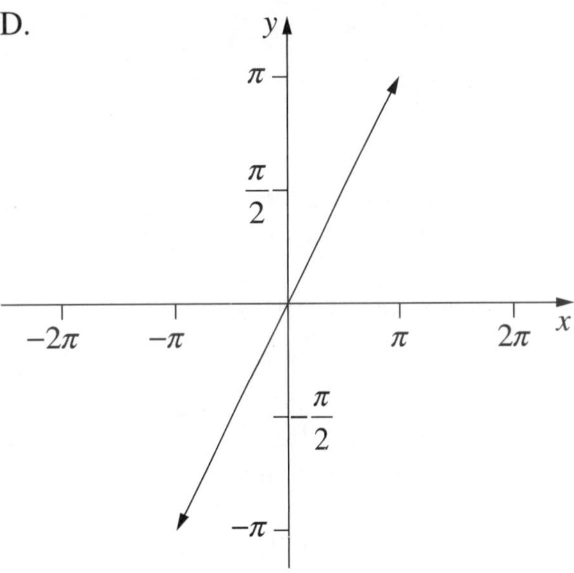

10 The members of a club voted for a new president. There were 15 candidates for the position of president and 3543 members voted. Each member voted for one candidate only.

One candidate received more votes than anyone else and so became the new president.

What is the smallest number of votes the new president could have received?

A. 236

B. 237

C. 238

D. 239

Section II

60 marks
Attempt Questions 11–14
Allow about 1 hour and 45 minutes for this section

Answer each question in the appropriate writing booklet. Extra writing booklets are available.

For questions in Section II, your responses should include relevant mathematical reasoning and/or calculations.

Question 11 (16 marks) Use the Question 11 Writing Booklet

(a) Find $\left(\underset{\sim}{i}+6\underset{\sim}{j}\right)+\left(2\underset{\sim}{i}-7\underset{\sim}{j}\right)$. **1**

(b) Expand and simplify $(2a-b)^4$. **2**

(c) Use the substitution $u = x+1$ to find $\int x\sqrt{x+1}\,dx$. **3**

(d) A committee containing 5 men and 3 women is to be formed from a group of 10 men and 8 women. **1**

In how many different ways can the committee be formed?

(e) A spherical bubble is moving up through a liquid. As it rises, the bubble gets bigger and its radius increases at the rate of 0.2 mm/s. **2**

At what rate is the volume of the bubble increasing when its radius reaches 0.6 mm? Express your answer in mm^3/s rounded to one decimal place.

(f) Evaluate $\displaystyle\int_0^{\sqrt{3}} \frac{1}{\sqrt{4-x^2}}\,dx$. **2**

(g) By factorising, or otherwise, solve $2\sin^3 x + 2\sin^2 x - \sin x - 1 = 0$ for $0 \le x \le 2\pi$. **3**

(h) The roots of $x^4 - 3x + 6 = 0$ are α, β, γ and δ. **2**

What is the value of $\dfrac{1}{\alpha}+\dfrac{1}{\beta}+\dfrac{1}{\gamma}+\dfrac{1}{\delta}$?

Question 12 (14 marks) Use the Question 12 Writing Booklet

(a) The direction field for a differential equation is given on page 1 of the Question 12 Writing Booklet [see the end of this paper]. **1**

The graph of a particular solution to the differential equation passes through the point P.

On the diagram provided in the writing booklet, sketch the graph of this particular solution.

(b) A bottle of water, with temperature 5°C, is placed on a table in a room. The temperature of the room remains constant at 25°C. After t minutes, the temperature of the water, in degrees Celsius, is T.

The temperature of the water can be modelled using the differential equation

$$\frac{dT}{dt} = k(T - 25) \qquad \text{(Do NOT prove this.)}$$

where k is the growth constant.

(i) After 8 minutes, the temperature of the water is 10°C. **3**

By solving the differential equation, find the value of t when the temperature of the water reaches 20°C. Give your answer to the nearest minute.

(ii) Sketch the graph of T as a function of t. **1**

(c) Use mathematical induction to prove that **3**

$$\frac{1}{1 \times 2 \times 3} + \frac{1}{2 \times 3 \times 4} + \cdots + \frac{1}{n(n+1)(n+2)} = \frac{1}{4} - \frac{1}{2(n+1)(n+2)}$$

for all integers $n \geq 1$.

(d) A function is defined by $f(x) = 4 - \left(1 - \frac{x}{2}\right)^2$ for x in the domain $(-\infty, 2]$.

(i) Sketch the graph of $y = f(x)$ showing the x- and y-intercepts. **2**

(ii) Find the equation of the inverse function, $f^{-1}(x)$, and state its domain. **3**

(iii) Sketch the graph of $y = f^{-1}(x)$. **1**

Question 13 (14 marks) Use the Question 13 Writing Booklet

(a) A 2-metre-high sculpture is to be made out of concrete. The sculpture is formed by rotating the region between $y = x^2$, $y = x^2 + 1$ and $y = 2$ around the y-axis. **3**

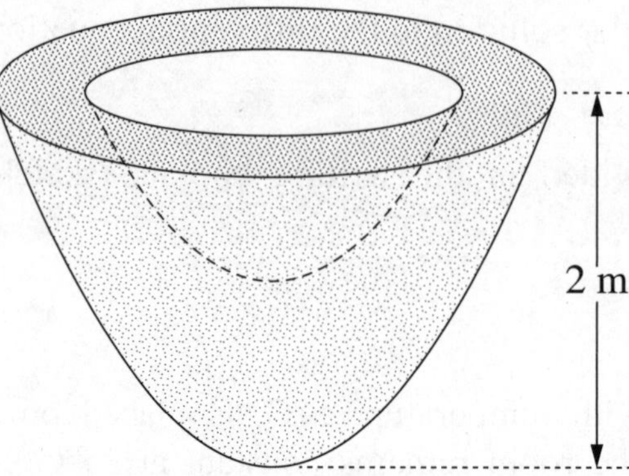

Find the volume of concrete needed to make the sculpture.

(b) When an object is projected from a point h metres above the origin with initial speed V m/s at an angle of $\theta°$ to the horizontal, its displacement vector, t seconds after projection, is **4**

$$\underset{\sim}{r}(t) = \left(Vt\cos\theta\right)\underset{\sim}{i} + \left(-5t^2 + Vt\sin\theta + h\right)\underset{\sim}{j}.$$

(Do NOT prove this.)

A person, standing in an empty room which is 3 m high, throws a ball at the far wall of the room. The ball leaves their hand 1 m above the floor and 10 m from the far wall. The initial velocity of the ball is 12 m/s at an angle of 30° to the horizontal.

Show that the ball will NOT hit the ceiling of the room but that it will hit the far wall without hitting the floor.

Question 13 continues on the following page

Question 13 (continued)

(c) The region enclosed by $y = 2 - |x|$ and $y = 1 - \dfrac{8}{4 + x^2}$ is shaded in the diagram. **3**

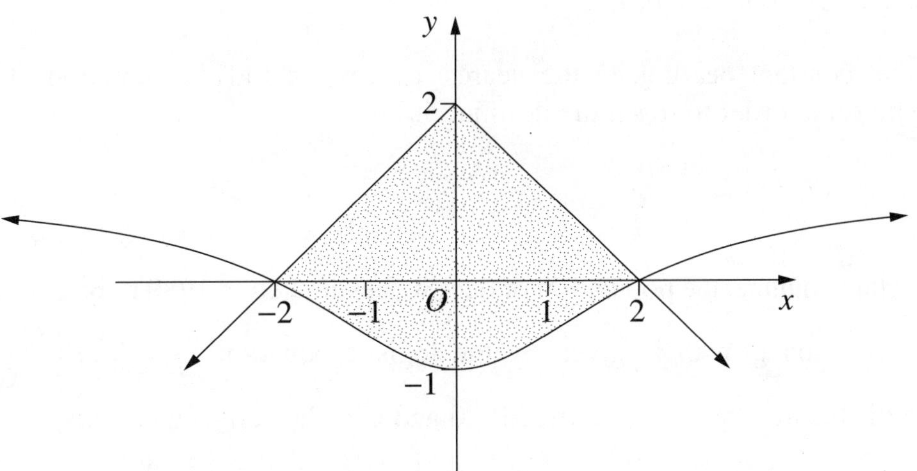

Find the exact value of the area of the shaded region.

(d) (i) The numbers A, B and C are related by the equations $A = B - d$ and $C = B + d$, where d is a constant. **2**

Show that $\dfrac{\sin A + \sin C}{\cos A + \cos C} = \tan B$.

(ii) Hence, or otherwise, solve $\dfrac{\sin\dfrac{5\theta}{7} + \sin\dfrac{6\theta}{7}}{\cos\dfrac{5\theta}{7} + \cos\dfrac{6\theta}{7}} = \sqrt{3}$, for $0 \le \theta \le 2\pi$. **2**

End of Question 13

Question 14 (16 marks) Use the Question 14 Writing Booklet

(a) A plane needs to travel to a destination that is on a bearing of 063°. The engine is set to fly at a constant 175 km/h. However, there is a wind from the south with a constant speed of 42 km/h. **3**

On what constant bearing, to the nearest degree, should the direction of the plane be set in order to reach the destination?

(b) In a certain country, the population of deer was estimated in 1980 to be 150 000. **4**

The population growth is given by the logistic equation $\dfrac{dP}{dt} = 0.1P\left(\dfrac{C-P}{C}\right)$ where t is the number of years after 1980 and C is the carrying capacity.

In the year 2000, the population of deer was estimated to be 600 000.

Use the fact that $\dfrac{C}{P(C-P)} = \dfrac{1}{P} + \dfrac{1}{C-P}$ to show that the carrying capacity is approximately 1 130 000.

(c) (i) For vector $\underset{\sim}{v}$, show that $\underset{\sim}{v} \cdot \underset{\sim}{v} = |\underset{\sim}{v}|^2$. **1**

(ii) In the trapezium $ABCD$, BC is parallel to AD and $\left|\overrightarrow{AC}\right| = \left|\overrightarrow{BD}\right|$. **3**

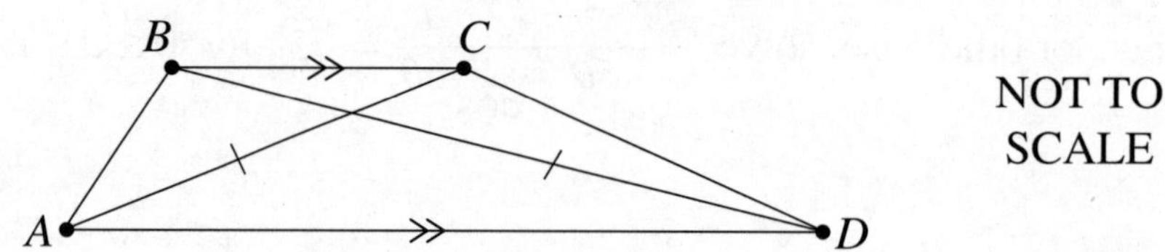

Let $\underset{\sim}{a} = \overrightarrow{AB}$, $\underset{\sim}{b} = \overrightarrow{BC}$ and $\overrightarrow{AD} = k\,\overrightarrow{BC}$, where $k > 0$.

Using part (i), or otherwise, show $2\underset{\sim}{a} \cdot \underset{\sim}{b} + (1-k)\,|\underset{\sim}{b}|^2 = 0$.

Question 14 continues on the following page

Question 14 (continued)

(d) At a certain factory, the proportion of faulty items produced by a machine is $p = \frac{3}{500}$, which is considered to be acceptable. To confirm that the machine is working to this standard, a sample of size n is taken and the sample proportion $\hat{p}$ is calculated. **3**

It is assumed that $\hat{p}$ is approximately normally distributed with $\mu = p$ and $\sigma^2 = \frac{p(1-p)}{n}$.

Production by this machine will be shut down if $\hat{p} \geq \frac{4}{500}$.

The sample size is to be chosen so that the chance of shutting down the machine unnecessarily is less than 2.5%.

Find the approximate sample size required, giving your answer to the nearest thousand.

(e) The polynomial $g(x) = x^3 + 4x - 2$ passes through the point $(1, 3)$. **2**

Find the gradient of the tangent to $f(x) = xg^{-1}(x)$ at the point where $x = 3$.

End of paper

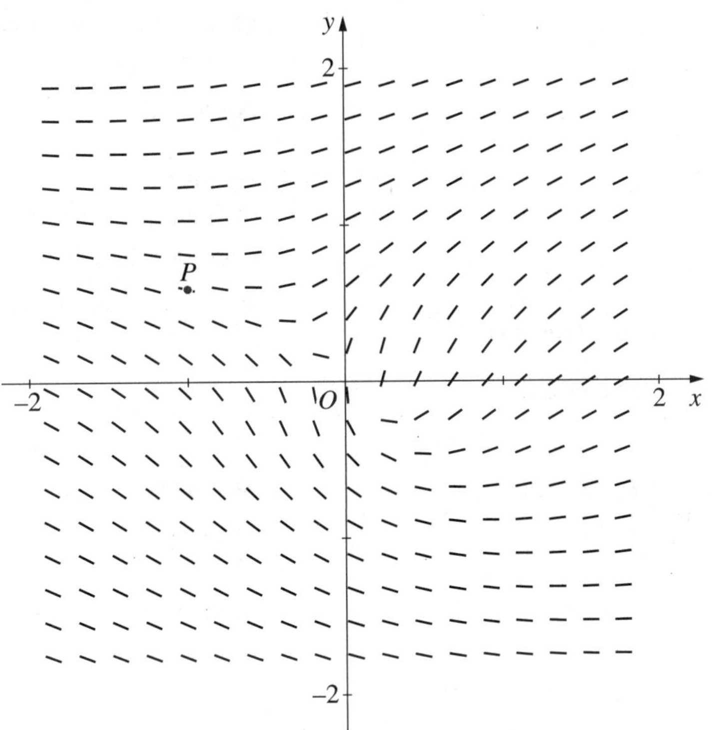

2021 Higher School Certificate
Worked answers

Section I

(Total 10 marks)

1 C	**2** B	**3** D	**4** A	**5** B
6 A	**7** D	**8** C	**9** A	**10** C

1 $\overrightarrow{OP}+\overrightarrow{PQ}=\overrightarrow{OQ}$

$$\overrightarrow{PQ}=\overrightarrow{OQ}-\overrightarrow{OP}$$

$$\overrightarrow{PQ}=\begin{pmatrix}2\\5\end{pmatrix}-\begin{pmatrix}-3\\1\end{pmatrix}$$

$$\overrightarrow{PQ}=\begin{pmatrix}5\\4\end{pmatrix}$$

Answer C

2 From the formula sheet

$\sin^2 nx=\frac{1}{2}(1-\cos 2nx)$, then

$$\sin^2 3x=\frac{1}{2}(1-\cos 6x)$$

$$\int \sin^2 3x\,dx=\int\frac{1-\cos 6x}{2}\,dx$$

Answer B

3 If $P(x)$ is divided by $x - k$, then the remainder is $P(k)$.

$$P(x)=-x^3-2x^2-3x+8$$

$$P(-2)=-(-2)^3-2(-2)^2-3(-2)+8$$

$$P(-2)=14$$

Answer D

4 $\frac{dy}{dx}=\frac{x}{y}$

$$y\,dy=x\,dx$$

$$\int y\,dy=\int x\,dx$$

$$\frac{y^2}{2}=\frac{x^2}{2}+D, \text{for some constant } D$$

$$y^2=x^2+c, \text{for some constant } c$$

Answer A

5 If $\overrightarrow{OA}\cdot\overrightarrow{OB}<0$, then

$$\left|\overrightarrow{OA}\right|\cdot\left|\overrightarrow{OB}\right|\cos\theta<0$$

Since $\left|\overrightarrow{OA}\right|\cdot\left|\overrightarrow{OB}\right|>0$, then $\cos\theta<0$

$\therefore \theta$ is obtuse.

Answer B

6 If $p = 0.9$, then $q = 0.1$.

$$r=P(X\geq 1)$$

$$=1-P(X=0)$$

$$=1-{}^{10}C_0\times 0.1^{10}$$

$$=0.999...$$

$\therefore r>0.9$

Answer A

7 $f(x)=-a\sin x+b\cos x$

$$f(0)=-a\sin(0)+b\cos(0)$$

$\therefore f(0)=b$ where $b>0$

$\therefore f(x)$ has a positive y-intercept.

$$f'(x)=-a\cos x-b\sin x$$

$$f'(0)=-a\cos(0)-b\sin(0)$$

$$f'(0)=-a$$

$\therefore$ Graph of $f(x)$ is decreasing at $x = 0$ and has a positive y-intercept.

Answer D

8 In the domain $-\frac{\pi}{2}\leq t\leq\frac{\pi}{2}$, $-1\leq\sin t\leq 1$ and $0\leq\cos t\leq 1$.

In the diagram, y is always less than the centre of the semicircle.

$\therefore y = 2 - \cos t$

Answer C

9 By substituting $x = \pi$ we see that options A or B are possible graphs of $y = \sin^{-1}(\sin x)$.

Consider the derivative function:

$$\frac{dy}{dx}=\frac{1}{\sqrt{1-(\sin x)^2}}\times\cos x$$

$$\frac{dy}{dx}=\frac{\cos x}{\sqrt{1-\sin^2 x}}$$

$$\frac{dy}{dx}=\frac{\cos x}{|\cos x|}$$

$$\frac{dy}{dx}=\pm 1$$

Neither $\sin^{-1} x$ nor $\sin x$ are discontinuous.

Answer A

10 n votes are to be given to m candidates.

If each candidate receives the same number of votes:

$3543 \div 15 = 236\frac{3}{15}$

$\therefore$ Three candidates could receive 237 votes and the others 236.

$\therefore$ The smallest number of votes for a majority is 238 votes.

Answer C

Section II

QUESTION 11

(a) $$\left(\underset{\sim}{i}+6\underset{\sim}{j}\right)+\left(2\underset{\sim}{i}-7\underset{\sim}{j}\right) = (1+2)\underset{\sim}{i}+(6-7)\underset{\sim}{j} = 3\underset{\sim}{i}-\underset{\sim}{j}$$

(1 mark)

(b) $$(2a-b)^4 = {}^4C_0(2a)^4 + {}^4C_1(2a)^3(-b) + {}^4C_2(2a)^2(-b)^2 + {}^4C_3(2a)^1(-b)^3 + {}^4C_4(-b)^4$$
$$= 16a^4 - 32a^3b + 24a^2b^2 - 8ab^3 + b^4$$

(2 marks)

(c) If $u = x + 1$, then $x = u - 1$.

$$\frac{du}{dx} = 1$$
$$du = dx$$
$$\int x\sqrt{x+1}\,dx = \int (u-1)\sqrt{u}\,du$$
$$= \int\left(u^{\frac{3}{2}} - u^{\frac{1}{2}}\right)du$$
$$= \frac{2u^{\frac{5}{2}}}{5} - \frac{2u^{\frac{3}{2}}}{3} + C$$
$$= \frac{2\sqrt{(x+1)^5}}{5} - \frac{2\sqrt{(x+1)^3}}{3} + C$$

(3 marks)

(d) Number of ways $= {}^{10}C_5 \times {}^8C_3$

$= 14112$

(1 mark)

(e) The bubble's radius increases at a rate of 0.2 mm/s.

So, $\frac{dr}{dt} = 0.2$

$$V = \frac{4}{3}\pi r^3$$
$$\frac{dV}{dr} = 4\pi r^2$$

Now, $\frac{dV}{dt} = \frac{dV}{dr} \times \frac{dr}{dt}$ (Chain Rule)

$$\frac{dV}{dt} = 4\pi r^2 \times 0.2$$
$$\frac{dV}{dt} = 0.8\pi r^2$$

For the rate when radius is 0.6 mm, substitute $r = 0.6$.

$$\therefore \frac{dV}{dt} = 0.8\pi(0.6)^2$$

$= 0.9$ mm^3/s, correct to one decimal place

(2 marks)

(f) $\int_0^{\sqrt{3}} \frac{1}{\sqrt{4-x^2}}\, dx$

$$=\left[\sin^{-1}\left(\frac{x}{2}\right)\right]_0^{\sqrt{3}}$$

$$=\left[\sin^{-1}\left(\frac{\sqrt{3}}{2}\right)-\sin^{-1}\left(\frac{0}{2}\right)\right]$$

$$=\frac{\pi}{3}$$

(2 marks)

(g) Factorising $2\sin^3 x + 2\sin^2 x - \sin x - 1 = 0$ by grouping in pairs:

$$2\sin^2 x\,(\sin x+1)-1\,(\sin x+1)=0$$

$$(2\sin^2 x-1)(\sin x+1)=0$$

$$2\sin^2 x-1=0 \quad \text{and} \quad \sin x+1=0$$

$$\sin x=\pm\frac{1}{\sqrt{2}} \quad \text{and} \quad \sin x=-1$$

$$\therefore x=\frac{\pi}{4},\frac{3\pi}{4},\frac{5\pi}{4},\frac{3\pi}{2},\frac{7\pi}{4}$$

(3 marks)

(h) For $x^4 - 3x + 6 = 0$, $a = 1$, $d = -3$ and $e = 6$.

$$\frac{1}{\alpha}+\frac{1}{\beta}+\frac{1}{\gamma}+\frac{1}{\delta}=\frac{\beta\gamma\delta}{\alpha\beta\gamma\delta}+\frac{\alpha\gamma\delta}{\alpha\beta\gamma\delta}+\frac{\alpha\beta\delta}{\alpha\beta\gamma\delta}+\frac{\alpha\beta\gamma}{\alpha\beta\gamma\delta}$$

$$=\frac{\alpha\beta\gamma+\beta\gamma\delta+\alpha\gamma\delta+\alpha\beta\delta}{\alpha\beta\gamma\delta}$$

$$=\frac{\frac{-d}{a}}{\frac{e}{a}}$$

$$=\frac{-d}{e}$$

$$=\frac{--3}{6}$$

$$=\frac{1}{2}$$

(2 marks)

QUESTION 12

(a) Two alternative solutions are drawn for the slope field.

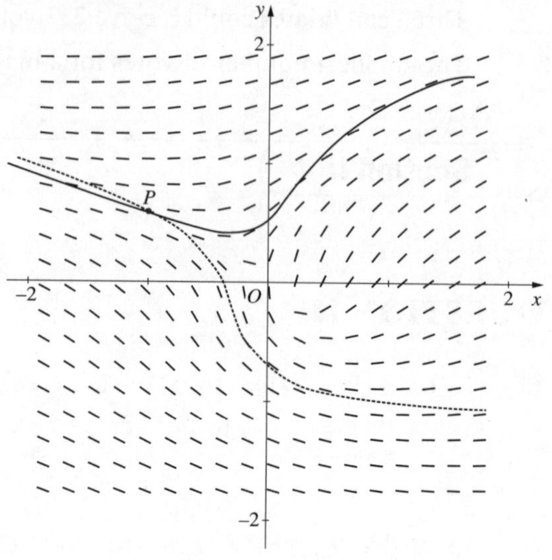

(1 mark)

(b) (i)

$$\frac{dT}{dt}=k(T-25)$$

$$\frac{1}{T-25}dT=k\,dt$$

$$\int\frac{1}{T-25}dT=\int k\ dt$$

$$\ln|T-25|=kt+C$$

When $t=0$, $T=5$.

$$\ln|5-25|=k(0)+C$$

$$\ln|-20|=C$$

$$\ln|T-25|=kt+\ln|-20|$$

$$kt=\ln|T-25|-\ln|-20|$$

$$kt=\ln\left|\frac{T-25}{-20}\right|$$

When $t=8$, $T=10$

$$8k=\ln\left|\frac{10-25}{-20}\right|$$

$$8k=\ln\left(\frac{3}{4}\right)$$

$$k=\frac{\ln\left(\frac{3}{4}\right)}{8}$$

$$k=-0.03596...$$

$$\frac{1}{8}\ln\left(\frac{3}{4}\right)t=\ln\left|\frac{T-25}{-20}\right|$$

When $T=20$,

$$t=\frac{8\ln\left|\frac{20-25}{-20}\right|}{\ln\left|\frac{3}{4}\right|}$$

$$t=38.55$$

∴ The bottle of water would reach 20 °C after 39 minutes.

(3 marks)

(ii)

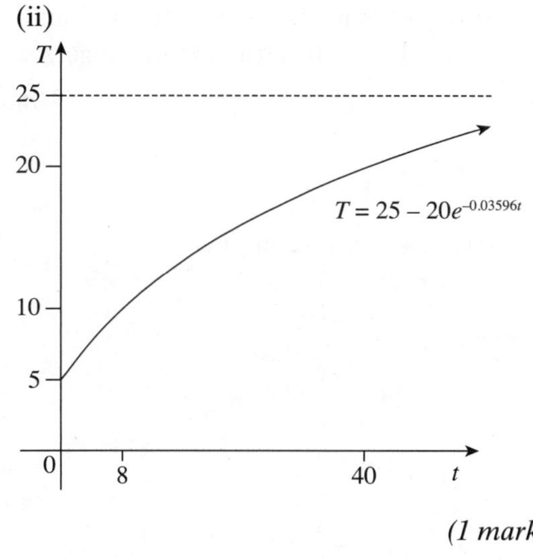

(1 mark)

(c) We want to prove that:

$$\frac{1}{1\times2\times3}+\frac{1}{2\times3\times4}+...+\frac{1}{n(n+1)(n+2)}=\frac{1}{4}-\frac{1}{2(n+1)(n+2)}$$

Prove true for $n = 1$.

$$\text{LHS}=\frac{1}{1\times2\times3}$$
$$=\frac{1}{6}$$

$$\text{RHS}=\frac{1}{4}-\frac{1}{2(2)(3)}$$
$$=\frac{1}{4}-\frac{1}{12}$$
$$=\frac{1}{6}$$

∴ the statement is true for $n = 1$.

Assume true for $n = k$,

i.e. assume

$$\frac{1}{1\times2\times3}+\frac{1}{2\times3\times4}+...+\frac{1}{k(k+1)(k+2)}=\frac{1}{4}-\frac{1}{2(k+1)(k+2)}$$

We want to prove true for $n = k + 1$.

If $n = k + 1$, then required to prove:

$$\frac{1}{1\times2\times3}+\frac{1}{2\times3\times4}+...+\frac{1}{k(k+1)(k+2)}+\frac{1}{(k+1)(k+2)(k+3)}=\frac{1}{4}-\frac{1}{2(k+2)(k+3)}$$

$$\text{LHS}=\frac{1}{4}-\frac{1}{2(k+1)(k+2)}+\frac{1}{(k+1)(k+2)(k+3)}$$
$$=\frac{1}{4}-\left(\frac{k+3}{2(k+1)(k+2)(k+3)}-\frac{2}{2(k+1)(k+2)(k+3)}\right)$$
$$=\frac{1}{4}-\left(\frac{k+1}{2(k+1)(k+2)(k+3)}\right)$$
$$=\frac{1}{4}-\frac{1}{2(k+2)(k+3)}$$
$$=\text{RHS}$$

The statement is true for $n = k + 1$.

Since the statement is true for $n = 1$ and $n = k + 1$, then it is true for all integers $n \geq 1$.

(3 marks)

(d) $f(x) = 4 - \left(1 - \frac{x}{2}\right)^2$

(i) For x-intercepts $f(x) = 0$.

$$4 - \left(1 - \frac{x}{2}\right)^2 = 0$$

$$\left(1 - \frac{x}{2}\right)^2 = 4$$

$$1 - \frac{x}{2} = \pm 2$$

$$-\frac{x}{2} = \pm 2 - 1$$

$-\frac{x}{2} = 1$ and $-\frac{x}{2} = -3$

$\therefore x = -2$ and $x = 6$

Since $f(x)$ has the domain $(-\infty, 2]$, the only x-intercept is at $(0, -2)$.

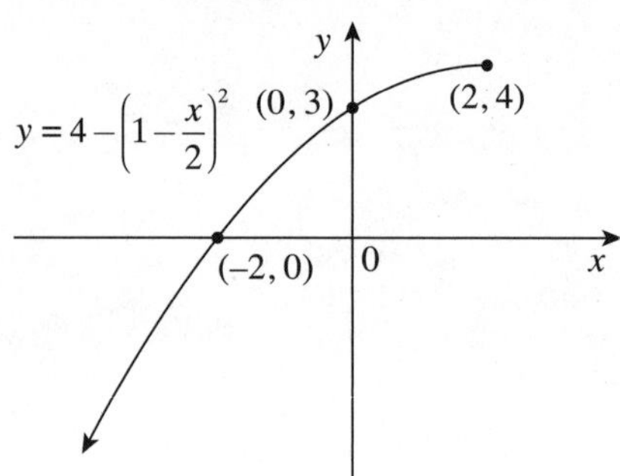

(2 marks)

(ii) $f(x) = 4 - \left(1 - \frac{x}{2}\right)^2$

$f^{-1}(x): x = 4 - \left(1 - \frac{y}{2}\right)^2$

$$\left(1 - \frac{y}{2}\right)^2 = 4 - x$$

$$1 - \frac{y}{2} = \pm\sqrt{4 - x}$$

$$-\frac{y}{2} = -1 \pm \sqrt{4 - x}$$

$$y = 2 \pm 2\sqrt{4 - x}$$

Since $f(x)$ has the domain $(-\infty, 2]$, then $f^{-1}(x)$ has the range of $(-\infty, 2]$.

$$2 = 2 \pm 2\sqrt{4 - x}$$

For $f^{-1}(x) = 2$, then $f^{-1}(x)$ must be the negative case.

$\therefore f^{-1}(x) = 2 - 2\sqrt{4 - x}$

(3 marks)

(iii) Sketch $f^{-1}(x) = 2 - 2\sqrt{4 - x}$

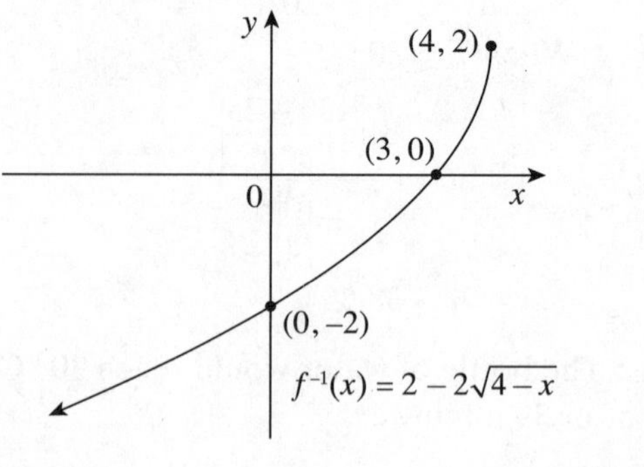

(1 mark)

QUESTION 13

(a) $V = \pi \int_0^2 x^2\, dy - \pi \int_1^2 x^2\, dy$

$$V = \pi \int_0^2 y\; dy - \pi \int_1^2 (y - 1)\; dy$$

$$V = \pi \left[\frac{y^2}{2}\right]_0^2 - \pi \left[\frac{y^2}{2} - y\right]_1^2$$

$$V = \pi \left[\frac{2^2}{2} - \frac{0^2}{2}\right] - \pi \left[\left(\frac{2^2}{2} - 2\right) - \left(\frac{1^2}{2} - 1\right)\right]$$

$$V = 2\pi - \pi \left[(0) - \left(-\frac{1}{2}\right)\right]$$

$$V = 2\pi - \frac{\pi}{2}$$

$$V = \frac{3\pi}{2}$$

$\therefore$ The volume is $\frac{3\pi}{2}$ units^3.

(3 marks)

(b) The ball is projected at an angle of 30° with initial velocity of 12 m/s and initial height of 1 m.

So, $\theta = \frac{\pi}{6}$, $V = 12$ and $h = 1$.

Hence

$$\underset{\sim}{r}(t) = \left(12t \cos\frac{\pi}{6}\right)\underset{\sim}{i} + \left(-5t^2 + 12t \sin\frac{\pi}{6} + 1\right)\underset{\sim}{j}$$

So in parametric form: $x(t) = 6\sqrt{3}\, t$ and $y(t) = -5t^2 + 6t + 1$.

To prove that the ball does not hit the roof, show that the max height < 3:

$$\dot{y}(t) = -10t + 6$$

$$0 = -10t + 6$$

$$t = \frac{3}{5}$$

$\therefore$ Since $\ddot{y} = -10$, the maximum height occurs when $t = \frac{3}{5}$ seconds.

Substitute $t=\frac{3}{5}$ into $y(t)$ gives the maximum height.

$$y=-5\left(\frac{3}{5}\right)^2+6\left(\frac{3}{5}\right)+1$$

$$y=\frac{14}{5}$$

$$<3$$

$\therefore$ The ball does not touch the ceiling.

To prove that the ball does not hit the floor, we can solve $x(t)=10$ and substitute the solution into $y(t)$.

$$6\sqrt{3}\,t=10$$

$$t=\frac{5}{3\sqrt{3}}$$

$\therefore$ the value of y when $x=10$ is:

$$y\left(\frac{5}{3\sqrt{3}}\right)=-5\left(\frac{5}{3\sqrt{3}}\right)^2+6\left(\frac{5}{3\sqrt{3}}\right)+1$$

$$\approx 2.144$$

$$>0$$

$\therefore$ When $t=\frac{5}{3\sqrt{3}}$, $x=10$ and $y\approx 2.144$, which shows the ball does not hit the floor.

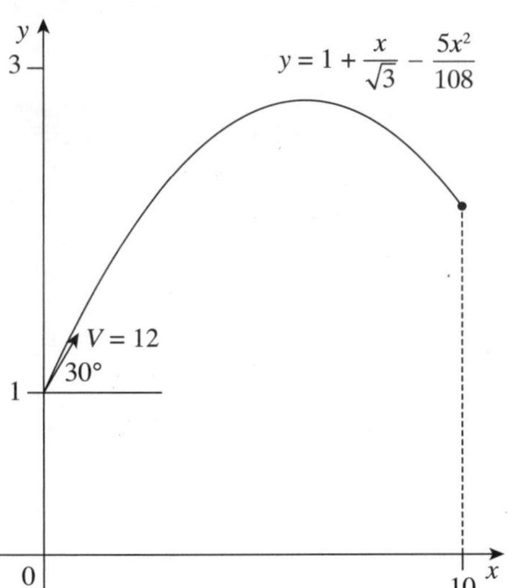

(4 marks)

(c) $A=\int_{-2}^{2}\left((2-|x|)-\left(1-\frac{8}{4+x^2}\right)\right)dx$

$$=2\int_0^2\left((2-|x|)-\left(1-\frac{8}{4+x^2}\right)\right)dx$$

$$=2\int_0^2 2-x-\left(1-\frac{8}{4+x^2}\right)dx$$

$$=2\int_0^2 1-x+\frac{8}{4+x^2}dx$$

$$=2\left[x-\frac{x^2}{2}+4\tan^{-1}\frac{x}{2}\right]_0^2$$

$$=2\left[\left(2-\frac{2^2}{2}+4\tan^{-1}\frac{2}{2}\right)-\left(0-\frac{0^2}{2}+4\tan^{-1}\frac{0}{2}\right)\right]$$

$$=2[\pi-0]$$

$$=2\pi$$

$\therefore$ The area is 2π units^2.

(3 marks)

(d) (i) As $A=B-d$ and $C=B+d$ then,

$$\frac{\sin A+\sin C}{\cos A+\cos C}=\frac{\sin(B-d)+\sin(B+d)}{\cos(B-d)+\cos(B+d)}$$

$$=\frac{2\sin B\cos d}{2\cos B\cos d}$$

(from the reference sheet)

$$=\frac{\sin B}{\cos B}$$

$$=\tan B$$

(2 marks)

(ii) Let $A=\frac{5\theta}{7}$ and $C=\frac{6\theta}{7}$.

Solving $\frac{5\theta}{7}=B-d$ and $\frac{6\theta}{7}=B+d$ simultaneously:

$$\frac{5\theta}{7}+\frac{6\theta}{7}=2B$$

$$B=\frac{11\theta}{14}$$

Since $\frac{\sin A+\sin C}{\cos A+\cos C}=\tan B$, then

$$\frac{\sin\frac{5\theta}{7}+\sin\frac{6\theta}{7}}{\cos\frac{5\theta}{7}+\cos\frac{6\theta}{7}}=\tan\frac{11\theta}{14}$$

$$\therefore \tan\frac{11\theta}{14}=\sqrt{3}$$

$$\frac{11\theta}{14}=\frac{\pi}{3},\frac{4\pi}{3}$$

$$\therefore \theta=\frac{14\pi}{33},\frac{56\pi}{33}$$

(2 marks)

QUESTION 14

(a) Let t = time, in hours.

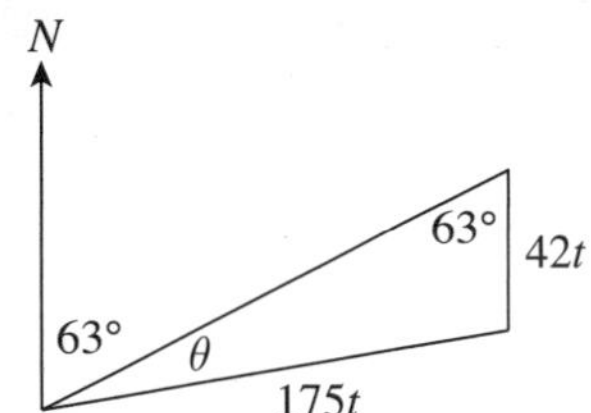

Using the sine rule:

$$\frac{\sin\theta}{42t} = \frac{\sin 63^\circ}{175t}$$

$$= \frac{42\sin 63^\circ}{175}$$

$$= 0.218\,34...$$

$$\therefore \theta = 12 \text{ (nearest degree)}$$

Bearing is 63 + 12 = 075.

(3 marks)

(b)
$$\frac{dP}{dt} = 0.1P\left(\frac{C-P}{C}\right)$$

$$\frac{C}{P(C-P)}dP = 0.1dt$$

$$\left(\frac{1}{P}+\frac{1}{C-P}\right)dP = \frac{1}{10}dt$$

$$\int \frac{1}{10}dt = \int\left(\frac{1}{P}+\frac{1}{C-P}\right)dP$$

$$\frac{t}{10} = \ln|P| - \ln|C-P| + K \text{ where } K \text{ is a constant}$$

$$\frac{t}{10} = \ln\left|\frac{P}{C-P}\right| + K$$

When $t = 0$, $P = 150\,000$:

$$0 = \ln\left|\frac{150\,000}{C-150\,000}\right| + K$$

$$K = -\ln\left|\frac{150\,000}{C-150\,000}\right|$$

$$\frac{t}{10} = \ln\left|\frac{P}{C-P}\right| - \ln\left|\frac{150\,000}{C-150\,000}\right|$$

When $t = 20$, $P = 600\,000$:

$$2 = \ln\left|\frac{600\,000}{C-600\,000}\right| - \ln\left|\frac{150\,000}{C-150\,000}\right|$$

$$2 = \ln\left|\frac{4(C-150\,000)}{C-600\,000}\right|$$

$$e^2 = \frac{4(C-150\,000)}{C-600\,000}$$

$$e^2(C-600\,000) = 4(C-150\,000)$$

$$e^2C - 600\,000e^2 = 4C - 600\,000$$

$$C(e^2-4) = 600\,000(e^2-1)$$

$$C = \frac{600\,000(e^2-1)}{e^2-4}$$

$$C = 1\,131\,121$$

$\therefore$ The carrying capacity is approximately 1 130 000.

(4 marks)

(c) (i) Let $\underset{\sim}{v} = x\underset{\sim}{i} + y\underset{\sim}{j}$, then

$$\underset{\sim}{v}\cdot\underset{\sim}{v} = x^2 + y^2$$
$$= \left(\sqrt{x^2 + y^2}\right)^2$$
$$= |\underset{\sim}{v}|^2$$

Alternatively, $\underset{\sim}{v}\cdot\underset{\sim}{v} = |\underset{\sim}{v}||\underset{\sim}{v}|\cos\theta$

Since $\theta = 0$, $\cos\theta = 1$.

$\underset{\sim}{v}\cdot\underset{\sim}{v} = |\underset{\sim}{v}|^2$ *(1 mark)*

(ii) $\overrightarrow{AC} = \overrightarrow{AB} + \overrightarrow{BC}$ and

$$\overrightarrow{BD} = \overrightarrow{BA} + \overrightarrow{AD}$$
$$= -\overrightarrow{AB} + \overrightarrow{AD}$$
$$= \overrightarrow{AD} - \overrightarrow{AB}$$
$$= k\underset{\sim}{b} - \underset{\sim}{a}$$

We know that $|\overrightarrow{AC}| = |\overrightarrow{BD}|$, so therefore:

$$|\overrightarrow{AC}|^2 = |\overrightarrow{BD}|^2$$
$$(\underset{\sim}{a}+\underset{\sim}{b})\cdot(\underset{\sim}{a}+\underset{\sim}{b}) = (k\underset{\sim}{b}-\underset{\sim}{a})\cdot(k\underset{\sim}{b}-\underset{\sim}{a})$$
$$\underset{\sim}{a}\cdot\underset{\sim}{a} + 2\underset{\sim}{a}\cdot\underset{\sim}{b} + \underset{\sim}{b}\cdot\underset{\sim}{b} = k^2\underset{\sim}{b}\cdot\underset{\sim}{b} - 2k\underset{\sim}{a}\cdot\underset{\sim}{b} + \underset{\sim}{a}\cdot\underset{\sim}{a}$$
$$2\underset{\sim}{a}\cdot\underset{\sim}{b} + |\underset{\sim}{b}|^2 = k^2|\underset{\sim}{b}|^2 - 2k\underset{\sim}{a}\cdot\underset{\sim}{b}$$
$$2\underset{\sim}{a}\cdot\underset{\sim}{b} + 2k\underset{\sim}{a}\cdot\underset{\sim}{b} = k^2|\underset{\sim}{b}|^2 - |\underset{\sim}{b}|^2$$
$$2(1+k)\underset{\sim}{a}\cdot\underset{\sim}{b} = (k^2-1)|\underset{\sim}{b}|^2$$
$$2(1+k)\underset{\sim}{a}\cdot\underset{\sim}{b} = -(1+k)(1-k)|\underset{\sim}{b}|^2$$
$$2\underset{\sim}{a}\cdot\underset{\sim}{b} = -(1-k)|\underset{\sim}{b}|^2$$
$$2\underset{\sim}{a}\cdot\underset{\sim}{b} + (1-k)|\underset{\sim}{b}|^2 = 0$$

(3 marks)

(d) $p = \dfrac{3}{500}$

$$\sigma^2 = \frac{p(1-p)}{n}$$
$$\sigma^2 = \frac{\frac{3}{500}\left(1-\frac{3}{500}\right)}{n}$$
$$= \frac{1491}{500^2 n}$$

$$P\left(\hat{p} \geq \frac{4}{500}\right) = P\left(Z \geq \frac{\frac{4}{500}-\mu}{\sigma}\right)$$
$$= P\left(Z \geq \frac{\frac{4}{500}-\frac{3}{500}}{\frac{1}{500}\sqrt{\frac{1491}{n}}}\right)$$
$$= P\left(Z \geq \frac{\sqrt{n}}{\sqrt{1491}}\right)$$

Hence we require $\dfrac{\sqrt{n}}{\sqrt{1491}} \geq 2$ (using the Empirical rule).

$$\sqrt{n} \geq 2\sqrt{1491}$$
$$n \geq 4\times1491$$
$$n \geq 5964$$

$\therefore$ The sample size required is 6000, to the nearest thousand.

(3 marks)

(e) If $g(x) = x^3 + 4x - 2$ passes through the point $(1, 3)$, then $g^{-1}(3) = 1$

From the product rule:

$$f'(x) = x\cdot\frac{d}{dx}(g^{-1}(x)) + 1\cdot g^{-1}(x)$$

Inverse functions have a gradient relationship where the gradient of $g(x)$ at (a, b) is the reciprocal of the gradient of $g^{-1}(x)$ at (b, a).

$\therefore$ The gradient of $g(x)$ at $x = 1$ is the reciprocal of the gradient of $g^{-1}(x)$ at $x = 3$

$$g'(x) = 3x^2 + 4$$
$$g'(1) = 3(1)^2 + 4$$
$$g'(1) = 7$$

$\therefore$ The gradient of $g^{-1}(x)$ at $x = 3$ is $\dfrac{1}{7}$

$$\therefore f'(3) = 3\cdot\frac{1}{7} + 1\cdot1$$
$$\therefore f'(3) = \frac{10}{3}$$

(2 marks)

NSW Education Standards Authority

2022 HIGHER SCHOOL CERTIFICATE EXAMINATION

Mathematics Extension 1

General Instructions

- Reading time – 10 minutes
- Working time – 2 hours
- Write using black pen
- Calculators approved by NESA may be used
- A reference sheet is provided at the back of this paper
- For questions in Section II, show relevant mathematical reasoning and/or calculations
- Write your Centre Number and Student Number on the Question 12 Writing Booklet attached

Total marks: 70

Section I – 10 marks

- Attempt Questions 1–10
- Allow about 15 minutes for this section

Section II – 60 marks

- Attempt Questions 11–14
- Allow about 1 hour and 45 minutes for this section

Section I

10 marks
Attempt Questions 1–10
Allow about 15 minutes for this section

Use the multiple-choice answer sheet for Questions 1–10.

---

1 It is given that $\cos\left(\dfrac{23\pi}{12}\right) = \dfrac{\sqrt{6}+\sqrt{2}}{4}$.

Which of the following is the value of $\cos^{-1}\left(\dfrac{\sqrt{6}+\sqrt{2}}{4}\right)$?

A. $\dfrac{23\pi}{12}$

B. $\dfrac{11\pi}{12}$

C. $\dfrac{\pi}{12}$

D. $-\dfrac{11\pi}{12}$

2 The graph of $f(x) = \dfrac{3}{x-1} + 2$ is shown.

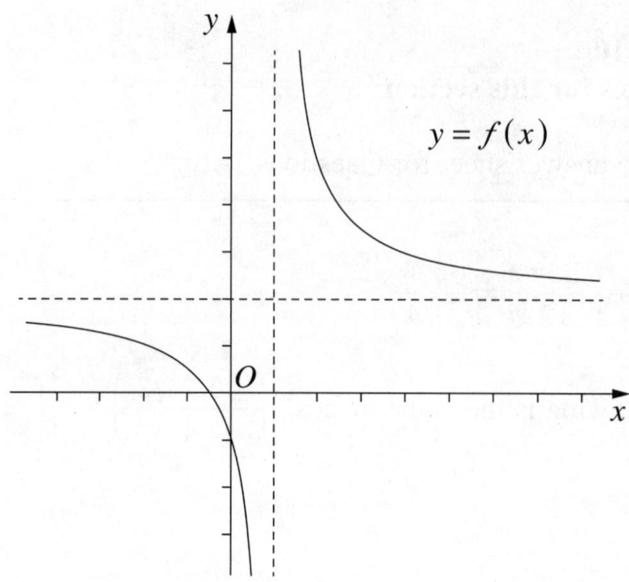

The graph of $f(x)$ was transformed to get the graph of $g(x)$ as shown.

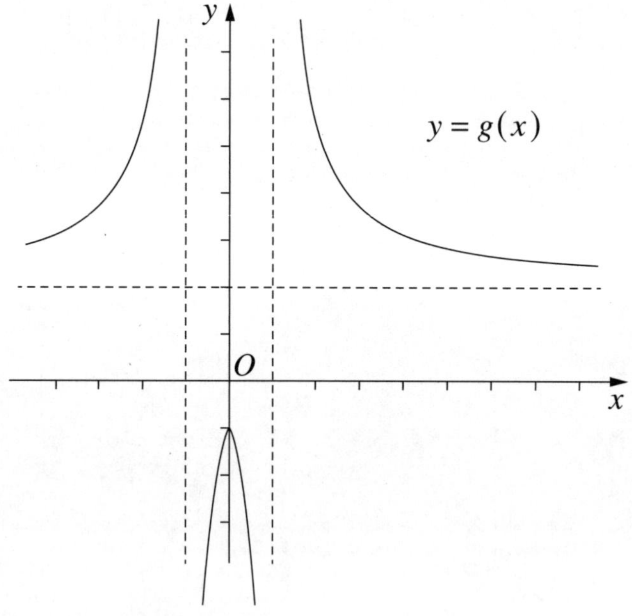

What transformation was applied?

A. $g(x) = f(|x|)$

B. $g(x) = \sqrt{f(x)}$

C. $g(x) = f(-x)$

D. $g(x) = \dfrac{1}{f(x)}$

3 Let $P(x)$ be a polynomial of degree 5. When $P(x)$ is divided by the polynomial $Q(x)$, the remainder is $2x + 5$.

Which of the following is true about the degree of Q?

A. The degree must be 1.

B. The degree could be 1.

C. The degree must be 2.

D. The degree could be 2.

4 The diagram shows the graph of the sum of the functions $f(x)$ and $g(x)$.

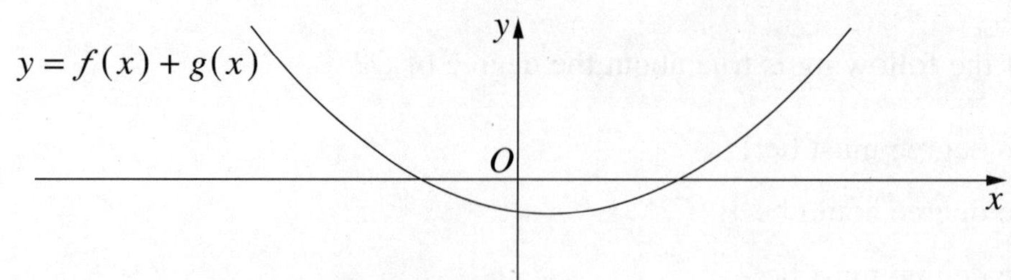

Which of the following best represents the graphs of both $f(x)$ and $g(x)$?

A.

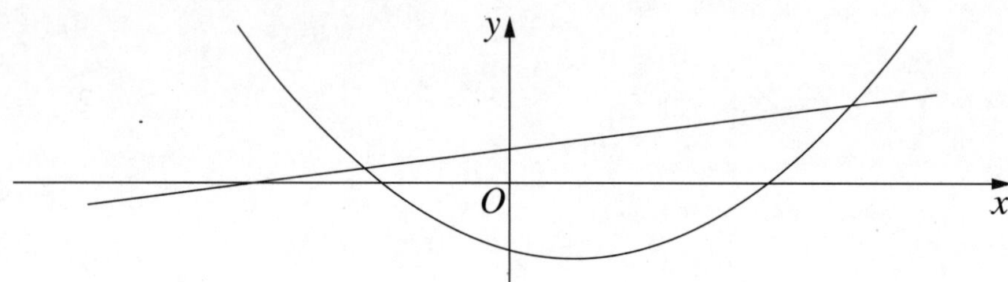

B.

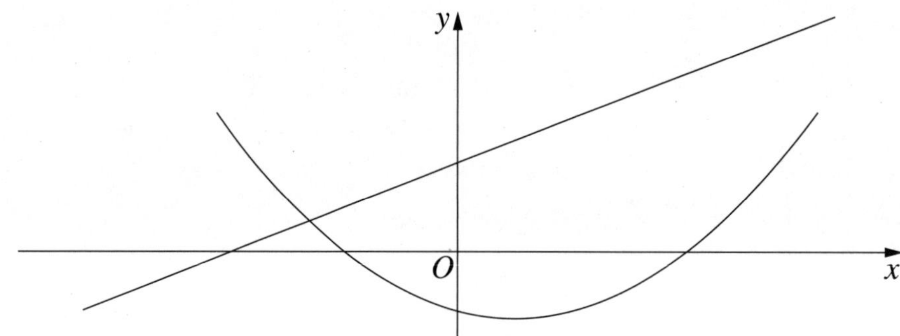

C.

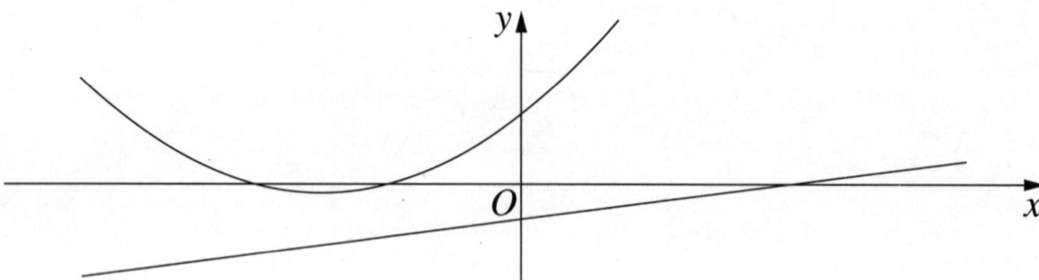

D.

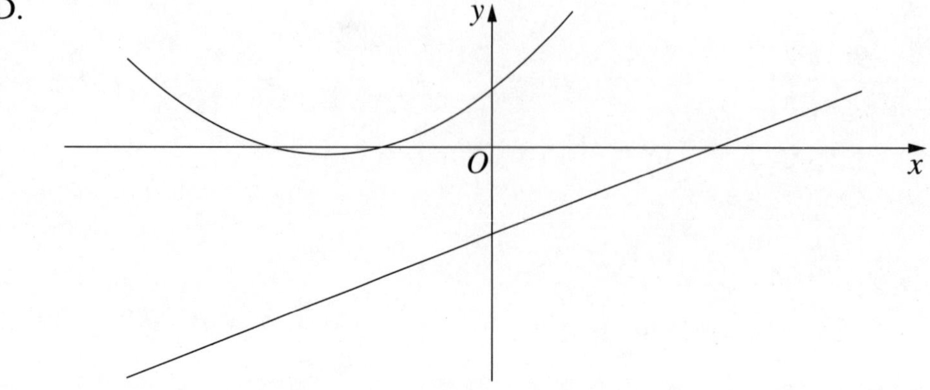

5 A curve is defined in parametric form by $x = 2 + t$ and $y = 3 - 2t^2$ for $-1 \le t \le 0$.

Which diagram best represents this curve?

A.

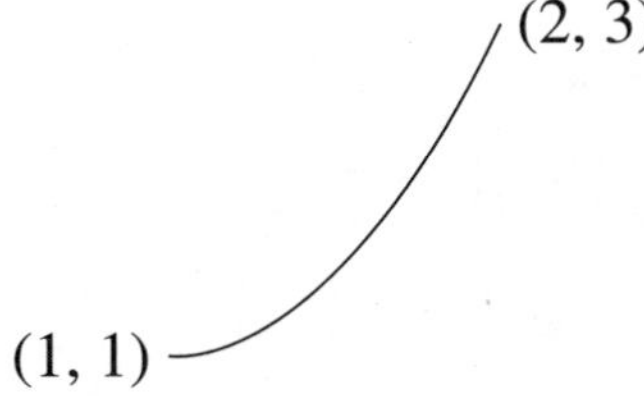

B.

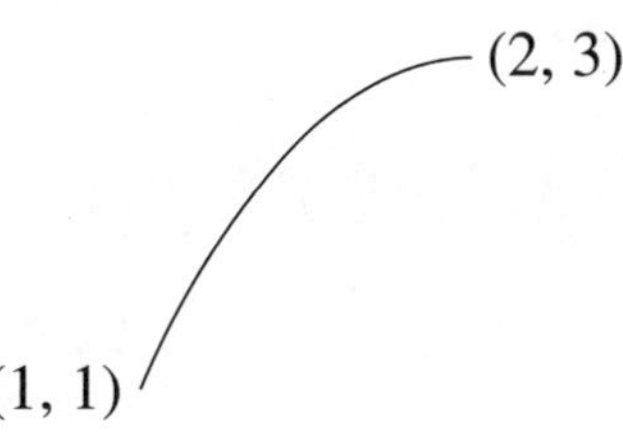

C.

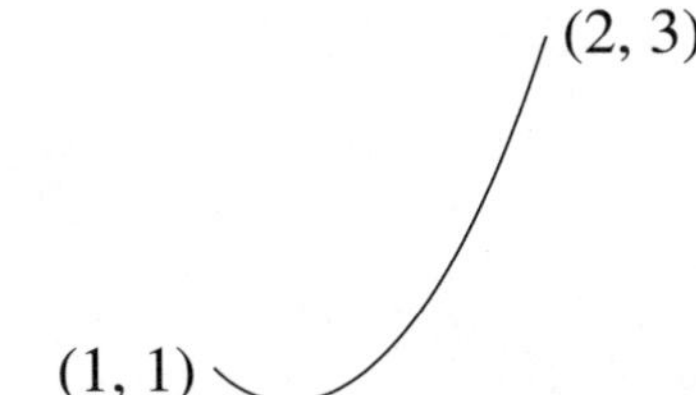

D.

6 The following diagram shows the vector $\underset{\sim}{u}$ and the vectors $\underset{\sim}{i} + \underset{\sim}{j}$, $-\underset{\sim}{i} + \underset{\sim}{j}$, $-\underset{\sim}{i} - \underset{\sim}{j}$ and $\underset{\sim}{i} - \underset{\sim}{j}$.

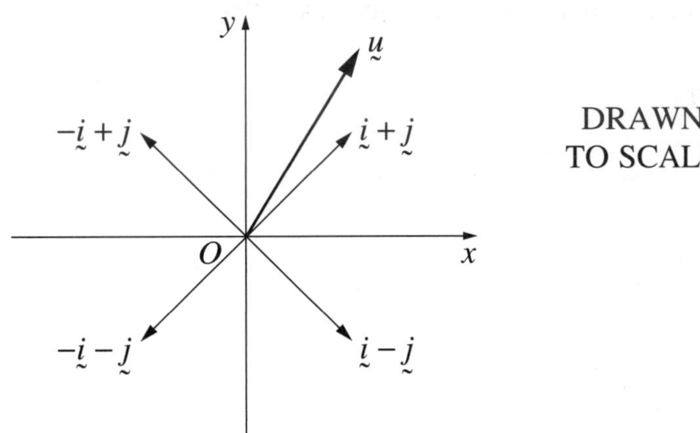

DRAWN TO SCALE

Which statement regarding this diagram could be true?

A. The projection of $\underset{\sim}{u}$ onto $\underset{\sim}{i} + \underset{\sim}{j}$ is the vector $1.1\underset{\sim}{i} + 1.8\underset{\sim}{j}$.

B. The projection of $\underset{\sim}{u}$ onto $-\underset{\sim}{i} + \underset{\sim}{j}$ is the vector $-0.4\underset{\sim}{i} + 0.4\underset{\sim}{j}$.

C. The projection of $\underset{\sim}{u}$ onto $-\underset{\sim}{i} - \underset{\sim}{j}$ is the vector $3.2\underset{\sim}{i} + 3.2\underset{\sim}{j}$.

D. The projection of $\underset{\sim}{u}$ onto $\underset{\sim}{i} - \underset{\sim}{j}$ is the vector $0.5\underset{\sim}{i} - 0.5\underset{\sim}{j}$.

7 The diagram shows triangle ABC with points chosen on each of the sides. On side AB, 3 points are chosen. On side AC, 4 points are chosen. On side BC, 5 points are chosen.

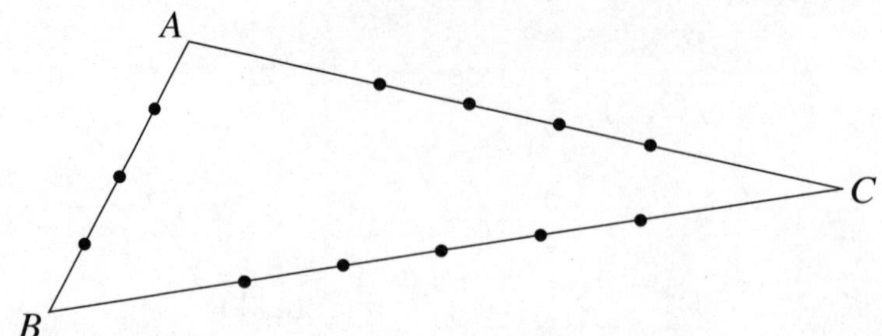

How many triangles can be formed using the chosen points as vertices?

A. 60

B. 145

C. 205

D. 220

8 The angle between two unit vectors $\underset{\sim}{a}$ and $\underset{\sim}{b}$ is θ and $|\underset{\sim}{a} + \underset{\sim}{b}| < 1$.

Which of the following best describes the possible range of values of θ?

A. $0 \le \theta < \dfrac{\pi}{3}$

B. $0 \le \theta < \dfrac{2\pi}{3}$

C. $\dfrac{\pi}{3} < \theta \le \pi$

D. $\dfrac{2\pi}{3} < \theta \le \pi$

9 A given function $f(x)$ has an inverse $f^{-1}(x)$.

The derivatives of $f(x)$ and $f^{-1}(x)$ exist for all real numbers x.

The graphs $y = f(x)$ and $y = f^{-1}(x)$ have at least one point of intersection.

Which statement is true for all points of intersection of these graphs?

A. All points of intersection lie on the line $y = x$.

B. None of the points of intersection lie on the line $y = x$.

C. At no point of intersection are the tangents to the graphs parallel.

D. At no point of intersection are the tangents to the graphs perpendicular.

10 Which of the following could be the graph of a solution to the differential equation

$$\frac{dy}{dx} = \sin y + 1?$$

A.

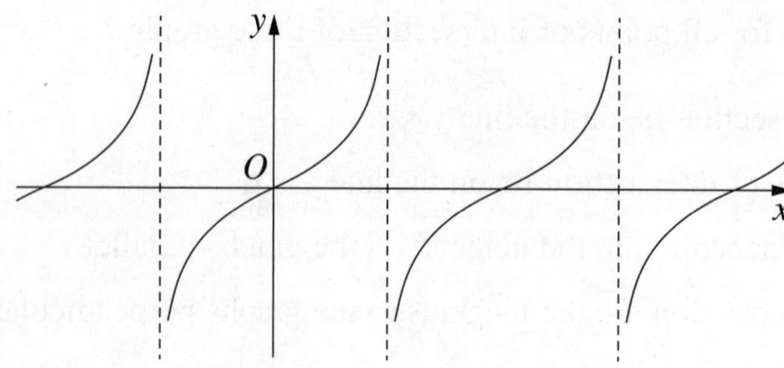

B.

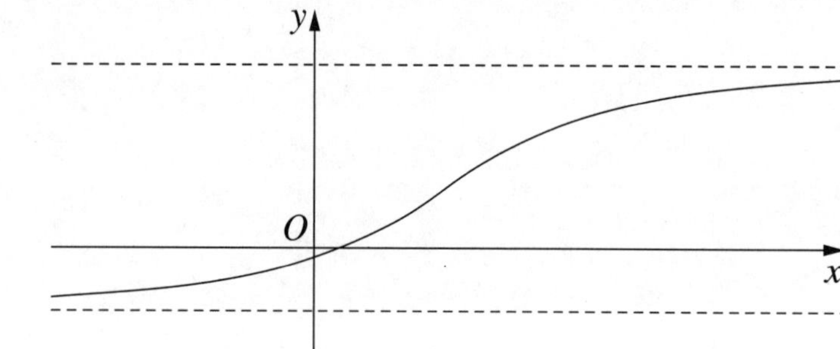

C.

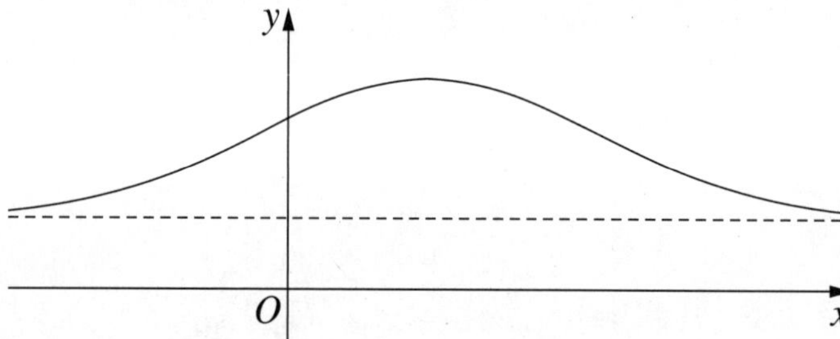

D.

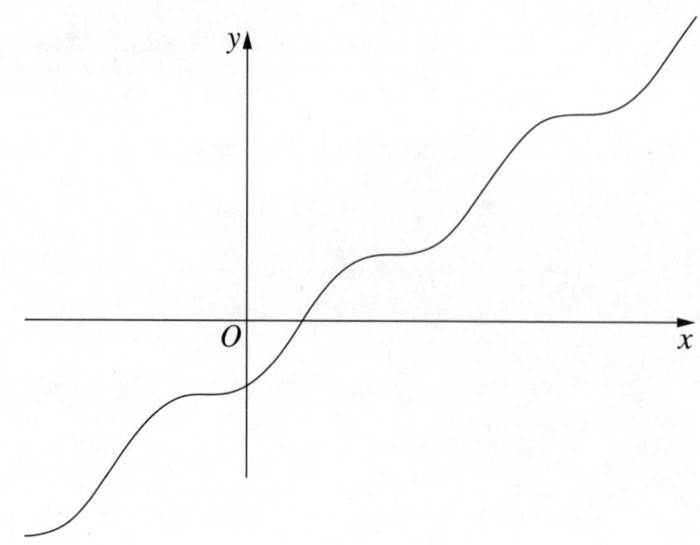

Section II

60 marks
Attempt Questions 11–14
Allow about 1 hour and 45 minutes for this section

Answer each question in the appropriate writing booklet. Extra writing booklets are available.

For questions in Section II, your responses should include relevant mathematical reasoning and/or calculations.

Question 11 (15 marks) Use the Question 11 Writing Booklet

(a) For the vectors $\underset{\sim}{u} = \underset{\sim}{i} - \underset{\sim}{j}$ and $\underset{\sim}{v} = 2\underset{\sim}{i} + \underset{\sim}{j}$, evaluate each of the following.

(i) $\underset{\sim}{u} + 3\underset{\sim}{v}$ **1**

(ii) $\underset{\sim}{u} \cdot \underset{\sim}{v}$ **1**

(b) Find the exact value of $\displaystyle\int_0^1 \frac{x}{\sqrt{x^2+4}}\,dx$ using the substitution $u = x^2 + 4$. **3**

(c) Find the coefficients of x^2 and x^3 in the expansion of $\left(1 - \dfrac{x}{2}\right)^8$. **2**

(d) The vectors $\underset{\sim}{u} = \begin{pmatrix} a \\ 2 \end{pmatrix}$ and $\underset{\sim}{v} = \begin{pmatrix} a-7 \\ 4a-1 \end{pmatrix}$ are perpendicular. **2**

What are the possible values of a?

(e) Express $\sqrt{3}\sin(x) - 3\cos(x)$ in the form $R\sin(x+\alpha)$. **3**

(f) Solve $\dfrac{x}{2-x} \geq 5$. **3**

Question 12 (16 marks) Use the Question 12 Writing Booklet

(a) A direction field is to be drawn for the differential equation **2**

$$\frac{dy}{dx} = \frac{x - 2y}{x^2 + y^2}.$$

On the diagram on page 1 of the Question 12 Writing Booklet,* clearly draw the correct slopes of the direction field at the points P, Q and R.

(b) A sports association manages 13 junior teams. It decides to check the age of all players. Any team that has more than 3 players above the age limit will be penalised. **2**

A total of 41 players are found to be above the age limit.

Will any team be penalised? Justify your answer.

(c) Find the equation of the tangent to the curve $y = x\arctan(x)$ at the point with coordinates $\left(1, \frac{\pi}{4}\right)$. Give your answer in the form $y = mx + c$. **3**

Question 12 continues on the following page

*See end of exam.

Question 12 (continued)

(d) In a room with temperature 12°C, coffee is poured into a cup. The temperature of the coffee when it is poured into the cup is 92°C, and it is far too hot to drink.

The temperature, T, in degrees Celsius, of the coffee, t minutes after it is made, can be modelled using the differential equation $\frac{dT}{dt} = k\left(T - T_1\right)$, where k is the constant of proportionality and T_1 is a constant.

(i) It takes 5 minutes for the coffee to cool to a temperature of 76°C. **3**

Using separation of variables, solve the given differential equation to show that $T = 12 + 80e^{\frac{t}{5}\ln\left(\frac{4}{5}\right)}$.

(ii) The optimal drinking temperature for a hot beverage is 57°C. **1**

Find the value of t when the coffee reaches this temperature, giving your answer to the nearest minute.

(e) A game consists of randomly selecting 4 balls from a bag. After each ball is selected it is replaced in the bag. The bag contains 3 red balls and 7 green balls. For each red ball selected, 10 points are earned and for each green ball selected, 5 points are deducted. For instance, if a player picks 3 red balls and 1 green ball, the score will be $3 \times 10 - 1 \times 5 = 25$ points. **2**

What is the expected score in the game?

(f) Use mathematical induction to prove that $15^n + 6^{2n+1}$ is divisible by 7 for all integers $n \geq 0$. **3**

End of Question 12

Question 13 (14 marks) Use the Question 13 Writing Booklet

(a) Three different points A, B and C are chosen on a circle centred at O. **3**

Let $\underset{\sim}{a} = \overrightarrow{OA}$, $\underset{\sim}{b} = \overrightarrow{OB}$ and $\underset{\sim}{c} = \overrightarrow{OC}$. Let $\underset{\sim}{h} = \underset{\sim}{a} + \underset{\sim}{b} + \underset{\sim}{c}$ and let H be the point such that $\overrightarrow{OH} = \underset{\sim}{h}$, as shown in the diagram.

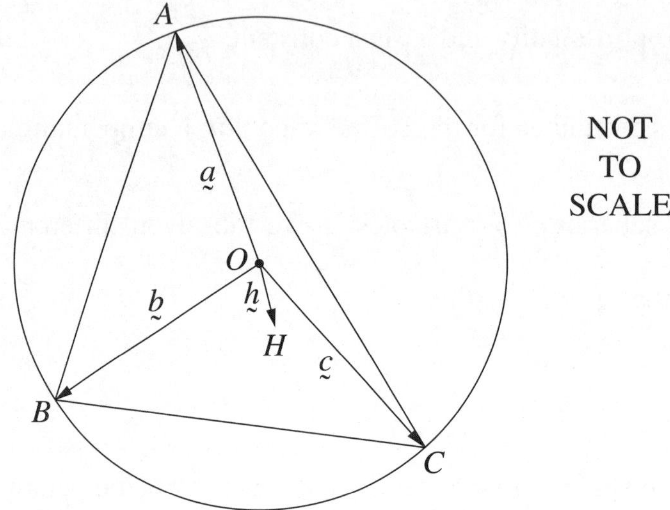

NOT TO SCALE

Show that $\overrightarrow{BH}$ and $\overrightarrow{CA}$ are perpendicular.

(b) A solid of revolution is to be found by rotating the region bounded by the x-axis and the curve $y = (k+1)\sin(kx)$, where $k > 0$, between $x = 0$ and $x = \dfrac{\pi}{2k}$ about the x-axis. **3**

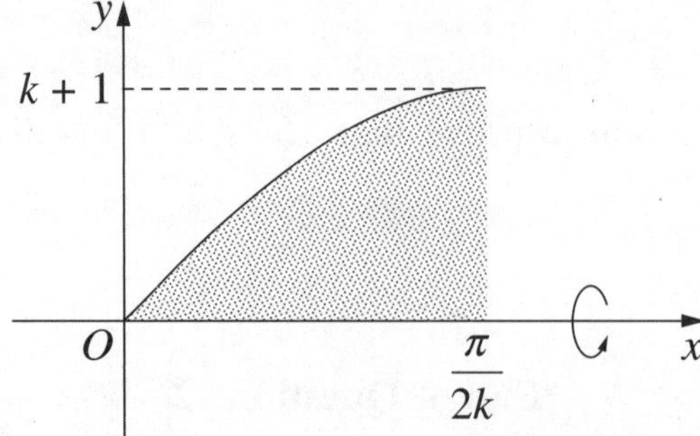

Find the value of k for which the volume is π^2.

Question 13 continues on the following page

Question 13 (continued)

(c) The function f is defined by $f(x) = \sin(x)$ for all real numbers x. Let g be the function defined on $[-1, 1]$ by $g(x) = \arcsin(x)$. **2**

Is g the inverse of f? Justify your answer.

(d) The monic polynomial, P, has degree 3 and roots α, β, γ. **3**

It is given that

$$\alpha^2 + \beta^2 + \gamma^2 = 85 \text{ and}$$

$$P'(\alpha) + P'(\beta) + P'(\gamma) = 87.$$

Find $\alpha\beta + \beta\gamma + \gamma\alpha$.

(e) *You may use the information on page 18* to answer this question.*

A chocolate factory sells 150-gram chocolate bars. There has been a complaint that the bars actually weigh less than 150 grams, so a team of inspectors was sent to the factory to check. They randomly selected 16 bars, weighed them and noted that 8 bars weighed less than 150 grams.

The factory manager claims 80% of the chocolate bars produced by the factory weigh 150 grams or more.

(i) The inspectors used the normal approximation to the binomial distribution to calculate the probability, $\mathcal{P}$, of having at least 8 bars weighing less than 150 grams in a random sample of 16, assuming the factory manager's claim is correct. **2**

Calculate the value of $\mathcal{P}$.

(ii) The factory manager disagrees with the method used by the inspectors as described in part (i). **1**

Explain why the method used by the inspectors might not be valid.

End of Question 13

* See end of exam for the table of values. (Page 18 is the original HSC page number)

Question 14 (15 marks) Use the Question 14 Writing Booklet

(a) Find the particular solution to the differential equation $(x-2)\frac{dy}{dx} = xy$ that passes through the point $(0, 1)$. **4**

(b) The vectors $\vec{u}$ and $\vec{v}$ are not parallel. The vector $\vec{p}$ is the projection of $\vec{u}$ onto the vector $\vec{v}$. **3**

The vector $\vec{p}$ is parallel to $\vec{v}$ so it can be written $\lambda_0\vec{v}$ for some real number λ_0. (Do NOT prove this.)

Prove that $|\vec{u} - \lambda\vec{v}|$ is smallest when $\lambda = \lambda_0$ by showing that, for all real numbers λ, $|\vec{u} - \lambda_0\vec{v}| \le |\vec{u} - \lambda\vec{v}|$.

Question 14 continues on the following page

Question 14 (continued)

(c) A video game designer wants to include an obstacle in the game they are developing. The player will reach one side of a pit and must shoot a projectile to hit a target on the other side of the pit in order to be able to cross. However, the instant the player shoots, the target begins to move away from the player at a constant speed that is half the initial speed of the projectile shot by the player, as shown in the diagram below. **4**

The initial distance between the player and the target is d, the initial speed of the projectile is $2u$ and it is launched at an angle of θ to the horizontal. The acceleration due to gravity is g. The launch angle is the ONLY parameter that the player can change.

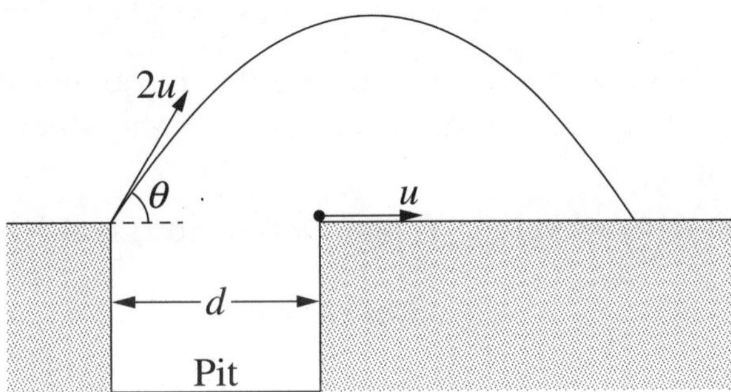

Taking the position of the player when the projectile is launched as the origin, the positions of the projectile and target at time t after the projectile is launched are as follows.

$$\vec{r}_P = \begin{pmatrix} 2ut\cos\theta \\ 2ut\sin\theta - \dfrac{g}{2}t^2 \end{pmatrix} \quad \text{Projectile}$$

$$\vec{r}_T = \begin{pmatrix} d + ut \\ 0 \end{pmatrix} \quad \text{Target}$$

(Do NOT prove these.)

Show that, for the player to have a chance of hitting the target, d must be less than 37% of the maximum possible range of the projectile (to 2 significant figures).

Question 14 continues on the following page

Question 14 (continued)

(d) *You may use the information on page 18* to answer this question.* **4**

An airline company that has empty seats on a flight is not maximising its profit.

An airline company has found that there is a probability of 5% that a passenger books a flight but misses it. The management of the airline company decides to allow for overbooking, which means selling more tickets than the number of seats available on each flight.

To protect their reputation, management makes the decision that no more than 1% of their flights should have more passengers showing up for the flight than available seats.

Given management's decision and using a suitable approximation, find the maximum number of tickets that can be sold for a flight which has 350 seats.

* See end of exam for the table of values. (Page 18 is the original HSC page number).

End of paper

Diagram for Question 12 (a)

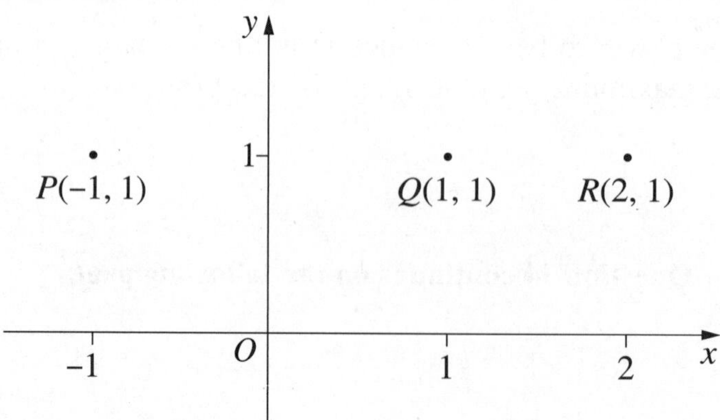

You may use the information below to answer Question 13 (e) and Question 14 (d).

Table of values $P(Z \leq z)$ for the normal distribution $N(0, 1)$

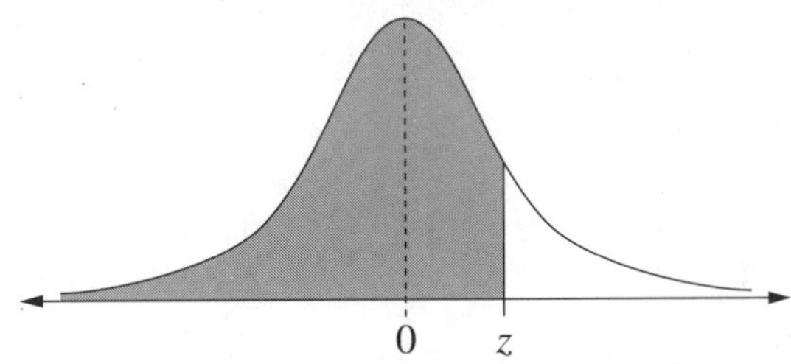

Z	0.00	0.01	0.02	0.03	0.04	0.05	0.06	0.07	0.08	0.09
0.0	0.5000	0.5040	0.5080	0.5120	0.5160	0.5199	0.5239	0.5279	0.5319	0.5359
0.1	0.5398	0.5438	0.5478	0.5517	0.5557	0.5596	0.5636	0.5675	0.5714	0.5753
0.2	0.5793	0.5832	0.5871	0.5910	0.5948	0.5987	0.6026	0.6064	0.6103	0.6141
0.3	0.6179	0.6217	0.6255	0.6293	0.6331	0.6368	0.6406	0.6443	0.6480	0.6517
0.4	0.6554	0.6591	0.6628	0.6664	0.6700	0.6736	0.6772	0.6808	0.6844	0.6879
0.5	0.6915	0.6950	0.6985	0.7019	0.7054	0.7088	0.7123	0.7157	0.7190	0.7224
0.6	0.7257	0.7291	0.7324	0.7357	0.7389	0.7422	0.7454	0.7486	0.7517	0.7549
0.7	0.7580	0.7611	0.7642	0.7673	0.7704	0.7734	0.7764	0.7794	0.7823	0.7852
0.8	0.7881	0.7910	0.7939	0.7967	0.7995	0.8023	0.8051	0.8078	0.8106	0.8133
0.9	0.8159	0.8186	0.8212	0.8238	0.8264	0.8289	0.8315	0.8340	0.8365	0.8389
1.0	0.8413	0.8438	0.8461	0.8485	0.8508	0.8531	0.8554	0.8577	0.8599	0.8621
1.1	0.8643	0.8665	0.8686	0.8708	0.8729	0.8749	0.8770	0.8790	0.8810	0.8830
1.2	0.8849	0.8869	0.8888	0.8907	0.8925	0.8944	0.8962	0.8980	0.8997	0.9015
1.3	0.9032	0.9049	0.9066	0.9082	0.9099	0.9115	0.9131	0.9147	0.9162	0.9177
1.4	0.9192	0.9207	0.9222	0.9236	0.9251	0.9265	0.9279	0.9292	0.9306	0.9319
1.5	0.9332	0.9345	0.9357	0.9370	0.9382	0.9394	0.9406	0.9418	0.9429	0.9441
1.6	0.9452	0.9463	0.9474	0.9484	0.9495	0.9505	0.9515	0.9525	0.9535	0.9545
1.7	0.9554	0.9564	0.9573	0.9582	0.9591	0.9599	0.9608	0.9616	0.9625	0.9633
1.8	0.9641	0.9649	0.9656	0.9664	0.9671	0.9678	0.9686	0.9693	0.9699	0.9706
1.9	0.9713	0.9719	0.9726	0.9732	0.9738	0.9744	0.9750	0.9756	0.9761	0.9767
2.0	0.9772	0.9778	0.9783	0.9788	0.9793	0.9798	0.9803	0.9808	0.9812	0.9817
2.1	0.9821	0.9826	0.9830	0.9834	0.9838	0.9842	0.9846	0.9850	0.9854	0.9857
2.2	0.9861	0.9864	0.9868	0.9871	0.9875	0.9878	0.9881	0.9884	0.9887	0.9890
2.3	0.9893	0.9896	0.9898	0.9901	0.9904	0.9906	0.9909	0.9911	0.9913	0.9916
2.4	0.9918	0.9920	0.9922	0.9925	0.9927	0.9929	0.9931	0.9932	0.9934	0.9936
2.5	0.9938	0.9940	0.9941	0.9943	0.9945	0.9946	0.9948	0.9949	0.9951	0.9952
2.6	0.9953	0.9955	0.9956	0.9957	0.9959	0.9960	0.9961	0.9962	0.9963	0.9964
2.7	0.9965	0.9966	0.9967	0.9968	0.9969	0.9970	0.9971	0.9972	0.9973	0.9974
2.8	0.9974	0.9975	0.9976	0.9977	0.9977	0.9978	0.9979	0.9979	0.9980	0.9981
2.9	0.9981	0.9982	0.9982	0.9983	0.9984	0.9984	0.9985	0.9985	0.9986	0.9986
3.0	0.9987	0.9987	0.9987	0.9988	0.9988	0.9989	0.9989	0.9989	0.9990	0.9990
3.1	0.9990	0.9991	0.9991	0.9991	0.9992	0.9992	0.9992	0.9992	0.9993	0.9993
3.2	0.9993	0.9993	0.9994	0.9994	0.9994	0.9994	0.9994	0.9995	0.9995	0.9995

2022 Higher School Certificate
Worked answers

Section I (*Total 10 marks*)

1	C	**2**	A	**3**	D	**4**	A	**5**	B
6	B	**7**	C	**8**	D	**9**	D	**10**	B

1 Since $\cos\left(\frac{23\pi}{12}\right) = \frac{\sqrt{6}+\sqrt{2}}{4}$ then $\cos\left(2\pi - \frac{23\pi}{12}\right) = \frac{\sqrt{6}+\sqrt{2}}{4}$.

As $y = \cos^{-1}(\theta)$ has a range of $[0, \pi]$, then $\cos^{-1}\left(\frac{\sqrt{6}+\sqrt{2}}{4}\right) = \frac{\pi}{12}$.

Answer C

2 Since the graph of $y = f(x)$ has been reflected about the y-axis to form $g(x)$, then $g(x) = f(|x|)$.

Answer A

3 If when $P(x)$ is divided by $Q(x)$ the remainder is $2x + 5$, which is degree 1, then $deg(Q)$ must be more than 1 but less than 5, such that $2 \leq deg(Q) \leq 4$.

Answer D

4 The roots of the function $f(x) + g(x)$ occur when $f(x) = a$ and $g(x) = -a$ such that $f(x) + g(x) = 0$.

Answer A

5 If $x = 2 + t$ and $y = 3 - 2t^2$, then

$x - 2 = t$

$\therefore y = 3 - 2(x - 2)^2$

The Cartesian equation shows a concave down parabola with vertex at $(2, 3)$.

Answer B

6 Let $\underset{\sim}{v} = -\underset{\sim}{i} + \underset{\sim}{j}$ such that the projection of $\underset{\sim}{u}$ onto $\underset{\sim}{v}$ is given by $\frac{(\underset{\sim}{u} \bullet \underset{\sim}{v})}{|\underset{\sim}{v}|}\underset{\sim}{v}$.

The length of v is given by $|\underset{\sim}{v}| = \sqrt{2}$ and $(\underset{\sim}{u} \bullet \underset{\sim}{v}) > 0$ since the angle θ between the vectors is acute.

Therefore the projection of $\underset{\sim}{u}$ onto $\underset{\sim}{v}$ would have the same direction as $\underset{\sim}{v}$ but a scalar of $\underset{\sim}{v}$.

$\therefore -0.4\underset{\sim}{i} + 0.4\underset{\sim}{j}$.

Answer B

7 Triangles are formed by choosing three vertices that are not all on the one line segment.

N = Total number of combinations without restrictions—three points on same line segment.

$N = {}^{12}C_3 - ({}^5C_3 + {}^4C_3 + {}^3C_3)$

$N = 205$

Answer C

8

$$|\underset{\sim}{a}+\underset{\sim}{b}| < 1$$

$$|\underset{\sim}{a}+\underset{\sim}{b}|^2 < 1$$

$$|\underset{\sim}{a}|^2 + |\underset{\sim}{b}|^2 + 2\underset{\sim}{a}\cdot\underset{\sim}{b} < 1$$

$$|\underset{\sim}{a}|^2 + |\underset{\sim}{b}|^2 + 2|\underset{\sim}{a}||\underset{\sim}{b}|\cos\theta < 1$$

Since $\underset{\sim}{a}$ and $\underset{\sim}{b}$ are unit vectors, then $|\underset{\sim}{a}| = |\underset{\sim}{b}| = 1$.

$2 + 2\cos\theta < 1$

$2\cos\theta < -1$

$\cos\theta < -\frac{1}{2}$

$\therefore \theta > \frac{2\pi}{3}$

Alternatively, the tail of the vector $\underset{\sim}{a} + \underset{\sim}{b}$ must remain inside the unit circle. Therefore the angle θ between $\underset{\sim}{a}$ and $\underset{\sim}{b}$ must not be less than $\frac{2\pi}{3}$.

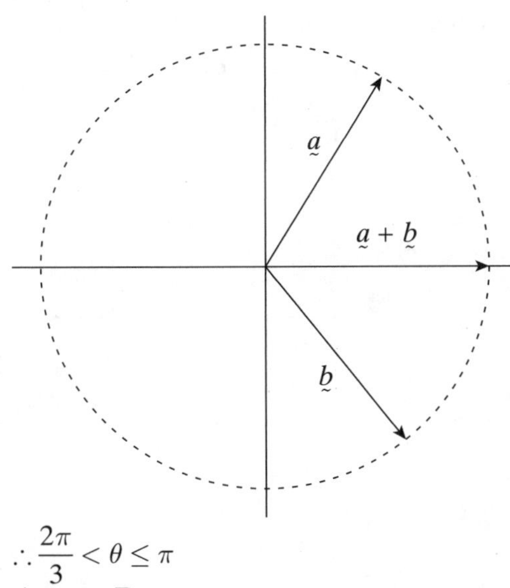

$\therefore \dfrac{2\pi}{3} < \theta \leq \pi$

Answer D

9 Points of intersection of function $f(x)$ and its $f^{-1}(x)$ inverse may have points of intersection on and off the line $y = x$. There are an infinite number of points with parallel tangents. The graph of $f(x) = \dfrac{1}{x}$ can be used as a counter example for A, B and C.

Answer D

10 Considering the gradient function, $\dfrac{dy}{dx} = \sin y + 1$, the graph of the differential equation must have a maximum gradient at $y = \dfrac{\pi}{2}$.

Also $0 < \dfrac{dy}{dx} \leq 2$. As y approaches $-\dfrac{\pi}{2}$ and $\dfrac{3\pi}{2}$, $\dfrac{dy}{dx}$ approaches 0, hence there are horizontal asymptotes at $y = -\dfrac{\pi}{2}$ and $y = \dfrac{3\pi}{2}$. This leaves B as the only option.

Alternatively, solving the differential equation gives:

$$\int \frac{dy}{1+\sin y} = \int dx$$

$$\begin{aligned}1 + \sin y &= 1+\cos\left(\frac{\pi}{2} - y\right)\\ &= 1+\cos\left(2\left(\frac{\pi}{4} - \frac{y}{2}\right)\right)\\ &= 1+2\cos^2\left(\frac{\pi}{4} - \frac{y}{2}\right) - 1\\ &= 2\cos^2\left(\frac{\pi}{4} - \frac{y}{2}\right)\end{aligned}$$

$$\therefore \int \frac{dy}{2\cos^2\left(\frac{\pi}{4} - \frac{y}{2}\right)} = \int dx$$

$$\frac{1}{2}\int \sec^2\left(\frac{\pi}{4} - \frac{y}{2}\right)dy = \int dx$$

$$\frac{1}{2}\left[-2\tan\left(\frac{\pi}{4} - \frac{y}{2}\right)\right] = x + C$$

$$\tan\left(\frac{\pi}{4} - \frac{y}{2}\right) = -x + C$$

$$\frac{\pi}{4} - \frac{y}{2} = \tan^{-1}(-x + C)$$

$$\frac{\pi}{4} - \tan^{-1}(-x+C) = \frac{y}{2}$$

$$\therefore y = \frac{\pi}{2} + 2\tan^{-1}(x - C)$$

Answer B

Section II

QUESTION 11

(a) (i) $\underset{\sim}{u} + \underset{\sim}{v} = (\underset{\sim}{i} - \underset{\sim}{j}) + 3(2\underset{\sim}{i} + \underset{\sim}{j})$
$= \underset{\sim}{i} - \underset{\sim}{j} + 6\underset{\sim}{i} + 3\underset{\sim}{j}$
$= 7\underset{\sim}{i} + 2\underset{\sim}{j}$

(1 mark)

(ii) $\underset{\sim}{u} \bullet \underset{\sim}{v} = 1 \times 2 + (-1) \times 1$
$= 1$

(1 mark)

(b) Let $u = x^2 + 4$

$\dfrac{du}{dx} = 2x$

$du = 2x\, dx$

When $x = 1, u = 5$ and $x = 0, u = 4$

$$\begin{aligned}\int_0^1 \frac{x}{\sqrt{x^2+4}}\,dx &= \frac{1}{2}\int_0^1 \frac{2x}{\sqrt{x^2+4}}\,dx\\ &= \frac{1}{2}\int_4^5 \frac{1}{\sqrt{u}}\,du\\ &= \frac{1}{2}\left[2\sqrt{u}\right]_4^5\\ &= \left[\sqrt{5} - \sqrt{4}\right]\\ &= \sqrt{5} - 2\end{aligned}$$

(3 marks)

(c) $\left(1-\frac{x}{2}\right)^8$

$= \binom{8}{0} + \binom{8}{1}\left(-\frac{x}{2}\right)^1 + \binom{8}{2}\left(-\frac{x}{2}\right)^2 + \binom{8}{3}\left(-\frac{x}{2}\right)^3 + \ldots$

$\therefore$ coefficient of x^2 is given by $\binom{8}{2}\left(\frac{1}{4}\right) = 7$, and

coefficient of x^3 is given by $\binom{8}{3}\left(-\frac{1}{8}\right) = -7$

(2 marks)

(d) Vectors perpendicular if $\underset{\sim}{u} \bullet \underset{\sim}{v} = 0$

$\underset{\sim}{u} \bullet \underset{\sim}{v} = a(a-7) + 2(4a-1)$

$0 = a^2 - 7a + 8a - 2$

$= a^2 + a - 2$

$= (a+2)(a-1)$

$\therefore a = -2$ or $a = 1$

(2 marks)

(e) $\sqrt{3}\sin(x) - 3\cos(x) \equiv R\sin(x + \alpha)$

$R = \sqrt{(\sqrt{3})^2 + (3)^2}$

$= \sqrt{12}$

$= 2\sqrt{3}$

$\tan\alpha = \frac{-3}{\sqrt{3}}$

$\alpha = \tan^{-1}\left(\frac{-3}{\sqrt{3}}\right)$

$= -\frac{\pi}{3}$ or $\frac{2\pi}{3}$

$\sqrt{3}\sin(x) - 3\cos(x) = 2\sqrt{3}\sin\left(x + \frac{2\pi}{3}\right)$

(3 marks)

(f) $\frac{x}{2-x} \geq 5$

$x \neq 2$

$\frac{x}{2-x} = 5$

$x = 5(2-x)$

$= 10 - 5x$

$6x = 10$

$x = \frac{5}{3}$

Testing regions:

When $x = \frac{2}{3}$

$\frac{\frac{2}{3}}{2-\frac{2}{3}} = \frac{1}{2}$

< 5

When $x = \frac{9}{5}$

$\frac{\frac{9}{5}}{2-\frac{9}{5}} = 9$

> 5

When $x = \frac{11}{5}$

$\frac{\frac{11}{5}}{2-\frac{11}{5}} = -11$

< 5

$\therefore \frac{5}{3} \leq x < 2$

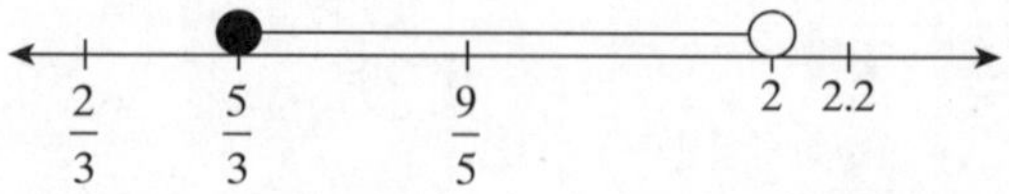

(3 marks)

QUESTION 12

(a) $\frac{dy}{dx} = \frac{x-2y}{x^2+y^2}$

At $(-1, 1)$

$\frac{dy}{dx} = \frac{(-1)-2(1)}{(-1)^2+(1)^2}$

$= \frac{-3}{2}$

At $(1, 1)$

$\frac{dy}{dx} = \frac{(1)-2(1)}{(1)^2+(1)^2}$

$= \frac{-1}{2}$

At $(2, 1)$

$\frac{dy}{dx} = \frac{(2)-2(1)}{(2)^2+(1)^2}$

$= 0$

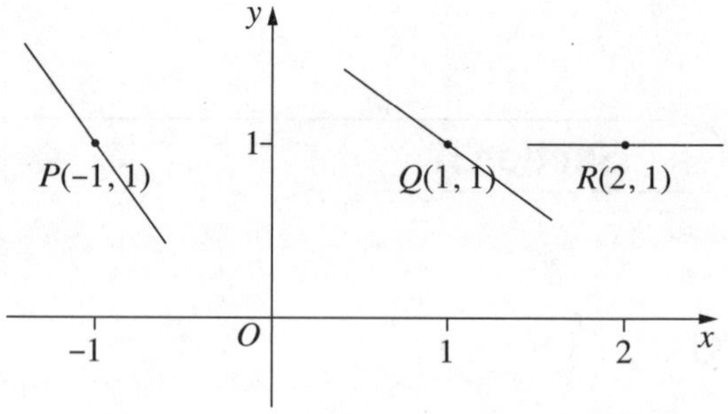

(2 marks)

(b) Let n = number of over-age players and k = number of teams

$\therefore \frac{n}{k} = \frac{41}{13}$

$= 3\frac{2}{13}$

$\therefore$ *Each* team can have 3 over-age players with 2 over-age players yet to be placed.

$\therefore$ At least one team will be penalised.

(2 marks)

(c) $y = x\arctan(x)$

Let $u = x$ and $v = \arctan(x)$

$$\frac{dy}{dx} = u\frac{dv}{dx} + v\frac{du}{dx}$$

$$= x\left(\frac{1}{1+x^2}\right) + \arctan(x)$$

At $\left(1, \frac{\pi}{4}\right)$, $m = (1)\left(\frac{1}{1+1^2}\right) + \arctan(1)$

$$\therefore m = \frac{1}{2} + \frac{\pi}{4}$$

$$y - y_1 = m(x - x_1)$$

$$y - \frac{\pi}{4} = \left(\frac{1}{2} + \frac{\pi}{4}\right)(x-1)$$

$$y - \frac{\pi}{4} = \frac{1}{2}x - \frac{1}{2} + \frac{\pi}{4}x - \frac{\pi}{4}$$

$$y = \frac{1}{2}x - \frac{1}{2} + \frac{\pi}{4}x$$

$$\therefore y = \left(\frac{1}{2} + \frac{\pi}{4}\right)x - \frac{1}{2}$$

(3 marks)

(d) (i) Newton's law of cooling states that $\frac{dT}{dt} = k(T - T_1)$ where $T_1 = 12$

$$\frac{dT}{dt} = k(T-12)$$

$$\frac{dT}{T-12} = k\,dt$$

$$\int \frac{1}{T-12}\,dT = \int k\,dt$$

$$\ln|T-12| = kt + C$$

When $t = 0$, $T = 92$

$$\ln|92 - 12| = k(0) + C$$

$$\therefore C = \ln 80$$

$$\ln|T-12| - \ln 80 = kt$$

$$\therefore \ln\left(\frac{T-12}{80}\right) = kt$$

When $t = 5$, $T = 76$

$$\ln\left(\frac{76-12}{80}\right) = 5k$$

$$\therefore k = \frac{1}{5}\ln\left(\frac{4}{5}\right)$$

$$\ln\left(\frac{T-12}{80}\right) = \frac{t}{5}\ln\left(\frac{4}{5}\right)$$

$$\frac{T-12}{80} = e^{\frac{t}{5}\ln\left(\frac{4}{5}\right)}$$

$$T - 12 = 80e^{\frac{t}{5}\ln\left(\frac{4}{5}\right)}$$

$$\therefore T = 12 + 80e^{\frac{t}{5}\ln\left(\frac{4}{5}\right)}$$

(3 marks)

(ii) Find when $T = 57$

$$T = 12 + 80e^{\frac{t}{5}\ln\left(\frac{4}{5}\right)}$$

$$57 = 12 + 80e^{\frac{t}{5}\ln\left(\frac{4}{5}\right)}$$

$$45 = 80e^{\frac{t}{5}\ln\left(\frac{4}{5}\right)}$$

$$\frac{9}{16} = e^{\frac{t}{5}\ln\left(\frac{4}{5}\right)}$$

$$\ln\left(\frac{9}{16}\right) = \frac{t}{5}\ln\left(\frac{4}{5}\right)$$

$$t = \frac{5\ln\left(\frac{9}{16}\right)}{\ln\left(\frac{4}{5}\right)}$$

$$t = 12.8922$$

$\therefore$ Coffee will reach 57 °C after 13 minutes.

(1 mark)

(e) Expected score from one draw:

$$E(X_1) = \frac{3}{10} \times 10 + \frac{7}{10} \times -5 = -\frac{1}{2}$$

Since events are independent, then the expected score from four draws:

$$E(X) = -\frac{1}{2} \times 4$$

$$= -2$$

(2 marks)

(f) Prove true for $n = 0$

$$15^0 + 6^{2(0)+1} = 1 + 6$$

$$= 7$$

Since 7 is divisible by 7, then true for $n = 0$.

Assume true for $n = k$.

$15^k + 6^{2k+1} = 7p$ where p is an integer

$$\therefore 6^{2k+1} = 7p - 15^k$$

Prove true for $n = k + 1$.

$15^{k+1} + 6^{2(k+1)+1} = 7q$ where q is an integer

$$\text{LHS} = 15^{k+1} + 6^{2k+2+1}$$

$$= 15^{k+1} + 6^{2k+1+2}$$

$$= 15 \times 15^k + 6^2 \times 6^{2k+1}$$

$$= 15 \times 15^k + 6^2(7p - 15^k)$$

from Assumption step

$$= 15 \times 15^k + 252p - 36 \times 15^k$$

$$= 252p - 21 \times 15^k$$

$$= 7(36p - 3 \times 15^k)$$

$= 7q$ for some integer q

(since p and k are integers)

$\therefore 15^{k+1} + 62^{(k+1)+1}$ is divisible by 7.

The statement is true for $n = 0$ and true for $n = k + 1$. So by mathematical induction it is true for all integers $n \geq 0$. *(3 marks)*

QUESTION 13

(a) $h = \underset{\sim}{a} + \underset{\sim}{b} + \underset{\sim}{c}$

$\therefore -\underset{\sim}{b} + \underset{\sim}{h} = \underset{\sim}{a} + \underset{\sim}{c}$

$\overrightarrow{BH} = -\underset{\sim}{b} + \underset{\sim}{h}$

$= \underset{\sim}{a} + \underset{\sim}{c}$

$\overrightarrow{CA} = -\underset{\sim}{c} + \underset{\sim}{a}$

Vectors are perpendicular if $\overrightarrow{BH} \bullet \overrightarrow{CA} = 0$

$\overrightarrow{BH} \bullet \overrightarrow{CA} = (\underset{\sim}{a} + \underset{\sim}{c})(-\underset{\sim}{c} + \underset{\sim}{a})$

$= -\underset{\sim}{a} \bullet c + \underset{\sim}{a} \bullet \underset{\sim}{a} - \underset{\sim}{c} \bullet \underset{\sim}{c} + \underset{\sim}{a} \bullet \underset{\sim}{c}$

$= \underset{\sim}{a} \bullet \underset{\sim}{a} - \underset{\sim}{c} \bullet \underset{\sim}{c}$

$= |\underset{\sim}{a}|^2 - |\underset{\sim}{c}|^2$

$= 0 \quad$ (since $|\underset{\sim}{a}|^2 = |\underset{\sim}{c}|^2$ as radii)

$\therefore \overrightarrow{BH}$ and $\overrightarrow{CA}$ are perpendicular.

(3 marks)

(b) $\int_0^{\frac{\pi}{2k}} (k+1)\sin(kx)\,dx$

$V = \pi \int_a^b y^2\,dx$

$\pi^2 = \pi \int_0^{\frac{\pi}{2k}} (k+1)^2 \sin^2(kx)\,dx$

$= \pi(k+1)^2 \int_0^{\frac{\pi}{2k}} \frac{1-\cos(2kx)}{2}\,dx$

$\pi = \frac{(k+1)^2}{2}\left[x - \frac{\sin(2kx)}{2k}\right]_0^{\frac{\pi}{2k}}$

$= \frac{(k+1)^2}{2}\left[\left(\frac{\pi}{2k} - \frac{\sin(\pi)}{2k}\right) - \left(0 - \frac{\sin(0)}{2k}\right)\right]$

$= \frac{(k+1)^2}{2}\left[\frac{\pi}{2k} - 0\right]$

$= \frac{\pi(k+1)^2}{4k}$

$1 = \frac{(k+1)^2}{4k}$

$4k = k^2 + 2k + 1$

$0 = k^2 - 2k + 1$

$= (k-1)^2$

$\therefore k = 1$

(3 marks)

(c) $f(x)$ is a many-to-one function.

$\therefore$ the inverse of $f(x)$ is not a function.

$g(x) = \sin^{-1}(x)$ is a function so it cannot be the inverse of f.

(2 marks)

(d) $P(x) = x^3 + bx^2 + cx + d$

$\alpha + \beta + \gamma = -b$

$\alpha\beta + \beta\gamma + \alpha\gamma = c$

$P'(x) = 3x^2 + 2bx + c$

Substituting α, β and γ into $P'(x)$ gives:

$P'(\alpha) = 3\alpha^2 + 2b\alpha + c$

$P'(\beta) = 3\beta^2 + 2b\beta + c$

$P'(\gamma) = 3\gamma^2 + 2b\gamma + c$

$P'(\alpha) + P'(\beta) + P'(\gamma) = 87$

$87 = 3\alpha^2 + 2b\alpha + c + 3\beta^2 + 2b\beta + c + 3\gamma^2 + 2b\gamma + c$

$= 3(\alpha^2 + \beta^2 + \gamma^2) + 2b(\alpha + \beta + \gamma) + 3c$

$= 3(\alpha^2 + \beta^2 + \gamma^2) - 2 \times -b(\alpha + \beta + \gamma) + 3(\alpha\beta + \beta\gamma + \alpha\gamma)$

$= 3(\alpha^2 + \beta^2 + \gamma^2) - 2(\alpha + \beta + \gamma)^2 + 3(\alpha\beta + \beta\gamma + \alpha\gamma)$

$= 3(\alpha^2 + \beta^2 + \gamma^2) - 2[\alpha^2 + \beta^2 + \gamma^2 + (2(\alpha\beta + \beta\gamma + \alpha\gamma))] + 3(\alpha\beta + \beta\gamma + \alpha\gamma)$

$= \alpha^2 + \beta^2 + \gamma^2 - (\alpha\beta + \beta\gamma + \alpha\gamma)$

$= 85 - (\alpha\beta + \beta\gamma + \alpha\gamma)$

$\therefore \alpha\beta + \beta\gamma + \alpha\gamma = -2$

(3 marks)

(e) (i) $n = 16, p = 0.2$

$$\begin{aligned}\mu &= np \\ &= 16 \times 0.2 \\ &= 3.2 \\ \sigma &= \sqrt{np(1-p)} \\ &= \sqrt{3.2(1-0.2)} \\ &= 1.6\end{aligned}$$

$$\begin{aligned}P(x \geq 8) &= P\left(Z \geq \frac{8-3.2}{1.6}\right) \\ &= P(Z \geq 3) \\ &= 1 - P(Z \leq 3) \\ &= 1 - 0.9987 \\ &= 0.0013\end{aligned}$$

(2 marks)

(ii) The method may not be valid as the sample size is not sufficient.

For approximation to be accurate $np \geq 10$.

$$\begin{aligned}np &= 3.2 \\ &< 10\end{aligned}$$

$\therefore$ Approximation is not considered to be valid.

(1 mark)

QUESTION 14

(a) $(x-2)\dfrac{dy}{dx} = xy$

$$\frac{dy}{y} = \frac{x}{x-2}dx$$

$$\int \frac{dy}{y} = \int \frac{x-2+2}{x-2}dx$$

$$\int \frac{dy}{y} = \int 1 + \frac{2}{x-2}dx$$

$$\ln y = x + 2\ln|x-2| + C$$

When $x = 0, y = 1$

$$\ln 1 = 0 + 2\ln|0-2| + C$$

$$0 = 2\ln 2 + C$$

$$\therefore C = -2\ln 2$$

$$\therefore \ln y = x + \ln(x-2)^2 - \ln 2^2$$

$$\ln y = x + \ln\left(\frac{x-2}{2}\right)^2$$

$$\ln y = \ln e^x + \ln\left(\frac{x-2}{2}\right)^2$$

$$\ln y = \ln\left[e^x\left(\frac{x-2}{2}\right)^2\right]$$

$$\therefore y = e^x\left(\frac{x-2}{2}\right)^2$$

(4 marks)

(b) The projection of $\underset{\sim}{u}$ onto $\underset{\sim}{v}$ is given by $\dfrac{\vec{u}\cdot\vec{v}}{|\underset{\sim}{v}|^2}\vec{\underset{\sim}{v}}$

$$\therefore \vec{p} = \frac{\vec{u}\cdot\vec{v}}{|\underset{\sim}{v}|^2}\vec{\underset{\sim}{v}}$$

$$\lambda_0\underset{\sim}{v} = \frac{\vec{u}\cdot\vec{v}}{|\underset{\sim}{v}|^2}\vec{\underset{\sim}{v}}$$

$$\therefore \lambda_0 = \frac{\vec{u}\cdot\vec{v}}{|\underset{\sim}{v}|^2}$$

$$\begin{aligned}|\vec{u} - \lambda\vec{v}|^2 &= (\vec{u} - \lambda\vec{v})\cdot(\vec{u} - \lambda\vec{v}) \\ &= |\vec{u}|^2 - 2\lambda\vec{u}\cdot\vec{v} + \lambda^2|\vec{u}|^2\end{aligned}$$

$\therefore$ The magnitude is given by a quadratic in terms of λ with minimum value at $\lambda = -\dfrac{b}{2a}$.

$$\begin{aligned}\lambda &= -\frac{b}{2a} \\ &= -\frac{-2\vec{u}\cdot\vec{v}}{2|\vec{v}|^2} \\ &= \frac{\vec{u}\cdot\vec{v}}{|\vec{v}|^2} \\ &= \lambda_0\end{aligned}$$

$$\therefore |\vec{u} - \lambda_0\vec{v}|^2 \leq |\vec{u} - \lambda\vec{v}|^2$$

$$\therefore |\vec{u} - \lambda_0\vec{v}| \leq |\vec{u} - \lambda\vec{v}|$$

(3 marks)

(c) Projectile will hit the ground when $2ut\sin\theta - \dfrac{g}{2}t^2 = 0$.

$$t\left(2u\sin\theta - \frac{g}{2}t\right) = 0$$

$$2u\sin\theta - \frac{g}{2}t = 0$$

$$2u\sin\theta = \frac{g}{2}t$$

$$\therefore t = \frac{4u\sin\theta}{g}$$

$$\begin{aligned}x_P &= 2ut\cos\theta \\ &= 2u\left(\frac{4u\sin\theta}{g}\right)\cos\theta \\ &= \frac{8u^2\sin\theta\cos\theta}{g} \\ &= \frac{4u^2\sin 2\theta}{g}\end{aligned}$$

The target's position is given by $x_T = d + ut$.

$$\begin{aligned}x_T &= d + u\left(\frac{4u\sin\theta}{g}\right) \\ &= d + \frac{4u^2\sin\theta}{g}\end{aligned}$$

For the player to hit the target, $x_T \leq x_P$

$$d+\frac{4u^2\sin\theta}{g} \leq \frac{4u^2\sin 2\theta}{g}$$

$$d \leq \frac{4u^2\sin 2\theta}{g} - \frac{4u^2\sin\theta}{g}$$

$$d \leq \frac{4u^2}{g}(\sin 2\theta - \sin\theta)$$

Let $f(\theta) = \sin 2\theta - \sin\theta$

$f'(\theta) = 2\cos 2\theta - \cos\theta$

$0 = 2\cos 2\theta - \cos\theta$

$0 = 2(2\cos^2\theta - 1) - \cos\theta$

$0 = 4\cos^2\theta - \cos\theta - 2$

$$\cos\theta = \frac{1 \pm \sqrt{(-1)^2 - 4\times 4\times -2}}{8}$$

$$\cos\theta = \frac{1\pm\sqrt{33}}{8}$$

$\cos\theta = 0.84$ or $\cos\theta = -0.59$

$\therefore \theta \approx 0.5678$

$\sin 2\theta - \sin\theta = \sin(2(0.5678)) - \sin(0.5678)$

$= 0.369$ (3 d.p.)

$$\therefore d \leq 0.37\frac{4u^2}{g}$$

$\therefore d \leq 0.37\ \text{Max}(x_P)$

(4 marks)

(d) $\mu = np$

$= 0.95n$ where p is Pr(not missing a flight)

$\sigma^2 = npq$

$= 0.05 \times 0.95n$

$= 0.0475n$

$$\Pr(X > 350) = \Pr\left(z > \frac{350 - 0.95n}{\sqrt{0.0475n}}\right) \leq 0.01$$

$$\Pr\left(z \leq \frac{350-0.95n}{\sqrt{0.0475n}}\right) \geq 0.99$$

From the z-score table provided:

$\Pr(z \leq 2.33) \geq 0.99$

$$\therefore \frac{350-0.95n}{\sqrt{0.0475n}} = 2.33$$

$350 - 0.95n = 2.33\sqrt{0.0475n}$

$(350 - 0.95n)^2 = 2.33^2 \times 0.0475n$

$122\,500 - 665n + 0.9025n^2 = 0.25\,787\ldots n$

$0.9025n^2 - 665.258n + 122\,500 = 0$

$$n = \frac{665.258 \pm \sqrt{(-665.258)^2 - 4\times 0.9025 \times 122\,500}}{2\times 0.9025}$$

$n = 358$ or 379

But since $\frac{350-0.95n}{\sqrt{0.0475n}}$, only $n = 358$ can be a possible solution.

$\therefore$ The maximum number of tickets that can be sold is 358.

(4 marks)

NSW Education Standards Authority

Centre Number

Student Number

2023 HIGHER SCHOOL CERTIFICATE EXAMINATION

Mathematics Extension 1

General Instructions

- Reading time – 10 minutes
- Working time – 2 hours
- Write using black pen
- Calculators approved by NESA may be used
- A reference sheet is provided at the back of this paper
- For questions in Section II, show relevant mathematical reasoning and/or calculations
- Write your Centre Number and Student Number at the top of this page

Total marks: 70

Section I – 10 marks

- Attempt Questions 1–10
- Allow about 15 minutes for this section

Section II – 60 marks

- Attempt Questions 11–14
- Allow about 1 hour and 45 minutes for this section

Section I

10 marks
Attempt Questions 1–10
Allow about 15 minutes for this section

Use the multiple-choice answer sheet for Questions 1–10.

---

1 The temperature $T(t)$°C of an object at time t seconds is modelled using Newton's Law of Cooling,

$$T(t) = 15 + 4e^{-3t}.$$

What is the initial temperature of the object?

A. -3

B. 4

C. 15

D. 19

2 A standard six-sided die is rolled 12 times.

Let $\hat{p}$ be the proportion of the rolls with an outcome of 2.

Which of the following expressions is the probability that at least 9 of the rolls have an outcome of 2?

A. $P\left(\hat{p} \geq \frac{3}{4}\right)$

B. $P\left(\hat{p} \geq \frac{1}{6}\right)$

C. $P\left(\hat{p} \leq \frac{3}{4}\right)$

D. $P\left(\hat{p} \leq \frac{1}{6}\right)$

3 The diagram shows the direction field of a differential equation. A particular solution to the differential equation passes through $(-2, 1)$.

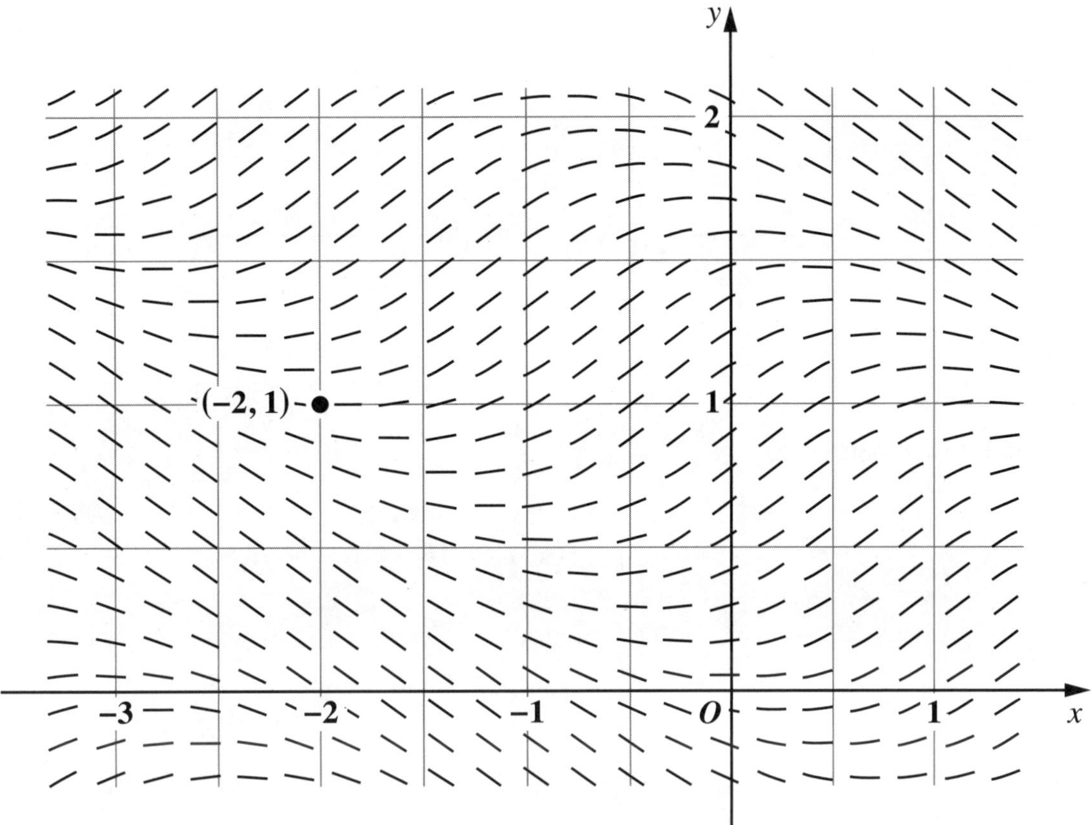

Where does the solution that passes through $(-2, 1)$ cross the y-axis?

A. $y = 1.12$

B. $y = 1.34$

C. $y = 1.56$

D. $y = 1.78$

4 The diagram shows the graphs of the functions $f(x)$ and $g(x)$.

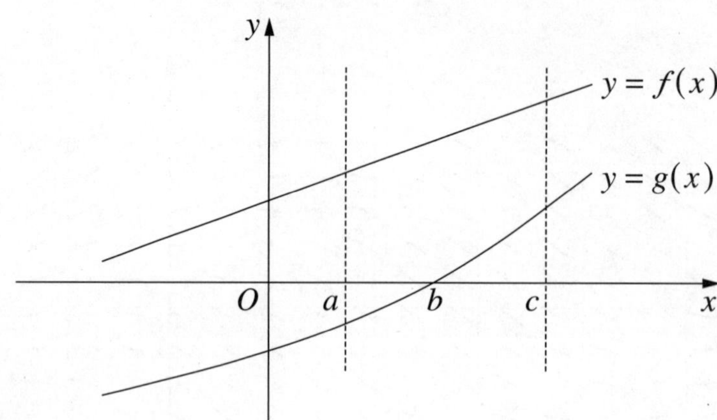

It is known that

$$\int_a^c f(x)\,dx = 10$$

$$\int_a^b g(x)\,dx = -2$$

$$\int_b^c g(x)\,dx = 3.$$

What is the area between the curves $y = f(x)$ and $y = g(x)$ between $x = a$ and $x = c$?

A. 5

B. 7

C. 9

D. 11

5 Which of the following is the value of $\sin^{-1}(\sin a)$ given that $\pi < a < \frac{3\pi}{2}$?

A. $a - \pi$

B. $\pi - a$

C. a

D. $-a$

6 Given the two non-zero vectors $\underset{\sim}{a}$ and $\underset{\sim}{b}$, let $\underset{\sim}{c}$ be the projection of $\underset{\sim}{a}$ onto $\underset{\sim}{b}$.

What is the projection of $10\underset{\sim}{a}$ onto $2\underset{\sim}{b}$?

A. $2\underset{\sim}{c}$

B. $5\underset{\sim}{c}$

C. $10\underset{\sim}{c}$

D. $20\underset{\sim}{c}$

7 Which statement is always true for real numbers a and b where $-1 \le a < b \le 1$?

A. $\sec a < \sec b$

B. $\sin^{-1} a < \sin^{-1} b$

C. $\arccos a < \arccos b$

D. $\cos^{-1} a + \sin^{-1} a < \cos^{-1} b + \sin^{-1} b$

8 The diagram shows the graph of a function.

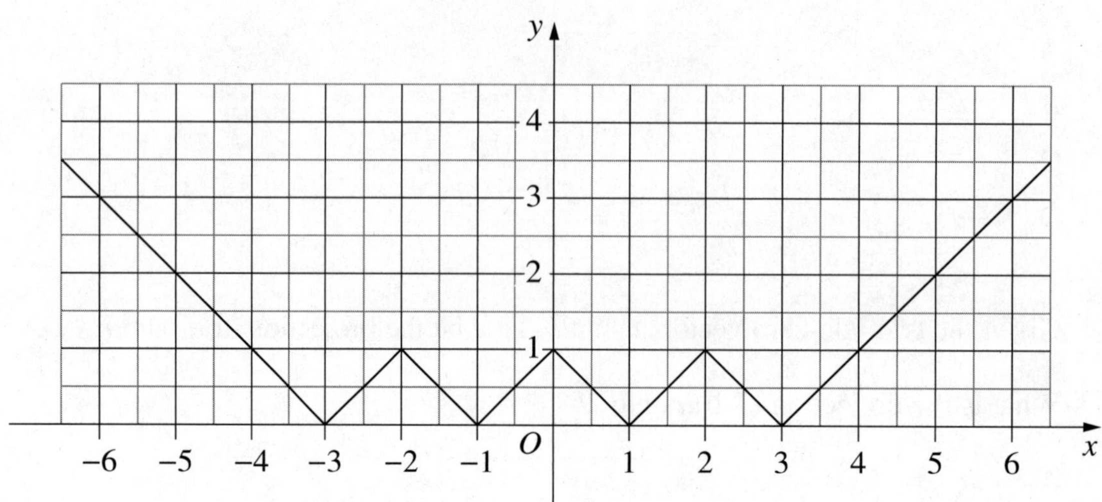

Which of the following is the equation of the function?

A. $y = \left|1 - \left||x| - 2\right|\right|$

B. $y = \left|2 - \left||x| - 1\right|\right|$

C. $y = \left|1 - |x - 2|\right|$

D. $y = \left|2 - |x - 1|\right|$

9 The graph of a cubic function, $y = f(x)$, is given below.

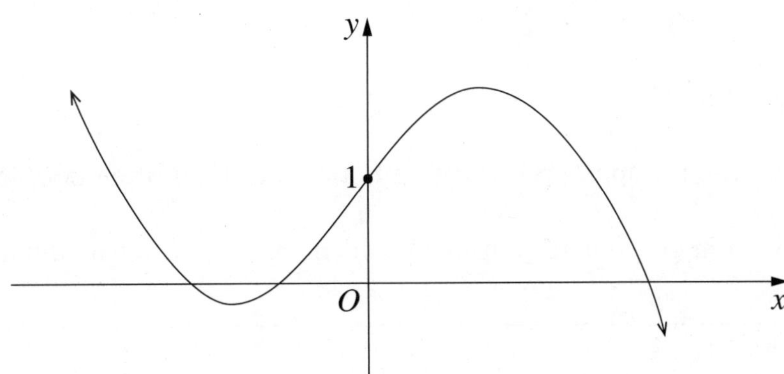

Which of the following functions has an inverse relation whose graph has more than 3 points with an x-coordinate of 1?

A. $y = \sqrt{f(x)}$

B. $y = \dfrac{1}{f(x)}$

C. $y = f(|x|)$

D. $y = |f(x)|$

10 A group with 5 students and 3 teachers is to be arranged in a circle.

In how many ways can this be done if no more than 2 students can sit together?

A. $4! \times 3!$

B. $5! \times 3!$

C. $2! \times 5! \times 3!$

D. $2! \times 2! \times 2! \times 3!$

Section II

60 marks
Attempt Questions 11–14
Allow about 1 hour and 45 minutes for this section

Answer each question in the appropriate writing booklet. Extra writing booklets are available.

For questions in Section II, your responses should include relevant mathematical reasoning and/or calculations.

Question 11 (16 marks) Use the Question 11 Writing Booklet

(a) The parametric equations of a line are given below. **2**

$$x = 1 + 3t$$
$$y = 4t$$

Find the Cartesian equation of this line in the form $y = mx + c$.

(b) In how many different ways can all the letters of the word CONDOBOLIN be arranged in a line? **2**

(c) Consider the polynomial **3**

$$P(x) = x^3 + ax^2 + bx - 12,$$

where a and b are real numbers.

It is given that $x + 1$ is a factor of $P(x)$ and that, when $P(x)$ is divided by $x - 2$, the remainder is -18.

Find a and b.

Question 11 continues on the following page

Question 11 (continued)

(d) Find $\int \frac{1}{\sqrt{4-9x^2}}\,dx$. **2**

(e) Solve $\cos\theta + \sin\theta = 1$ for $0 \le \theta \le 2\pi$. **3**

(f) A recent census found that 30% of Australians were born overseas.

A sample of 900 randomly selected Australians was surveyed.

Let $\hat{p}$ be the sample proportion of surveyed people who were born overseas.

A normal distribution is to be used to approximate $P(\hat{p} \le 0.31)$.

(i) Show that the variance of the random variable $\hat{p}$ is $\frac{7}{30\,000}$. **2**

(ii) Use the standard normal distribution and the information on page 16 to approximate $P(\hat{p} \le 0.31)$, giving your answer correct to two decimal places. **2**

End of Question 11

Question 12 (15 marks) Use the Question 12 Writing Booklet

(a) Evaluate $\int_3^4 (x+2)\sqrt{x-3}\,dx$ using the substitution $u = x - 3$. **3**

(b) Use mathematical induction to prove that **3**

$$\left(1 \times 2\right) + \left(2 \times 2^2\right) + \left(3 \times 2^3\right) + \cdots + \left(n \times 2^n\right) = 2 + (n-1)2^{n+1}$$

for all integers $n \geq 1$.

(c) A gym has 9 pieces of equipment: 5 treadmills and 4 rowing machines.

On average, each treadmill is used 65% of the time and each rowing machine is used 40% of the time.

(i) Find an expression for the probability that, at a particular time, exactly 3 of the 5 treadmills are in use. **2**

(ii) Find an expression for the probability that, at a particular time, exactly 3 of the 5 treadmills are in use and no rowing machines are in use. **1**

(d) It is known that ${}^{n}C_r = {}^{n-1}C_{r-1} + {}^{n-1}C_r$ for all integers such that $1 \leq r \leq n-1$. (Do NOT prove this.) **2**

Find ONE possible set of values for p and q such that

$${}^{2022}C_{80} + {}^{2022}C_{81} + {}^{2023}C_{1943} = {}^{p}C_q.$$

Question 12 continues on the following page

Question 12 (continued)

(e) The region, R, bounded by the hyperbola $y = \dfrac{60}{x+5}$, the line $x = 10$ and the coordinate axes is shown. **4**

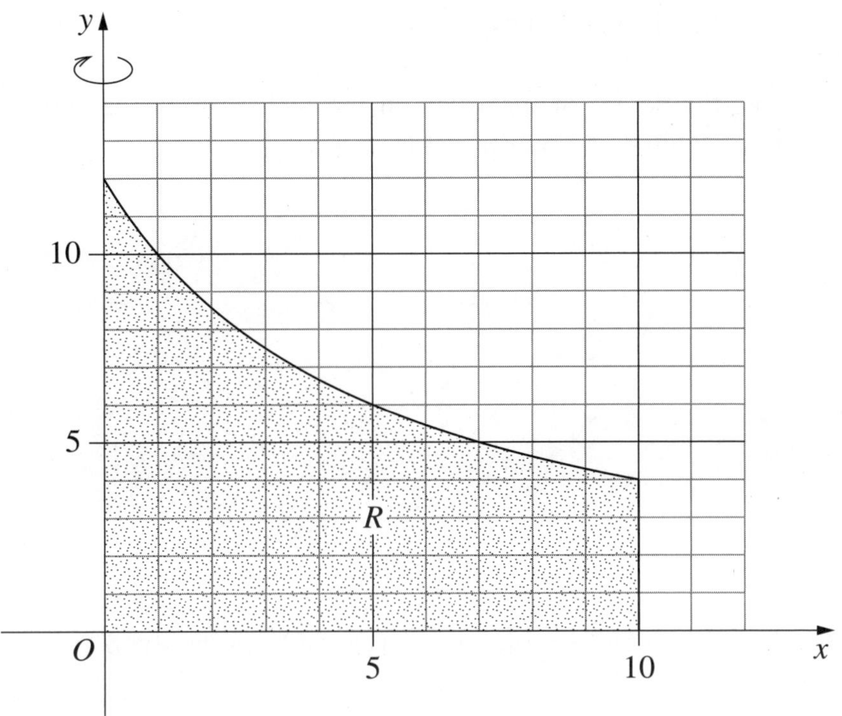

Find the volume of the solid of revolution formed when the region R is rotated about the y-axis. Leave your answer in exact form.

End of Question 12

Question 13 (15 marks) Use the Question 13 Writing Booklet

(a) A hemispherical water tank has radius R cm. The tank has a hole at the bottom which allows water to drain out.

Initially the tank is empty. Water is poured into the tank at a constant rate of $2kR$ cm^3 s^{-1}, where k is a positive constant.

After t seconds, the height of the water in the tank is h cm, as shown in the diagram, and the volume of water in the tank is V cm^3.

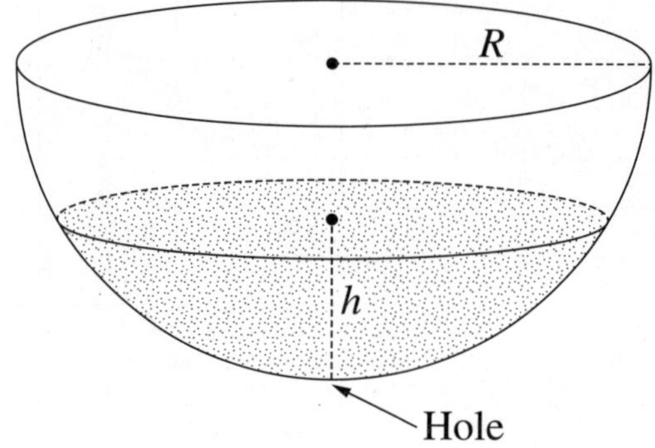

It is known that $V = \pi\left(Rh^2 - \dfrac{h^3}{3}\right)$. (Do NOT prove this.)

While water flows into the tank and also drains out of the bottom, the rate of change of the volume of water in the tank is given by $\dfrac{dV}{dt} = k(2R - h)$.

(i) Show that $\dfrac{dh}{dt} = \dfrac{k}{\pi h}$. **2**

(ii) Show that the tank is full of water after $T = \dfrac{\pi R^2}{2k}$ seconds. **2**

(iii) The instant the tank is full, water stops flowing into the tank, but it continues to drain out of the hole at the bottom as before. **3**

Show that the tank takes 3 times as long to empty as it did to fill.

Question 13 continues on the following page

Question 13 (continued)

(b) Particle A is projected from the origin with initial speed v m s^{-1} at an angle θ with the horizontal plane. At the same time, particle B is projected horizontally with initial speed u m s^{-1} from a point that is H metres above the origin, as shown in the diagram.

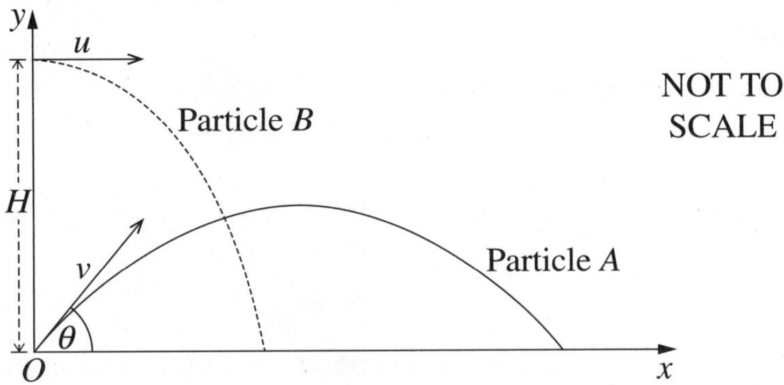

The position vector of particle A, t seconds after it is projected, is given by

$$\mathbf{r}_A(t) = \begin{pmatrix} vt\cos\theta \\ vt\sin\theta - \frac{1}{2}gt^2 \end{pmatrix}. \qquad \text{(Do NOT prove this.)}$$

The position vector of particle B, t seconds after it is projected, is given by

$$\mathbf{r}_B(t) = \begin{pmatrix} ut \\ H - \frac{1}{2}gt^2 \end{pmatrix}. \qquad \text{(Do NOT prove this.)}$$

The angle θ is chosen so that $\tan\theta = 2$.

The two particles collide.

(i) By first showing that $\cos\theta = \frac{1}{\sqrt{5}}$, verify that $v = \sqrt{5}u$. **2**

(ii) Show that the particles collide at time $T = \frac{H}{2u}$. **1**

When the particles collide, their velocity vectors are perpendicular.

(iii) Show that $H = \frac{2u^2}{g}$. **3**

(iv) Prior to the collision, the trajectory of particle A was a parabola. (Do NOT prove this.) **2**

Find the height of the vertex of that parabola above the horizontal plane. Give your answer in terms of H.

End of Question 13

Question 14 (14 marks) Use the Question 14 Writing Booklet

(a) Let $f(x) = 2x + \ln x$, for $x > 0$.

(i) Explain why the inverse of $f(x)$ is a function. **1**

(ii) Let $g(x) = f^{-1}(x)$. By considering the value of $f(1)$, or otherwise, evaluate $g'(2)$. **2**

(b) Consider the hyperbola $y = \dfrac{1}{x}$ and the circle $(x - c)^2 + y^2 = c^2$, where c is a constant.

(i) Show that the x-coordinates of any points of intersection of the hyperbola and circle are zeros of the polynomial $P(x) = x^4 - 2cx^3 + 1$. **1**

(ii) The graphs of $y = x^4 - 2cx^3 + 1$ for $c = 0.8$ and $c = 1$ are shown. **3**

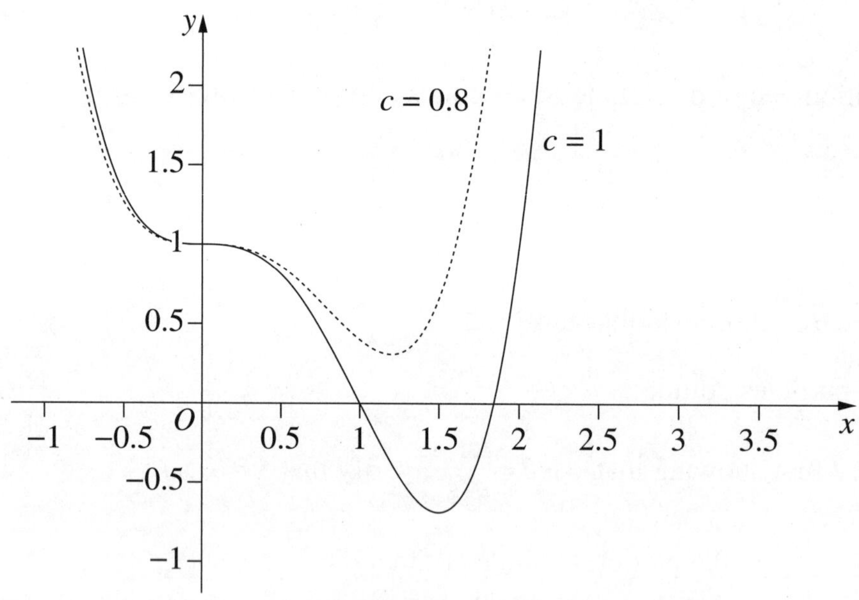

By considering the given graphs, or otherwise, find the exact value of $c > 0$ such that the hyperbola $y = \dfrac{1}{x}$ and the circle $(x - c)^2 + y^2 = c^2$ intersect at only one point.

Question 14 continues on the following page

Question 14 (continued)

(c) (i) Given a non-zero vector $\begin{pmatrix} p \\ q \end{pmatrix}$, it is known that the vector $\begin{pmatrix} q \\ -p \end{pmatrix}$ is perpendicular to $\begin{pmatrix} p \\ q \end{pmatrix}$ and has the same magnitude. (Do NOT prove this.) **3**

Points A and B have position vectors $\overrightarrow{OA} = \begin{pmatrix} a_1 \\ a_2 \end{pmatrix}$ and $\overrightarrow{OB} = \begin{pmatrix} b_1 \\ b_2 \end{pmatrix}$, respectively.

Using the given information, or otherwise, show that the area of triangle OAB is $\frac{1}{2}\left|a_1b_2 - a_2b_1\right|$.

(ii) The point P lies on the circle centred at $I(r, 0)$ with radius $r > 0$, such that $\overrightarrow{IP}$ makes an angle of t to the horizontal. **4**

The point Q lies on the circle centred at $J(-R, 0)$ with radius $R > 0$, such that $\overrightarrow{JQ}$ makes an angle of $2t$ to the horizontal.

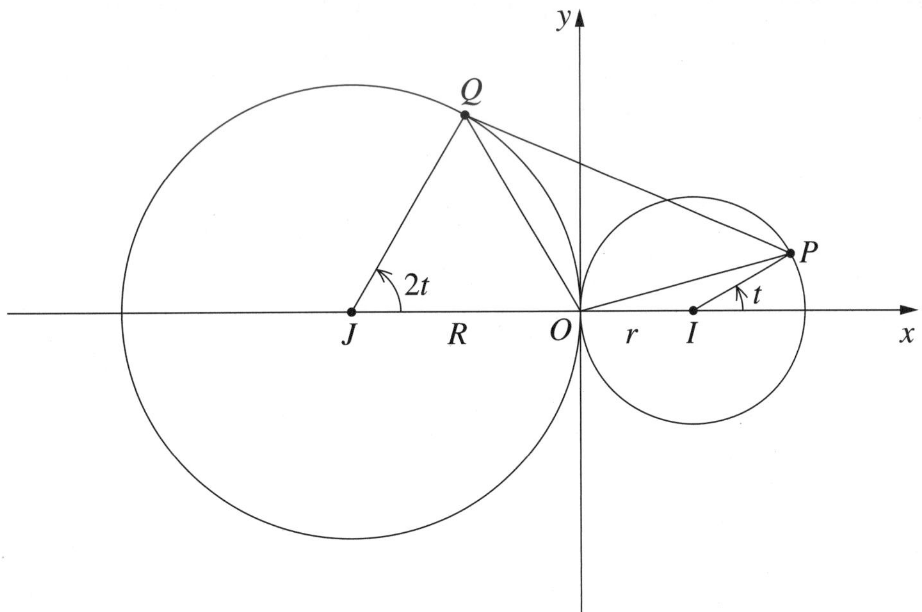

Note that $\overrightarrow{OP} = \overrightarrow{OI} + \overrightarrow{IP}$ and $\overrightarrow{OQ} = \overrightarrow{OJ} + \overrightarrow{JQ}$.

Using part (i), or otherwise, find the values of t, where $-\pi \le t \le \pi$, that maximise the area of triangle OPQ.

End of paper

2023 Higher School Certificate
Worked answers

Section I

(*Total 10 marks*)

1 D	**2** A	**3** C	**4** C	**5** B
6 B	**7** B	**8** A	**9** D	**10** B

1 $T(0) = 15 + 4e^{-3(0)} = 19$

Answer D

2 $P\left(\hat{p} \geq \frac{9}{12}\right) = P\left(\hat{p} \geq \frac{3}{4}\right)$

Answer A

3

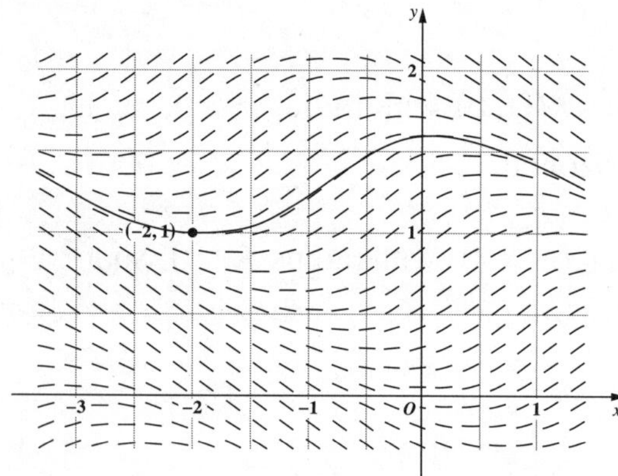

The solution curve will cross the y-axis just above $y = 1.5$.

Answer C

4

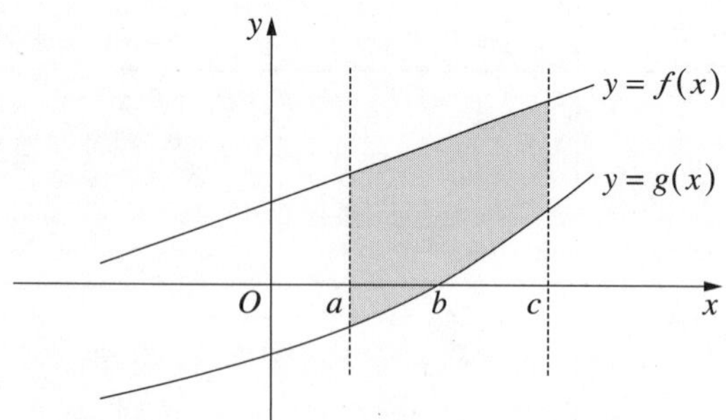

The grey area represents the area between the curves $y = f(x)$ and $y = g(x)$ between $x = a$ and $x = c$.

This area is equal to

$\int_a^c f(x)dx + \left|\int_a^b g(x)dx\right| - \int_b^c g(x)dx$

$= 10 + 2 - 3 = 9.$

Answer C

5 The range of $\sin^{-1} x$ is $\left[-\frac{\pi}{2}, \frac{\pi}{2}\right]$.

$\sin a = \sin(\pi - a)$

If $\pi < a < \frac{3\pi}{2}$, then $-\frac{\pi}{2} < \pi - a < 0$

$\therefore \sin^{-1}(\sin a) = \pi - a$

Answer B

6 $\underset{\sim}{c} = \text{proj}_{\underset{\sim}{b}}\underset{\sim}{a} = \frac{\underset{\sim}{a} \cdot \underset{\sim}{b}}{|\underset{\sim}{b}|^2}\underset{\sim}{b}$

$$\text{proj}_{2\underset{\sim}{b}} 10\underset{\sim}{a} = \frac{(10\underset{\sim}{a}) \cdot (2\underset{\sim}{b})}{|2\underset{\sim}{b}|^2}(2\underset{\sim}{b})$$

$$= \frac{20(\underset{\sim}{a} \cdot \underset{\sim}{b})}{4|\underset{\sim}{b}|^2}2\underset{\sim}{b}$$

$$= 5\frac{\underset{\sim}{a} \cdot \underset{\sim}{b}}{|\underset{\sim}{b}|^2}\underset{\sim}{b}$$

$$= 5\underset{\sim}{c}$$

Answer B

7

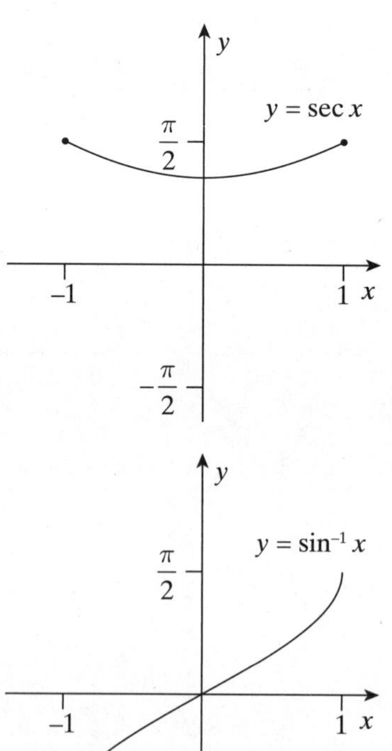

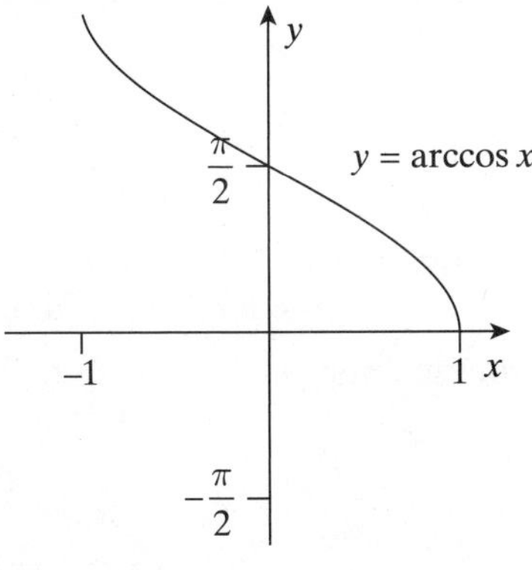

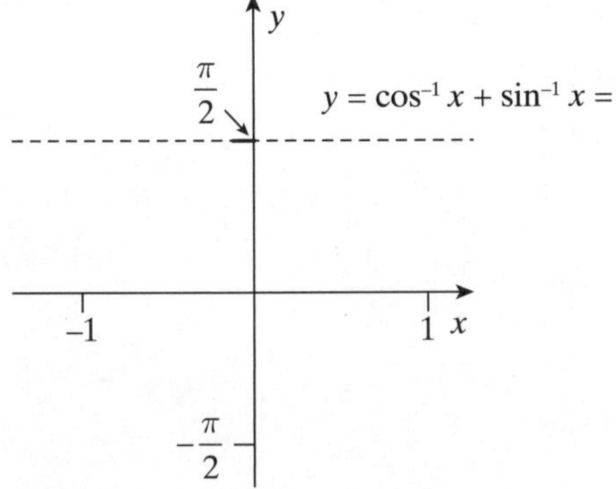

As the function $y = \sin^{-1} x$ is always increasing on the domain $[-1,1]$, $\sin^{-1} a < \sin^{-1} b$ if $-1 \leq a < b \leq 1$.

Answer B

8 The graph is symmetrical over the y-axis. Thus the answer is either Option A or Option B.

Test x-intercepts, $x = \pm1, \pm3$, in the even functions in Option A and Option B:
A: $|1 - ||\pm1| - 2|| = 0, |1 - ||\pm3| - 2|| = 0$
B: $|2 - ||\pm1| - 1|| = 2, |2 - ||\pm3| - 1|| = 0$
Only option A is even and has the correct x-intercepts.

Answer A

9 When $f(x)$ has an x-coordinate of 1, its inverse relation will have a y-coordinate of 1.

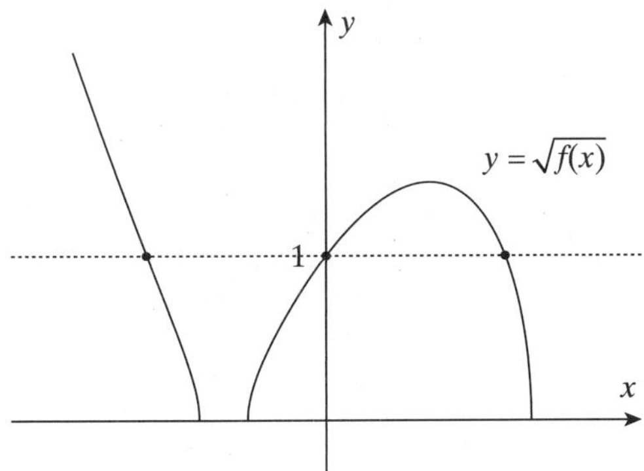

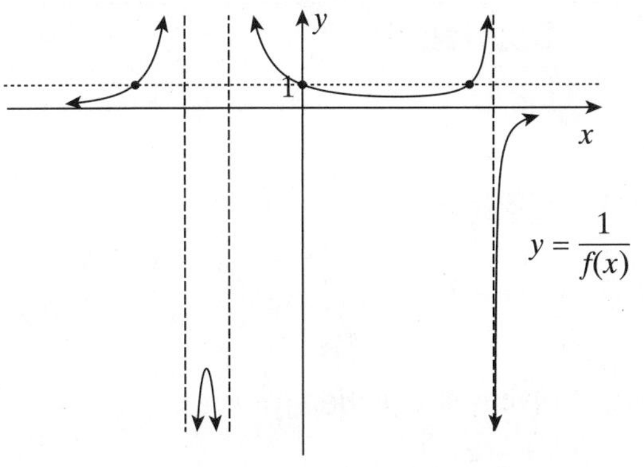

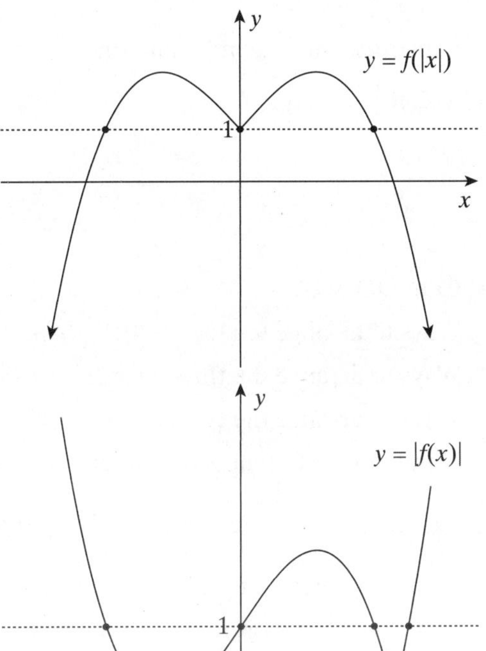

Only $y = |f(x)|$ has more than three points with a y-coordinate of 1 so its inverse is the only one that has more than 3 points with an x-coordinate of 1.

Answer D

10 There are $(5 - 1)! = 4!$ ways to arrange five students in a circle. The only configuration for the teachers such that no more than three students sit together is as follows:

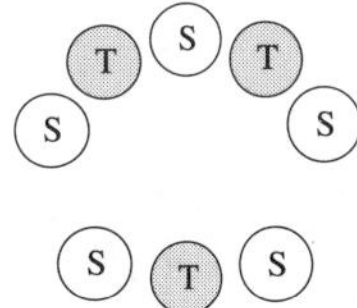

There are five ways to rotate the configuration of teachers among the five students and there are 3! ways to arrange the three teachers into the three places.

Total arrangements $= 4! \times 5 \times 3! = 5! \times 3!$

Answer B

Section II

QUESTION 11

(a) $x = 1 + 3t$

$y = 4t$

Solve first equation for t:

$t = \frac{1}{3}(x-1)$

Substitute into second equation:

$y = 4\left(\frac{1}{3}(x-1)\right)$

$y = \frac{4}{3}x - \frac{4}{3}$

(2 marks)

(b) CONDOBOLIN

Ways to arrange ten letters: 10!

Ways to arrange the three O's: 3!

Ways to arrange the two N's: 2!

Number of different arrangements: $\frac{10!}{3!2!}$

(2 marks)

(c) $P(x) = x^3 + ax^2 + bx - 12$

By the factor theorem if $x + 1$ is a factor of $P(x)$, $P(-1) = 0$.

$(-1)^3 + a(-1)^2 + b(-1) - 12 = 0$

$a - b = 13$ ①

By the remainder theorem if dividing $P(x)$ by $x - 2$ leaves a remainder of -18, $P(2) = -18$.

$(2)^3 + a(2)^2 + b(2) - 12 = -18$

$2a + b = -7$ ②

Solve ① and ② simultaneously:

① + ②: $3a = 6 \rightarrow a = 2$

Substitute $a = 2$ into ①:

$2 - b = 13, b = -11$

(3 marks)

(d) $$\int \frac{1}{\sqrt{4-9x^2}}dx = \frac{1}{3}\int \frac{3}{\sqrt{(2)^2-(3x)^2}}dx$$

$$= \frac{1}{3}\sin^{-1}\frac{3x}{2} + c$$

(2 marks)

(e) $\cos\theta + \sin\theta = 1, 0 \le \theta \le 2\pi$

Using the auxiliary angle method, let $\cos\theta + \sin\theta = R\sin(\theta + \alpha)$ for some $R > 0$ and $0 < \alpha < \frac{\pi}{2}$.

Then $\cos\theta + \sin\theta = R(\sin\theta\cos\alpha + \cos\theta\sin\alpha)$ by the sine angle sum identity.

Equating coefficients:

$R\cos\alpha = 1, \cos\alpha = \frac{1}{R}$

$R\sin\alpha = 1, \sin\alpha = \frac{1}{R}$

$R = \sqrt{1^2 + 1^2} = \sqrt{2}$

$\tan\alpha = 1 \rightarrow \alpha = \frac{\pi}{4}$

Solve for θ:

$\sqrt{2}\sin\left(\theta + \frac{\pi}{4}\right) = 1, \frac{\pi}{4} \le \theta + \frac{\pi}{4} \le \frac{9\pi}{4}$

$\sin\left(\theta + \frac{\pi}{4}\right) = \frac{1}{\sqrt{2}}$

$\theta + \frac{\pi}{4} = \frac{\pi}{4}, \frac{3\pi}{4}, \frac{9\pi}{4}$

$\theta = 0, \frac{\pi}{2}, 2\pi$

(3 marks)

(f) (i) $p = 0.3$

$$\text{Var}(\hat{p}) = \frac{p(1-p)}{n} = \frac{0.3 \times 0.7}{900} = \frac{0.21}{900} = \frac{7}{30\,000}$$

(2 marks)

(ii) Let $Y \sim N(\text{E}(\hat{p}), \text{Var}(\hat{p}))$

$= N\left(0.3, \frac{7}{30\,000}\right)$

Find the z-score for $Y = 0.31$:

$$z = \frac{0.31 - 0.3}{\sqrt{\frac{7}{30\,000}}} \approx 0.65$$

$P(\hat{p} \le 0.31) \approx P(Y \le 0.65) = 0.7422$
$= 74.22\%$ (from table at the end of the HSC)

(2 marks)

QUESTION 12

(a) Let $u = x - 3$.
Then $x = u + 3$
and $du = dx$.

Redefine the limits of integration:
When $x = 4, u = 1$ and when $x = 3, u = 0$.

$$\int_3^4 (x+2)\sqrt{x-3}\,dx = \int_0^1 (u+3+2)\sqrt{u}\,du$$
$$= \int_0^1 (u+5)u^{\frac{1}{2}}\,du$$
$$= \int_0^1 \left(u^{\frac{3}{2}} + 5u^{\frac{1}{2}}\right)du$$
$$= \left[\frac{2}{5}u^{\frac{5}{2}} + 5\times\frac{2}{3}u^{\frac{3}{2}}\right]_0^1$$
$$= \left[\frac{2}{5}(1)^{\frac{5}{2}} + \frac{10}{3}(1)^{\frac{3}{2}}\right] - \left[\frac{2}{5}(0)^{\frac{5}{2}} + \frac{10}{3}(0)^{\frac{3}{2}}\right]$$
$$= \frac{2}{5} + \frac{10}{3} = \frac{56}{15}$$

(3 marks)

(b) Prove $(1 \times 2) + (2 \times 2^2) + (3 \times 2^3) + \ldots + (n \times 2^n) = 2 + (n-1)2^{n+1}$ for $n \geq 1$.
Let $n = 1$.
LHS $= (1 \times 2) = 2$
RHS $= 2 + (1-1)2^{1+1} = 2$
LHS = RHS so it is true for $n = 1$.
Assume it is true for $n = k$. That is,
$(1 \times 2) + (2 \times 2^2) + (3 \times 2^3) + \ldots + (k \times 2^k) = 2 + (k-1)2^{k+1}$
Let $n = k + 1$. Need to show:
$(1 \times 2) + (2 \times 2^2) + (3 \times 2^3) + \ldots + (k \times 2^k) + ((k+1) \times 2^{k+1}) = 2 + (k)2^{k+2}$
By the assumption, LHS $= 2 + (k-1)2^{k+1} + ((k+1) \times 2^{k+1})$
$= 2 + ((k-1) + (k+1))2^{k+1}$
$= 2 + (2k)2^{k+1}$
$= 2 + (k)2^{k+2}$
= RHS
Thus it is true for $n = k + 1$ if it is true for $n = k$.
Therefore, by the principle of mathematical induction, it is true for $n \geq 1$.

(3 marks)

(c) (i) Let random variable X be the number of treadmills in use.
$X \sim \text{Bin}(5, 0.65)$
$P(X = 3) = {}^5C_3(0.65)^3(1-0.65)^{5-3}$
$= {}^5C_3(0.65)^3(0.35)^2$

(2 marks)

(ii) Let random variable Y be the number of rowing machines in use.
$Y \sim \text{Bin}(4, 0.4)$
$P(Y = 0) = {}^4C_0(0.4)^0(1-0.4)^{4-0}$
$= (0.6)^4$

$P(X = 3 \text{ and } Y = 0) = {}^5C_3(0.65)^3(0.35)^2(0.6)^4$

(1 mark)

(d) It is given that ${}^{n}C_r = {}^{n-1}C_{r-1} + {}^{n-1}C_r$

Method 1:

${}^{2022}C_{80} + {}^{2022}C_{81} + {}^{2023}C_{1943}$

$= {}^{2023}C_{81} + {}^{2023}C_{1943}$ by the given identity

$= {}^{2023}C_{81} + {}^{2023}C_{80}$ by the symmetry of Pascal's triangle

$= {}^{2024}C_{81}$ by the given identity

Thus $p = 2024$ and $q = 81$.

Method 2:

${}^{2022}C_{80} + {}^{2022}C_{81} + {}^{2023}C_{1943}$

$= {}^{2023}C_{81} + {}^{2023}C_{1943}$ by the given identity

$= {}^{2023}C_{1942} + {}^{2023}C_{1943}$ by the symmetry of Pascal's triangle

$= {}^{2024}C_{1943}$ by the given identity

Thus $p = 2024$ and $q = 1943$.

(2 marks)

(e)

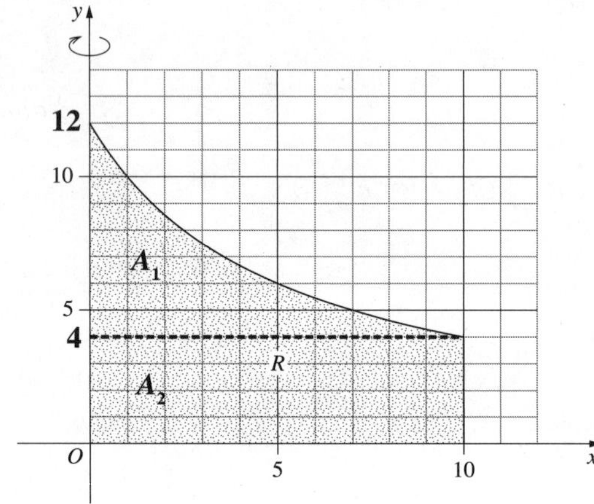

As we must integrate along the y-axis, rearrange the function to make x the subject.

$y = \dfrac{60}{x+5} \rightarrow x = \dfrac{60}{y} - 5$

Volume of A_1 rotated about the y-axis:

$$V_1 = \int_4^{12} \pi \left(\frac{60}{y} - 5\right)^2 dy$$
$$= \pi \int_4^{12} \left(\frac{60^2}{y^2} - \frac{600}{y} + 25\right) dy$$
$$= \pi \left[-\frac{60^2}{y} - 600 \ln y + 25y\right]_4^{12}$$
$$= \pi[(-300 - 600 \ln 12 + 300) - (-900 - 600 \ln 4 + 100)]$$
$$= \pi\left(800 + 600 \ln \frac{1}{3}\right)$$
$$= \pi(800 - 600 \ln 3)$$

Volume of A_2 rotated about the y-axis is a cylinder: $V_2 = \pi(10)^2(4) = 400\pi$

Total volume: $V = V_1 + V_2 = \pi(800 - 600 \ln 3) + 400\pi = 600\pi\,(2 - \ln 3)$

(4 marks)

QUESTION 13

(a) (i) It is given that $V = \pi\left(Rh^2 - \frac{h^3}{3}\right)$ and $\frac{dV}{dt} = k(2R-h)$.

Differentiate V with respect to t, applying the chain rule as h is a function of t:

$$\frac{dV}{dt} = \pi\left(2Rh\frac{dh}{dt} - \frac{3h^2}{3}\frac{dh}{dt}\right)$$
$$= \frac{dh}{dt}\pi h(2R-h)$$
$$\frac{dh}{dt} = \frac{\frac{dV}{dt}}{\pi h(2R-h)}$$

Substitute the value of $\frac{dV}{dt}$ given:

$$\frac{dh}{dt} = \frac{k(2R-h)}{\pi h(2R-h)}$$
$$= \frac{k}{\pi h}$$

(2 marks)

(ii) Find h as a function of t by solving the differential equation found in part (i):

$$\frac{dh}{dt} = \frac{k}{\pi h}$$
$$h\ dh = \frac{k}{\pi}dt$$
$$\int h\ dh = \int \frac{k}{\pi}dt$$
$$\frac{1}{2}h^2 = \frac{k}{\pi}t + c_1$$

Solve for c_1 given $h = 0$ (tank is empty) when $t = 0$:

$$\frac{1}{2}(0)^2 = \frac{k}{\pi}(0) + c_1 \rightarrow c_1 = 0$$
$$\therefore \frac{1}{2}h^2 = \frac{k}{\pi}t$$

The tank is full when $h = R$ (radius of the tank).

$$\frac{1}{2}R^2 = \frac{k}{\pi}T$$
$$T = \frac{\pi R^2}{2k}$$

(2 marks)

(iii) Consider the rate of change of the volume:

$$\frac{dV}{dt} = k(2R - h) = 2kR - kh$$

It was given that water is flowing into the tank at a rate of $2kR$. Thus the water is flowing out at a rate of kh.

As the water stops flowing in when the tank is full, we now have $\frac{dV}{dt} = -kh$.

By part (i), $\frac{dh}{dt} = \frac{\frac{dV}{dt}}{\pi h(2R-h)}$.

Substituting the new value of $\frac{dV}{dt}$:

$$\frac{dh}{dt} = \frac{\frac{dV}{dt}}{\pi h(2R-h)}$$
$$= \frac{-kh}{\pi h(2R-h)}$$
$$= \frac{-k}{\pi(2R-h)}$$

Find h as a function of t by solving this differential equation:

$$\frac{dh}{dt} = \frac{-k}{\pi(2R-h)}$$
$$(2R-h)\ dh = -\frac{k}{\pi}dt$$
$$\int(2R-h)\ dh = \int -\frac{k}{\pi}dt$$
$$2Rh - \frac{1}{2}h^2 = -\frac{k}{\pi}t + c_2$$

Solve for c_2 given $h = R$ (tank is full) when $t = 0$:

$$2R^2 - \frac{1}{2}R^2 = -\frac{k}{\pi}(0) + c_2 \rightarrow c_2 = \frac{3}{2}R^2$$
$$\therefore 2Rh - \frac{1}{2}h^2 = -\frac{k}{\pi}t + \frac{3}{2}R^2$$

The tank is empty when $h = 0$.

$$2R(0) - \frac{1}{2}(0)^2 = -\frac{k}{\pi}t + \frac{3}{2}R^2$$
$$0 = -\frac{k}{\pi}t + \frac{3}{2}R^2$$
$$t = \frac{3\pi R^2}{2k} = 3 \times \frac{\pi R^2}{2k} = 3T$$

Therefore the tank takes three times as long to empty as it does to fill.

(3 marks)

(b) (i) It is given that $\tan\theta = 2 = \frac{2}{1}$

hypotenuse: $\sqrt{1^2 + 2^2} = \sqrt{5}$

$$\therefore \cos\theta = \frac{1}{\sqrt{5}}$$

Since the particles collide,

$$vt\cos\theta = ut$$
$$v \times \frac{1}{\sqrt{5}} = u$$
$$v = \sqrt{5}u$$

(2 marks)

(ii) Particles collide when

$$vT\sin\theta - \frac{1}{2}gT^2 = H - \frac{1}{2}gT^2$$

$vT \sin \theta = H$

By part (i), substitute $\sin \theta = \frac{2}{\sqrt{5}}$ and $v = \sqrt{5}u$.

$$\sqrt{5}uT \times \frac{2}{\sqrt{5}} = H$$

$$T = \frac{H}{2u}$$

(1 mark)

(iii) When two vectors are perpendicular, their dot product equals zero. Thus when particles collide at $t = T$, $r'_A \cdot r'_B = 0$.

$$r'_A = \begin{pmatrix} v\cos\theta \\ v\sin\theta - gt \end{pmatrix} \text{ and } r'_B = \begin{pmatrix} u \\ -gt \end{pmatrix}$$

$$r'_A \cdot r'_B = uv \cos \theta + (v \sin \theta - gT)(-gT) = 0$$

$$uv \cos \theta - vgT \sin \theta + g^2T^2 = 0$$

By part (i), substitute $\cos \theta = \frac{1}{\sqrt{5}}$, $\sin \theta = \frac{2}{\sqrt{5}}$ and $v = \sqrt{5}u$.

$$u\left(\sqrt{5}u\right) \times \frac{1}{\sqrt{5}} - \sqrt{5}ugT \times \frac{2}{\sqrt{5}} + g^2T^2 = 0$$

$$u^2 - 2ugT + g^2T^2 = 0$$

By part (ii), $T = \frac{H}{2u}$.

$$u^2 - 2ug\left(\frac{H}{2u}\right) + g^2\left(\frac{H}{2u}\right)^2 = 0$$

$$\frac{g^2}{4u^2}H^2 - gH + u^2 = 0$$

This equation is quadratic in terms of H. Solve using the quadratic formula:

$$H = \frac{g \pm \sqrt{g^2 - 4\frac{g^2}{4u^2} \times u^2}}{2 \times \frac{g^2}{4u^2}} = \frac{g}{\frac{g^2}{2u^2}} = \frac{2u^2}{g}$$

(3 marks)

(iv) Maximum height occurs when vertical component of velocity is zero. By part (iii), vertical component of velocity is $v \sin \theta - gt$.

$$v\sin\theta - gt = 0 \rightarrow t = \frac{v\sin\theta}{g}$$

Determine vertical component of position at this time:

$$y_{\max} = vt\sin\theta - \frac{1}{2}gt^2 \text{ when } t = \frac{v\sin\theta}{g}:$$

$$y_{\max} = v\left(\frac{v\sin\theta}{g}\right)\sin\theta - \frac{1}{2}g\left(\frac{v\sin\theta}{g}\right)^2 = \frac{v^2\sin^2\theta}{g} - \frac{v^2\sin^2\theta}{2g} = \frac{v^2}{2g}\sin^2\theta$$

By part (i), substitute $\theta = \frac{2}{\sqrt{5}}$ and $v = \sqrt{5}u$.

$$y_{\max} = \frac{\left(\sqrt{5}u\right)^2}{2g} \times \left(\frac{2}{\sqrt{5}}\right)^2 = \frac{2u^2}{g} = H \text{ by part (iii).}$$

Thus the height of the vertex of the parabola is H.

(2 marks)

QUESTION 14

(a) (i) $f(x) = 2x + \ln x, x > 0$
The inverse of $f(x)$ is a function if and only if $f(x)$ is one-to-one.
Consider $f'(x) = 2 + \frac{1}{x}$.
On the domain $x > 0, f'(x) > 0$, thus $f(x)$ is always increasing.
$\therefore f(x)$ is one-to-one and its inverse is a function.

(1 mark)

(ii) $g(x) = f^{-1}(x)$
Since $f(1) = 2(1) + \ln(1) = 2, g(2) = 1$.

Thus $g'(2) = \dfrac{1}{f'(1)} = \dfrac{1}{2 + \frac{1}{1}} = \dfrac{1}{3}$

(2 marks)

(b) (i) $y = \frac{1}{x}$ and $(x - c)^2 + y^2 = c^2$
Solve simultaneously. Substitute equation of hyperbola into equation of circle:

$$(x-c)^2 + \left(\frac{1}{x}\right)^2 = c^2$$

$$x^2 - 2cx + c^2 + \frac{1}{x^2} = c^2$$

$$x^2 - 2cx + \frac{1}{x^2} = 0$$

Multiply both sides by x^2:
$x^4 - 2cx^3 + 1 = 0$
Thus the x-coordinates of the points of intersection are the zeros of the polynomial
$P(x) = x^4 - 2cx^3 + 1$.

(1 mark)

(ii) For the graphs to have only one intersection, the polynomial must have only one zero. This will occur at the turning point seen on the graph for a value of c, $0.8 < c < 1$, such that the turning point is tangent to the x-axis (a double root).
Find the x-coordinate of the turning point in terms of c:

$$P'(x) = 4x^3 - 6cx^2$$
$$4x^3 - 6cx^2 = 0$$
$$2x^2(2x - 3c) = 0$$

Stationary points occur when $x = 0$ or $x = \frac{3c}{2}$
$x = 0$ is the horizontal point of inflection seen on the graph.

$x = \frac{3c}{2}$ is the turning point.
If $x = \frac{3c}{2}$ is a zero of $P(x)$,

$$\left(\frac{3c}{2}\right)^4 - 2c\left(\frac{3c}{2}\right)^3 + 1 = 0$$

$$\frac{81c^4}{16} - \frac{27c^4}{4} + 1 = 0$$

$$\frac{27}{16}c^4 = 1$$

$$c = \sqrt[4]{\frac{16}{27}}$$

$$= \frac{2}{\sqrt[4]{27}} \quad (c > 0)$$

(3 marks)

(c) (i) Consider any two vectors $\overrightarrow{OA} = \begin{pmatrix} a_1 \\ a_2 \end{pmatrix}$ and $\overrightarrow{OB} = \begin{pmatrix} b_1 \\ b_2 \end{pmatrix}$ and the triangle formed.

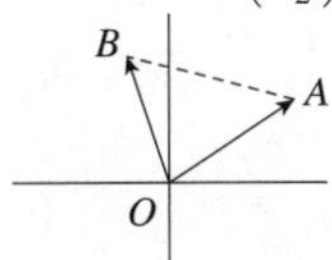

Let $\overrightarrow{OA}$ be the base of the triangle. Let $\overrightarrow{OA'} = \begin{pmatrix} a_2 \\ -a_1 \end{pmatrix}$, which is perpendicular and equal in length to $\overrightarrow{OA}$, as given.

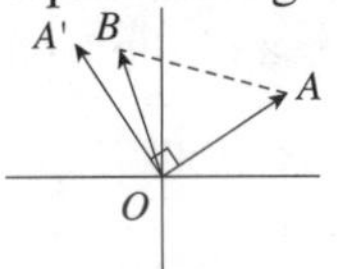

To scale $\overrightarrow{OA'}$ to the height of the triangle, consider the projection of $\overrightarrow{OB}$ onto $\overrightarrow{OA'}$.

$$\text{proj}_{\overrightarrow{OA'}}\overrightarrow{OB} = \frac{\overrightarrow{OB} \cdot \overrightarrow{OA'}}{\left|\overrightarrow{OA'}\right|^2}\overrightarrow{OA'}$$

$$= \frac{b_1a_2 + b_2(-a_1)}{a_1^2 + a_2^2}\overrightarrow{OA'}$$

$$= \frac{b_1a_2 - b_2a_1}{a_1^2 + a_2^2}\overrightarrow{OA'}$$

The area of the triangle is $A = \frac{1}{2}bh$:

$$A = \frac{1}{2}\left|\overrightarrow{OA}\right|\left|\frac{b_1a_2 - b_2a_1}{a_1^2 + a_2^2}\overrightarrow{OA'}\right|$$

$$= \frac{1}{2}\left|\overrightarrow{OA}\right|\left|\frac{b_1a_2 - b_2a_1}{a_1^2 + a_2^2}\right|\left|\overrightarrow{OA'}\right|$$

$$= \frac{1}{2}\sqrt{a_1^2 + a_2^2} \times \frac{|-(a_1b_2 - a_2b_1)|}{a_1^2 + a_2^2}\sqrt{a_1^2 + a_2^2}$$

$$= \frac{1}{2}|a_1b_2 - a_2b_1|$$

(3 marks)

(ii)

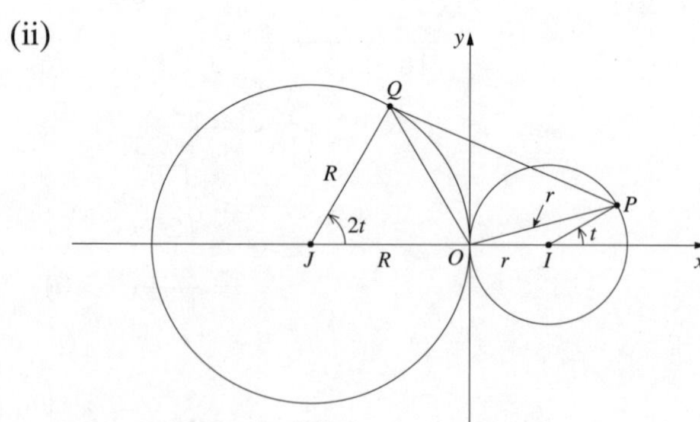

As $\overrightarrow{IP}$ and $\overrightarrow{JQ}$ are radii of the circles centred at I and J, their lengths are r and R respectively.

Thus $\overrightarrow{IP} = r\begin{pmatrix}\cos t\\ \sin t\end{pmatrix}$ and $\overrightarrow{JQ} = R\begin{pmatrix}\cos 2t\\ \sin 2t\end{pmatrix}$

Also $\overrightarrow{OI} = \begin{pmatrix}r\\ 0\end{pmatrix}$ and $\overrightarrow{OJ} = \begin{pmatrix}-R\\ 0\end{pmatrix}$

As given, $\overrightarrow{OP} = \overrightarrow{OI} + \overrightarrow{IP} = \begin{pmatrix}r\\ 0\end{pmatrix} + r\begin{pmatrix}\cos t\\ \sin t\end{pmatrix} = r\begin{pmatrix}\cos t + 1\\ \sin t\end{pmatrix}$

and $\overrightarrow{OQ} = \overrightarrow{OJ} + \overrightarrow{JQ} = \begin{pmatrix}-R\\ 0\end{pmatrix} + R\begin{pmatrix}\cos 2t\\ \sin 2t\end{pmatrix} = R\begin{pmatrix}\cos 2t - 1\\ \sin 2t\end{pmatrix}$

By part (i), the area of the triangle OPQ is

$$A = \frac{1}{2}|r(\cos t + 1) \times R\sin 2t - r\sin t \times R(\cos 2t - 1)|$$

$$= \frac{Rr}{2}|\sin 2t\cos t + \sin 2t - \cos 2t\sin t + \sin t| \quad (R > 0, r > 0)$$

$$= \frac{Rr}{2}|\sin 2t\cos t - \cos 2t\sin t + \sin 2t + \sin t|$$

$$= \frac{Rr}{2}|\sin(2t - t) + \sin 2t + \sin t| \quad \text{(sine angle sum identity)}$$

$$= \frac{Rr}{2}|2\sin t + \sin 2t|$$

The area is maximised when $|2\sin t + \sin 2t|$ is maximised. Find the derivative of the inner function and find stationary points:

$$\frac{d}{dt}(2\sin t + \sin 2t) = 2\cos t + 2\cos 2t = 0$$

$$\cos t + \cos 2t = 0$$

$$\cos t + 2\cos^2 t - 1 = 0 \quad \text{(cosine double angle identity)}$$

$$2\cos^2 t + \cos t - 1 = 0$$

$$(2\cos t - 1)(\cos t + 1) = 0$$

$$\cos t = \frac{1}{2} \text{ or } \cos t = -1$$

$$t = \pm\frac{\pi}{3}, \pm\pi \qquad (-\pi \le t \le \pi)$$

The endpoints are stationary points, and $2\sin t + \sin 2t = 0$ when $t = \pm\pi$.

When $t = \pm\frac{\pi}{3}$, $2\sin t + \sin 2t$ has maximum and minimum values which are equal in magnitude since sine is odd. Both become equal maximums when absolute value is taken.

Thus $A = \frac{Rr}{2}|2\sin t + \sin 2t|$ is maximised when $t = \pm\frac{\pi}{3}$. *(4 marks)*

NSW Education Standards Authority

Centre Number

Student Number

2024 **HIGHER SCHOOL CERTIFICATE EXAMINATION**

Mathematics Extension 1

General Instructions

- Reading time – 10 minutes
- Working time – 2 hours
- Write using black pen
- Calculators approved by NESA may be used
- A reference sheet is provided at the back of this paper
- For questions in Section II, show relevant mathematical reasoning and/or calculations
- Write your Centre Number and Student Number at the top of this page and on the Question 13 Writing Booklet attached

Total marks: 70

Section I – 10 marks

- Attempt Questions 1–10
- Allow about 15 minutes for this section

Section II – 60 marks

- Attempt Questions 11–14
- Allow about 1 hour and 45 minutes for this section

Section I

10 marks
Attempt Questions 1–10
Allow about 15 minutes for this section

Use the multiple-choice answer sheet for Questions 1–10.

---

1 The polynomial $x^3 + 2x^2 - 5x - 6$ has zeros -1, -3 and α.

What is the value of α?

A. -2

B. 2

C. 3

D. 6

2 Consider the functions $y = f(x)$ and $y = g(x)$, and the regions shaded in the diagram below.

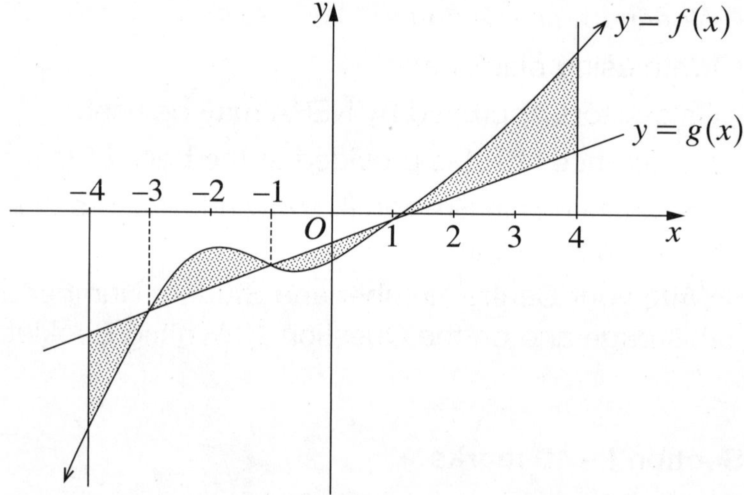

Which of the following gives the total area of the shaded regions?

A. $\int_{-4}^{4} f(x) - g(x)\,dx$

B. $\left|\int_{-4}^{4} f(x) - g(x)\,dx\right|$

C. $\int_{-4}^{-3} f(x) - g(x)\,dx + \int_{-3}^{-1} f(x) - g(x)\,dx + \int_{-1}^{1} f(x) - g(x)\,dx + \int_{1}^{4} f(x) - g(x)\,dx$

D. $-\int_{-4}^{-3} f(x) - g(x)\,dx + \int_{-3}^{-1} f(x) - g(x)\,dx - \int_{-1}^{1} f(x) - g(x)\,dx + \int_{1}^{4} f(x) - g(x)\,dx$

3 Students from 4 different schools come together to form a choir.

What is the minimum size of the choir to know that there must be at least 20 students in the choir from one of the schools?

A. 76

B. 77

C. 80

D. 81

4 What are the domain and range of the function $y = 2\cos^{-1}(2x) + 2\sin^{-1}(2x)$?

A. Domain: $[-0.5, 0.5]$ and Range: $\{\pi\}$

B. Domain: $[-0.5, 0.5]$ and Range: $[-\pi, 3\pi]$

C. Domain: $[-2, 2]$ and Range: $\{\pi\}$

D. Domain: $[-2, 2]$ and Range: $[-\pi, 3\pi]$

5 Consider the function $g(x) = 2\sin^{-1}(3x)$.

Which transformations have been applied to $f(x) = \sin^{-1}(x)$ to obtain $g(x)$?

A. Vertical dilation by a factor of $\frac{1}{2}$ and a horizontal dilation by a factor of $\frac{1}{3}$

B. Vertical dilation by a factor of $\frac{1}{2}$ and a horizontal dilation by a factor of 3

C. Vertical dilation by a factor of 2 and a horizontal dilation by a factor of $\frac{1}{3}$

D. Vertical dilation by a factor of 2 and a horizontal dilation by a factor of 3

6 How many real value(s) of x satisfy the equation

$$|b| = |b\sin(4x)|,$$

where $x \in [0, 2\pi]$ and b is not zero?

A. 1

B. 2

C. 4

D. 8

7 A driver's knowledge test contains 30 multiple-choice questions, each with 4 options. An applicant must get at least 29 correct to pass.

If an applicant correctly answers the first 25 questions and randomly guesses the last 5 questions, what is the probability that the applicant will pass the test?

A. $\frac{1}{256}$

B. $\frac{15}{1024}$

C. $\frac{1}{64}$

D. $\frac{21}{256}$

8 A local council is proposing to ban dog-walking on the beach. It is known that the proportion of households that have a dog is $\frac{7}{12}$.

The local council wishes to poll n households about this proposal.

Let $\hat{p}$ be the random variable representing the proportion of households polled that have a dog.

What is the smallest sample size, n, for which the standard deviation of $\hat{p}$ is less than 0.06?

A. 67

B. 68

C. 94

D. 95

9 A bag contains n metal coins, $n \geq 3$, that are made from either silver or bronze.

There are k silver coins in the bag and the rest are bronze.

Two coins are to be drawn at random from the bag, with the first coin drawn not being replaced before the second coin is drawn.

Which of the following expressions will give the probability that the two coins drawn are made of the same metal?

A. $\dfrac{k(k-1)+(n-k)(n-k-1)}{n(n-1)}$

B. $\binom{n}{2}\binom{n}{k}\left(1-\dfrac{k}{n}\right)^{n-2}$

C. $\dfrac{\binom{k}{2}+\binom{n-k}{2}}{n(n-1)}$

D. $\dfrac{k^2+(n-k)^2}{n^2}$

10 For real numbers a and b, where $a \neq 0$ and $b \neq 0$, we can find numbers $\alpha, \beta, \gamma, \delta$ and R such that $a\cos x + b\sin x$ can be written in the following 4 forms:

$$R\sin(x+\alpha)$$
$$R\sin(x-\beta)$$
$$R\cos(x+\gamma)$$
$$R\cos(x-\delta)$$

where $R > 0$ and $0 < \alpha, \beta, \gamma, \delta < 2\pi$.

What is the value of $\alpha + \beta + \gamma + \delta$?

A. 0

B. π

C. 2π

D. 4π

Section II

60 marks
Attempt Questions 11–14
Allow about 1 hour and 45 minutes for this section

Answer each question in the appropriate writing booklet. Extra writing booklets are available.

For questions in Section II, your responses should include relevant mathematical reasoning and/or calculations.

Question 11 (15 marks) Use the Question 11 Writing Booklet

(a) Consider the vectors $\underset{\sim}{a} = 3\underset{\sim}{i} + 2\underset{\sim}{j}$ and $\underset{\sim}{b} = -\underset{\sim}{i} + 4\underset{\sim}{j}$.

(i) Find $2\underset{\sim}{a} - \underset{\sim}{b}$. **1**

(ii) Find $\underset{\sim}{a} \cdot \underset{\sim}{b}$. **1**

(b) Solve $x^2 - 8x - 9 \leq 0$. **2**

(c) Using the substitution $u = x - 1$, find $\int x\sqrt{x-1}\,dx$. **3**

(d) Solve the differential equation $\dfrac{dy}{dx} = xy$, given $y > 0$. Express your answer in the form $y = e^{f(x)}$. **2**

(e) Differentiate the function $f(x) = \arcsin\left(x^5\right)$. **1**

Question 11 continues on the following page

Question 11 (continued)

(f) The volume of a sphere of radius r cm, is given by $V = \frac{4}{3}\pi r^3$, and the volume of the sphere is increasing at a rate of $10 \text{ cm}^3 \text{ s}^{-1}$. **2**

Show that the rate of increase of the radius is given by $\frac{dr}{dt} = \frac{5}{2\pi r^2} \text{ cm s}^{-1}$.

(g) The region, R, is bounded by the curves $y = \sin x$, $y = x$ and the line $x = \frac{\pi}{2}$ as shown in the diagram. **3**

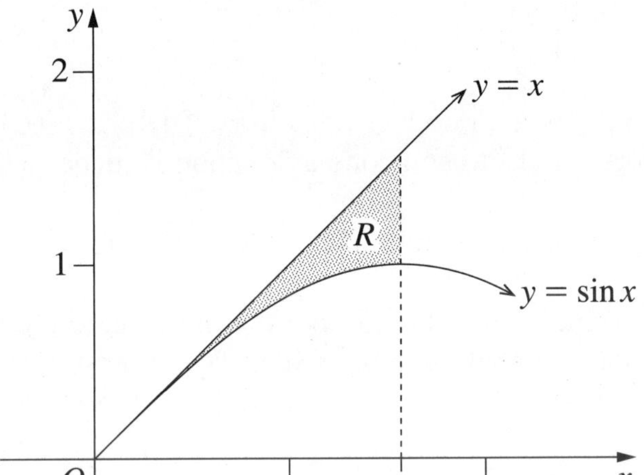

Find the area of the region R.

End of Question 11

Question 12 (15 marks) Use the Question 12 Writing Booklet

(a) The vectors $\begin{pmatrix} a^2 \\ 2 \end{pmatrix}$ and $\begin{pmatrix} a+5 \\ a-4 \end{pmatrix}$ are perpendicular. **3**

Find the possible values of a.

(b) The region, R, is bounded by the function, $y = x^3$, the x-axis and the lines $x = 1$ and $x = 2$. **3**

What is the volume of the solid of revolution obtained when the region R is rotated about the x-axis?

(c) A charity employs a worker to collect donations. There is a 0.31 chance that when the charity worker talks to someone a donation is made to the charity. **3**

Each day the charity worker must talk to exactly 100 people.

Use the standard normal distribution and the information on page 13 to approximate the probability that, on a particular day, at least 35% of the people talked to made a donation.

(d) Use mathematical induction to prove that $2^{3n} + 13$ is divisible by 7 for all integers $n \geq 1$. **3**

Question 12 continues on the following page

Question 12 (continued)

(e) The diagram shows the graph of $y = \dfrac{1}{|x-5|}$. **3**

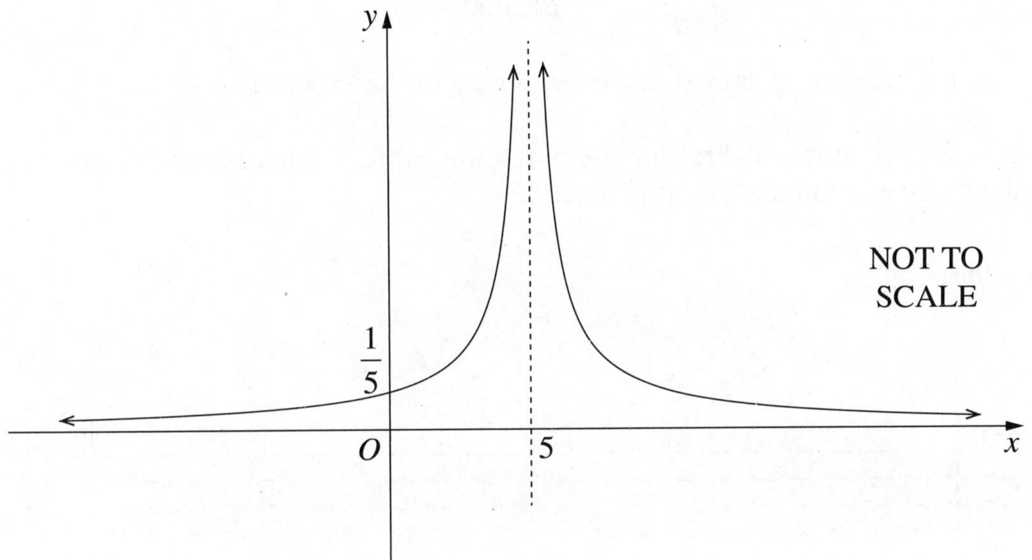

For what values of x is $\dfrac{x}{6} \geq \dfrac{1}{|x-5|}$?

End of Question 12

Question 13 (16 marks) Use the Question 13 Writing Booklet

(a) In an experiment, the population of insects, $P(t)$, was modelled by the logistic differential equation

$$\frac{dP}{dt} = P(2000 - P)$$

where t is the time in days after the beginning of the experiment.

The diagram shows a direction field for this differential equation, with the point S representing the initial population.

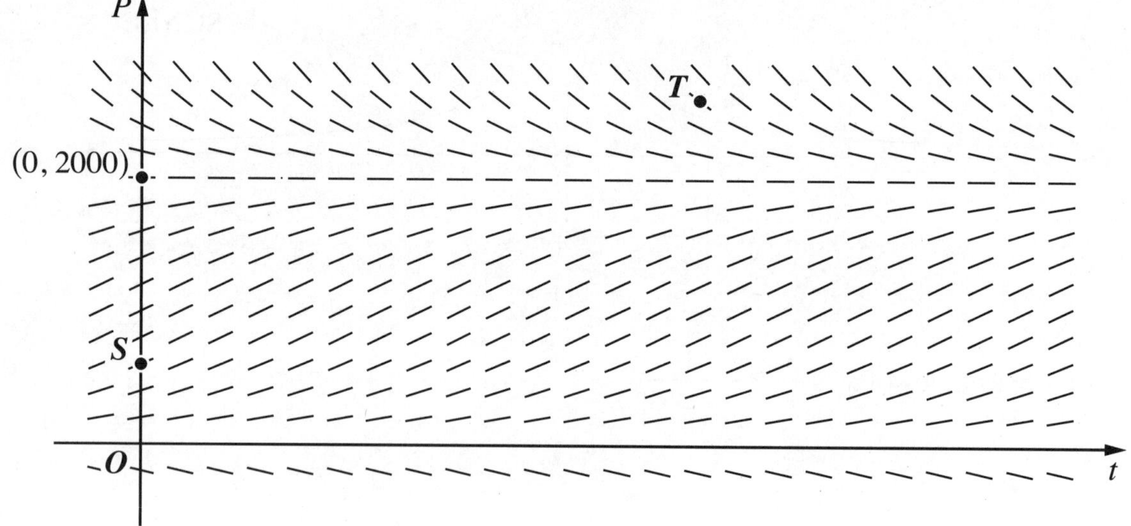

(i) Explain why the graph of the solution that passes through the point S cannot also pass through the point T. **1**

(ii) On the diagram provided on page 1 of the Question 13 Writing Booklet, clearly sketch the graph of the solution that passes through the point S. **1**

(iii) Find the predicted value of the population, $P(t)$, at which the rate of growth of the population is largest. **2**

Question 13 continues on the following page

Question 13 (continued)

(b) (i) Show that $\cos^4 x + \sin^4 x = \dfrac{1 + \cos^2 2x}{2}$. **2**

(ii) Hence, or otherwise, evaluate $\displaystyle\int_0^{\frac{\pi}{4}} \left(\cos^4 x + \sin^4 x\right) dx$. **3**

(c) The vector $\underset{\sim}{a}$ is $\begin{pmatrix} 1 \\ 3 \end{pmatrix}$ and the vector $\underset{\sim}{b}$ is $\begin{pmatrix} 2 \\ -1 \end{pmatrix}$. **4**

The projection of a vector $\underset{\sim}{x}$ onto the vector $\underset{\sim}{a}$ is $k\underset{\sim}{a}$, where k is a real number.

The projection of the vector $\underset{\sim}{x}$ onto the vector $\underset{\sim}{b}$ is $p\underset{\sim}{b}$, where p is a real number.

Find the vector $\underset{\sim}{x}$ in terms of k and p.

(d) Using the substitution $u = e^x + 2e^{-x}$, and considering u^2, find **3**

$$\int \frac{e^{3x} - 2e^x}{4 + 8e^{2x} + e^{4x}}\, dx.$$

End of Question 13

Question 14 (14 marks) Use the Question 14 Writing Booklet

(a) Find the domain and range of the function that is the solution to the differential equation **4**

$$\frac{dy}{dx} = e^{x+y}$$

and whose graph passes through the origin.

(b) For what values of the constant k would the function $f(x) = \dfrac{kx}{1+x^2} + \arctan x$ have an inverse? **3**

(c) (i) Explain why the equation $\tan^{-1}(3x) + \tan^{-1}(10x) = \theta$, where $-\pi < \theta < \pi$, has exactly one solution. **1**

(ii) Solve $\tan^{-1}(3x) + \tan^{-1}(10x) = \dfrac{3\pi}{4}$. **2**

(d) A particle is projected from the origin, with initial speed V at an angle of θ to the horizontal. The position vector of the particle, $\underset{\sim}{r}(t)$, where t is the time after projection and g is the acceleration due to gravity, is given by **4**

$$\underset{\sim}{r}(t) = \begin{pmatrix} Vt\cos\theta \\ Vt\sin\theta - \dfrac{gt^2}{2} \end{pmatrix}. \qquad \text{(Do NOT prove this.)}$$

Let $D(t)$ be the distance of the particle from the origin at time t, so $D(t) = |\underset{\sim}{r}(t)|$.

Show that for $\theta < \sin^{-1}\left(\sqrt{\dfrac{8}{9}}\right)$ the distance, $D(t)$, is increasing for all $t > 0$.

End of paper

Use the information below to answer Question 12 (c).

Table of values $P(Z \le z)$ for the normal distribution $N(0,1)$

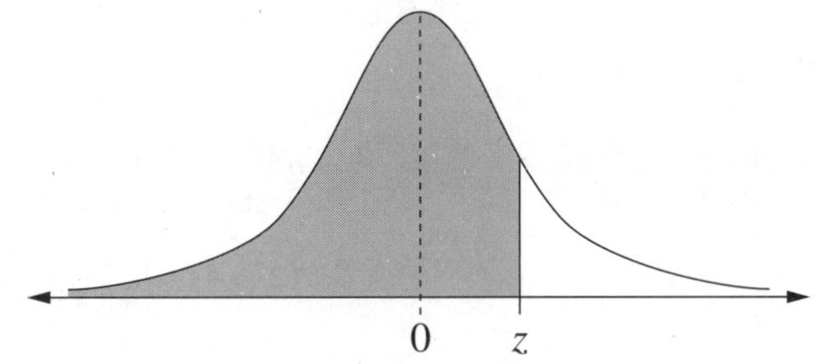

Z	0.00	0.01	0.02	0.03	0.04	0.05	0.06	0.07	0.08	0.09
0.0	0.5000	0.5040	0.5080	0.5120	0.5160	0.5199	0.5239	0.5279	0.5319	0.5359
0.1	0.5398	0.5438	0.5478	0.5517	0.5557	0.5596	0.5636	0.5675	0.5714	0.5753
0.2	0.5793	0.5832	0.5871	0.5910	0.5948	0.5987	0.6026	0.6064	0.6103	0.6141
0.3	0.6179	0.6217	0.6255	0.6293	0.6331	0.6368	0.6406	0.6443	0.6480	0.6517
0.4	0.6554	0.6591	0.6628	0.6664	0.6700	0.6736	0.6772	0.6808	0.6844	0.6879
0.5	0.6915	0.6950	0.6985	0.7019	0.7054	0.7088	0.7123	0.7157	0.7190	0.7224
0.6	0.7257	0.7291	0.7324	0.7357	0.7389	0.7422	0.7454	0.7486	0.7517	0.7549
0.7	0.7580	0.7611	0.7642	0.7673	0.7704	0.7734	0.7764	0.7794	0.7823	0.7852
0.8	0.7881	0.7910	0.7939	0.7967	0.7995	0.8023	0.8051	0.8078	0.8106	0.8133
0.9	0.8159	0.8186	0.8212	0.8238	0.8264	0.8289	0.8315	0.8340	0.8365	0.8389
1.0	0.8413	0.8438	0.8461	0.8485	0.8508	0.8531	0.8554	0.8577	0.8599	0.8621
1.1	0.8643	0.8665	0.8686	0.8708	0.8729	0.8749	0.8770	0.8790	0.8810	0.8830
1.2	0.8849	0.8869	0.8888	0.8907	0.8925	0.8944	0.8962	0.8980	0.8997	0.9015
1.3	0.9032	0.9049	0.9066	0.9082	0.9099	0.9115	0.9131	0.9147	0.9162	0.9177
1.4	0.9192	0.9207	0.9222	0.9236	0.9251	0.9265	0.9279	0.9292	0.9306	0.9319
1.5	0.9332	0.9345	0.9357	0.9370	0.9382	0.9394	0.9406	0.9418	0.9429	0.9441
1.6	0.9452	0.9463	0.9474	0.9484	0.9495	0.9505	0.9515	0.9525	0.9535	0.9545
1.7	0.9554	0.9564	0.9573	0.9582	0.9591	0.9599	0.9608	0.9616	0.9625	0.9633
1.8	0.9641	0.9649	0.9656	0.9664	0.9671	0.9678	0.9686	0.9693	0.9699	0.9706
1.9	0.9713	0.9719	0.9726	0.9732	0.9738	0.9744	0.9750	0.9756	0.9761	0.9767
2.0	0.9772	0.9778	0.9783	0.9788	0.9793	0.9798	0.9803	0.9808	0.9812	0.9817
2.1	0.9821	0.9826	0.9830	0.9834	0.9838	0.9842	0.9846	0.9850	0.9854	0.9857
2.2	0.9861	0.9864	0.9868	0.9871	0.9875	0.9878	0.9881	0.9884	0.9887	0.9890
2.3	0.9893	0.9896	0.9898	0.9901	0.9904	0.9906	0.9909	0.9911	0.9913	0.9916
2.4	0.9918	0.9920	0.9922	0.9925	0.9927	0.9929	0.9931	0.9932	0.9934	0.9936
2.5	0.9938	0.9940	0.9941	0.9943	0.9945	0.9946	0.9948	0.9949	0.9951	0.9952
2.6	0.9953	0.9955	0.9956	0.9957	0.9959	0.9960	0.9961	0.9962	0.9963	0.9964
2.7	0.9965	0.9966	0.9967	0.9968	0.9969	0.9970	0.9971	0.9972	0.9973	0.9974
2.8	0.9974	0.9975	0.9976	0.9977	0.9977	0.9978	0.9979	0.9979	0.9980	0.9981
2.9	0.9981	0.9982	0.9982	0.9983	0.9984	0.9984	0.9985	0.9985	0.9986	0.9986
3.0	0.9987	0.9987	0.9987	0.9988	0.9988	0.9989	0.9989	0.9989	0.9990	0.9990
3.1	0.9990	0.9991	0.9991	0.9991	0.9992	0.9992	0.9992	0.9992	0.9993	0.9993
3.2	0.9993	0.9993	0.9994	0.9994	0.9994	0.9994	0.9994	0.9995	0.9995	0.9995

2024 Higher School Certificate
Worked answers

Section I
(Total 10 marks)

1 B	**2** D	**3** B	**4** A	**5** C
6 D	**7** C	**8** B	**9** A	**10** D

1 $\alpha\beta\gamma = -\frac{d}{a}$

$\alpha(-1)(-3) = -\frac{-6}{1}$

$3\alpha = 6$

$\alpha = 2$

Answer B

2 $\int_{-4}^{-3} g(x) - f(x)dx = -\int_{-4}^{-3} f(x) - g(x)dx$

$\int_{-1}^{1} g(x) - f(x)dx = -\int_{-1}^{1} f(x) - g(x)dx$

Answer D

3 19 from each of the four schools, plus one more: $19 \times 4 + 1 = 77$

Answer B

4 Domain:

$-1 \le 2x \le 1$

$-\frac{1}{2} \le x \le \frac{1}{2}$

Range:

$y = 2(\cos^{-1} 2x + \sin^{-1} 2x) = 2\left(\frac{\pi}{2}\right) = \pi$

Answer A

5 $g(x) = a \sin^{-1}(bx)$

Vertical dilation: $a = 2$

Horizontal dilation: $\frac{1}{b} = \frac{1}{3}$

Answer C

6 $|b| = |b \sin(4x)|$

$\pm b = b \sin(4x)$

$\sin(4x) = \pm 1$

$4x = \frac{\pi}{2}, \frac{3\pi}{2} \dots$

Consider solutions on domain

$0 \le x \le 2\pi$

$0 \le 4x \le 8\pi$

Four cycles from 0 to 2π, two solutions per cycle, for a total of 8 solutions

Answer D

7 Applicant must get 4 or 5 out of the last 5 questions correct to pass.

$P(X = 4) = \binom{5}{4}\left(\frac{1}{4}\right)^4\left(\frac{3}{4}\right)^1$

$= 5 \times \frac{1}{256} \times \frac{3}{4}$

$= \frac{15}{1024}$

$P(X = 5) = \binom{5}{5}\left(\frac{1}{4}\right)^5\left(\frac{3}{4}\right)^0$

$= 1 \times \frac{1}{1024} \times 1$

$= \frac{1}{1024}$

$P(X = 4 \text{ or } X = 5) = \frac{15}{1024} + \frac{1}{1024}$

$= \frac{1}{64}$

Answer C

8 $\text{Var}(X) = \frac{p(1-p)}{n}, \sigma = \sqrt{\text{Var}(X)}$

$\sqrt{\frac{\left(\frac{7}{12}\right)\left(\frac{5}{12}\right)}{n}} < 0.06$

$\frac{35}{144n} < 0.0036$

$\frac{1}{n} < 0.014811\dots$

$n > 67.5154...$

Thus, smallest sample size is 68.

Answer B

9 P(two silver)

$= P$(first choice silver)P(second choice silver|first choice silver)

$= \frac{k}{n} \times \frac{k-1}{n-1}$

P(two bronze)

$= P$(first choice bronze)P(second choice bronze|first choice bronze)

$= \frac{n-k}{n} \times \frac{n-k-1}{n-1}$

$P(\text{two same metal})$
$= \frac{k}{n} \times \frac{k-1}{n-1} + \frac{n-k}{n} \times \frac{n-k-1}{n-1}$
$= \frac{k(k-1)+(n-k)(n-k-1)}{n(n-1)}$
Answer A

10 For $R\sin(x+\alpha) = R\sin(x-\beta)$, $\alpha = 2\pi - \beta$
$\therefore \alpha + \beta = 2\pi$
For $R\cos(x+\gamma) = R\cos(x-\delta)$, $\gamma = 2\pi - \delta$
$\therefore \gamma + \delta = 2\pi$
Thus $\alpha + \beta + \gamma + \delta = 4\pi$
Answer D

Section II

QUESTION 11

(a) (i) $2\underset{\sim}{a} - \underset{\sim}{b}$
$= 2(3\underset{\sim}{i} + 2\underset{\sim}{j}) - (-\underset{\sim}{i} + 4\underset{\sim}{j})$
$= 6\underset{\sim}{i} + 4\underset{\sim}{j} + \underset{\sim}{i} - 4\underset{\sim}{j}$
$= 7\underset{\sim}{i}$

(1 mark)

(ii) $\underset{\sim}{a} \cdot \underset{\sim}{b}$
$= 3 \times -1 + 2 \times 4$
$= 5$

(1 mark)

(b) $x^2 - 8x - 9 \leq 0$
$(x-9)(x+1) \leq 0$
The concave-up parabola is below the x-axis between the x-intercepts of $x = 9$ and $x = -1$.
Solution: $-1 \leq x \leq 9$

(2 marks)

(c) $u = x - 1 \rightarrow x = u + 1$
$du = dx$
The integral becomes
$\int (u+1)u^{\frac{1}{2}}\,du$
$= \int (u^{\frac{3}{2}} + u^{\frac{1}{2}})du$
$= \frac{2u^{\frac{5}{2}}}{5} + \frac{2u^{\frac{3}{2}}}{3} + C$
$= \frac{2}{5}(x-1)^{\frac{5}{2}} + \frac{2}{3}(x-1)^{\frac{3}{2}} + C$

(3 marks)

(d) $\frac{dy}{dx} = xy, y > 0$
Separable differential equation:
$\frac{1}{y}dy = x\,dx$
$\int \frac{1}{y}dy = \int x\,dx$
$\ln y = \frac{x^2}{2} + C \; (y > 0)$
$y = e^{\frac{x^2}{2}+C}$

(2 marks)

(e) $f(x) = \arcsin(x^5)$
$f'(x) = \frac{5x^4}{\sqrt{1-(x^5)^2}}$
$= \frac{5x^4}{\sqrt{1-x^{10}}}$

(1 mark)

(f) $V = \frac{4}{3}\pi r^3, \frac{dV}{dt} = 10$
$\frac{dr}{dt} = \frac{dr}{dV} \times \frac{dV}{dt}$
$\frac{dV}{dr} = 4\pi r^2 \rightarrow \frac{dr}{dV} = \frac{1}{4\pi r^2}$
$\therefore \frac{dr}{dt} = \frac{1}{4\pi r^2} \times 10 = \frac{5}{2\pi r^2}$ cms^{-1}

(2 marks)

(g) $A = \int_0^{\frac{\pi}{2}} (x - \sin x)\,dx$
$= \left[\frac{x^2}{2} + \cos x\right]_0^{\frac{\pi}{2}}$
$= \left[\frac{1}{2}\left(\frac{\pi}{2}\right)^2 + \cos\left(\frac{\pi}{2}\right)\right] - \left[\frac{1}{2}(0)^2 + \cos(0)\right]$
$= \frac{\pi^2}{8} - 1$ units^2

(3 marks)

QUESTION 12

(a) When vectors are perpendicular, their dot product is equal to zero.

$$\begin{pmatrix} a^2 \\ 2 \end{pmatrix} \cdot \begin{pmatrix} a+5 \\ a-4 \end{pmatrix} = 0$$

$a^2(a + 5) + 2(a - 4) = 0$

$a^3 + 5a^2 + 2a - 8 = 0$

By inspection, $a = 1$ is a root:

$(1)^3 + 5(1)^2 + 2(1) - 8 = 0$

$$\begin{array}{r} a^2 + 6a + 8 \\ a-1\overline{)a^3 + 5a^2 + 2a - 8} \\ \underline{-(a^3 - a^2)} \quad \downarrow \quad \downarrow \\ 6a^2 + 2a \\ \underline{-(6a^2 - 6a)} \\ 8a - 8 \\ \underline{-(8a - 8)} \\ 0 \end{array}$$

$(a - 1)(a^2 + 6a + 8) = 0$

$(a - 1)(a + 4)(a + 2) = 0$

The possible values of a are $a = 1, a = -4, a = -2$.

(3 marks)

(b)
$$\begin{aligned} V &= \int_a^b \pi[r(x)]^2\,dx \\ &= \pi\int_1^2 (x^3)^2\,dx \\ &= \pi\int_1^2 x^6\,dx \\ &= \pi\left[\frac{x^7}{7}\right]_1^2 \\ &= \pi\left(\frac{2^7}{7} - \frac{1^7}{7}\right) \\ &= \frac{127\pi}{7} \text{ units}^3 \end{aligned}$$

(3 marks)

(c) $p = 0.31, n = 100$

Normal approximation to population proportion:

$\mu = E(X) = p = 0.31$

$\text{Var}(X) = \dfrac{p(1-p)}{n} = \dfrac{0.31 \times 0.69}{100} = 0.002\,139$

$\sigma = \sqrt{\text{Var}(X)} = \sqrt{0.002\,139}$

$z = \dfrac{x - \mu}{\sigma} = \dfrac{0.35 - 0.31}{\sqrt{0.002\,139}} = 0.86 \text{ (2 d.p.)}$

$$\begin{aligned} P(X \geq 0.35) &\approx P(Z \geq 0.86) \\ &= 1 - P(Z < 0.86) \\ &= 1 - 0.8051 \\ &= 0.1949 \end{aligned}$$

(0.8051 is obtained from the table at the end of the test.)

There is an approximately 19.49% chance at least 35% make a donation.

(3 marks)

(d) Need to prove $2^{3n} + 13 = 7p$ for some integer p

Prove base case:

Let $n = 1$.

$2^{3(1)} + 13 = 21$, which is divisible by 7.

Assume true for some $n = k$:

$2^{3k} + 13 = 7p$ for some integer p.

Let $n = k + 1$.

Need to show $2^{3(k+1)} + 13 = 7q$ for some integer q:

LHS $= 2^{3k+3} + 13 = 2^{3k} \times 2^3 + 13$

By the assumption, $2^{3k} = 7p - 13$.

$$\begin{aligned} \text{So LHS} &= (7p - 13) \times 2^3 + 13 \\ &= 7p \times 2^3 - 104 + 13 \\ &= 7p \times 2^3 - 91 \\ &= 7(p \times 2^3 - 13) \\ &= 7q \end{aligned}$$

$\therefore$ by the principle of mathematical induction, the statement is true for $n \geq 1$.

(3 marks)

(e) Consider the graphs of $y = \dfrac{1}{|x-5|}$ and $y = \dfrac{x}{6}$.

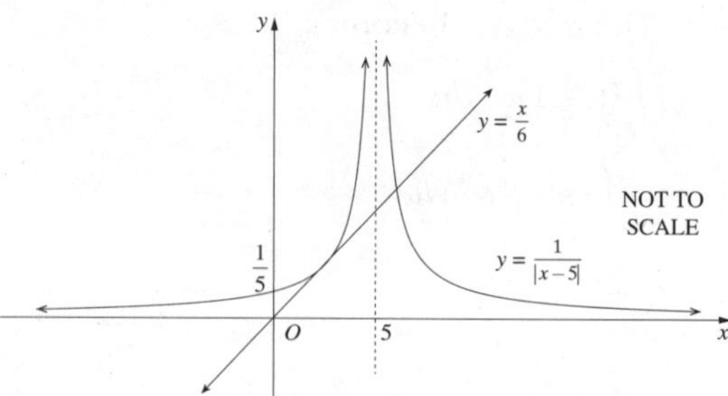

$\dfrac{x}{6} \geq \dfrac{1}{|x-5|}$ when the line $y = \dfrac{x}{6}$ is intersecting or above the graph of $y = \dfrac{1}{|x-5|}$.

Check for intersection(s) to the left of the vertical asymptote:

$$\begin{aligned} \frac{x}{6} &= -\frac{1}{x-5} \\ x(x - 5) &= -6 \\ x^2 - 5x + 6 &= 0 \\ (x - 3)(x - 2) &= 0 \\ x &= 2, x = 3 \end{aligned}$$

The line is above the graph between $x = 2$ and $x = 3$.
Find intersection to the right of the vertical asymptote:

$$\frac{x}{6} = \frac{1}{x-5}$$
$$x(x-5) = 6$$
$$x^2 - 5x - 6 = 0$$
$$(x-6)(x+1) = 0$$

$x = -1$ and $x = 6$
Only $x = 6$ is to the right of the vertical asymptote.
The line is above the graph from $x = 6$.
Solution: $[2, 3] \cup [6, \infty)$ *(3 marks)*

QUESTION 13

(a) (i) The points S and T are separated by a line of horizontal gradients at $y = 2000$, which represents a horizontal asymptote. Any solution that passes through S cannot pass $y = 2000$ to reach the point T.

(1 mark)

(ii)

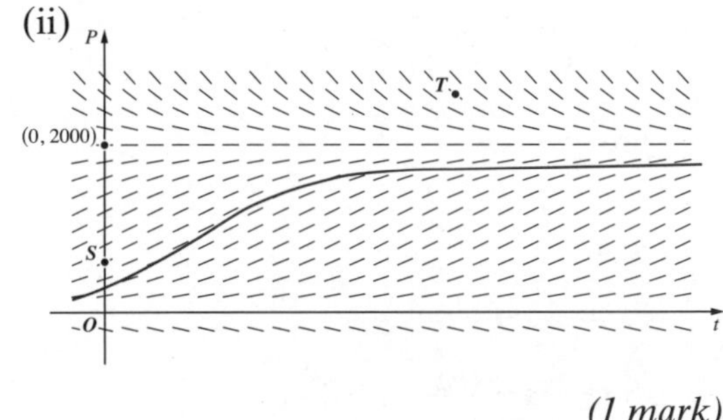

(1 mark)

(iii) Find when $\frac{dP}{dt}$ has a maximum:

$\frac{dP}{dt} = 2000P - P^2$ is quadratic in terms of P with leading coefficient negative.
Therefore, the vertex at

$$P = -\frac{b}{2a}$$
$$= -\frac{2000}{2(-1)}$$
$$= 1000$$

is a global maximum.
The rate at which the growth of the population is largest occurs when the population is 1000.

(2 marks)

(b) (i)
$$\begin{aligned}\text{LHS} &= \cos^4 x + \sin^4 x \\ &= (\cos^2 x)^2 + (\sin^2 x)^2 \\ &= \left(\frac{1}{2}(1-\cos(2x))\right)^2 + \left(\frac{1}{2}(1+\cos(2x))\right)^2 \\ &= \frac{1}{4}(1 - 2\cos(2x) + \cos^2(2x) + 1 + 2\cos(2x) + \cos^2(2x)) \\ &= \frac{1}{4}(2 + 2\cos^2(2x)) \\ &= \frac{1+\cos^2(2x)}{2} \\ &= \text{RHS}\end{aligned}$$

(2 marks)

(ii)
$$\begin{aligned}&\int_0^{\frac{\pi}{4}} (\cos^4 x + \sin^4 x)\,dx \\ &= \int_0^{\frac{\pi}{4}} \frac{1+\cos^2(2x)}{2}\,dx \text{ (by part (i))} \\ &= \frac{1}{2}\int_0^{\frac{\pi}{4}} \left(1 + \frac{1}{2}(1+\cos(4x))\right)dx \\ &= \frac{1}{4}\int_0^{\frac{\pi}{4}} (3+\cos(4x))\,dx \\ &= \frac{1}{4}\left[3x + \frac{1}{4}\sin(4x)\right]_0^{\frac{\pi}{4}} \\ &= \frac{1}{4}\left\{\left[3\left(\frac{\pi}{4}\right) + \frac{1}{4}\sin(\pi)\right] - \left[3(0) + \frac{1}{4}\sin(0)\right]\right\} \\ &= \frac{3\pi}{16}\end{aligned}$$

(3 marks)

(c) $\underset{\sim}{a} = \begin{pmatrix} 1 \\ 3 \end{pmatrix}, \underset{\sim}{b} = \begin{pmatrix} 2 \\ -1 \end{pmatrix}$, and let $\underset{\sim}{x} = \begin{pmatrix} x_1 \\ x_2 \end{pmatrix}$

$\text{proj}_{\underset{\sim}{a}}\underset{\sim}{x} = k\underset{\sim}{a}$

$\dfrac{\underset{\sim}{x} \cdot \underset{\sim}{a}}{\underset{\sim}{a} \cdot \underset{\sim}{a}} = k$

$\dfrac{x_1 + 3x_2}{10} = k$

$x_1 + 3x_2 = 10k$ ①

$\text{proj}_{\underset{\sim}{b}}\underset{\sim}{x} = p\underset{\sim}{b}$

$\dfrac{\underset{\sim}{x} \cdot \underset{\sim}{b}}{\underset{\sim}{b} \cdot \underset{\sim}{b}} = p$

$\dfrac{2x_1 - x_2}{5} = p$

$2x_1 - x_2 = 5p$ ②

$2 \times$ ① – ②:

$7x_2 = 20k - 5p$

$x_2 = \dfrac{20}{7}k - \dfrac{5}{7}p$

Substitute x_2 into ①:

$x_1 + 3\left(\dfrac{20}{7}k - \dfrac{5}{7}p\right) = 10\text{k}$

$x_1 + \dfrac{60}{7}k - \dfrac{15}{7}p = 10k$

$x_1 = \dfrac{10}{7}k + \dfrac{15}{7}p$

$\underset{\sim}{x} = \begin{pmatrix} \dfrac{10}{7}k + \dfrac{15}{7}p \\ \dfrac{20}{7}k - \dfrac{5}{7}p \end{pmatrix}$ *(4 marks)*

(d) $u = e^x + 2e^{-x}$

$du = (e^x - 2e^{-x})dx$

$u^2 = (e^x + 2e^{-x})^2$

$\int \dfrac{e^{3x} - 2e^x}{4 + 8e^{2x} + e^{4x}}dx$

$= \int \dfrac{e^{2x}(e^x - 2e^{-x})}{4 + 8e^{2x} + e^{4x}}dx$

$= \int \dfrac{e^x - 2e^{-x}}{4e^{-2x} + 8 + e^{2x}}dx$

$= \int \dfrac{e^x - 2e^{-x}}{4e^{-2x} + 4 + e^{2x} + 4}dx$

$= \int \dfrac{e^x - 2e^{-x}}{(2e^{-x} + e^x)^2 + 4}dx$

The integral becomes:

$\int \dfrac{1}{u^2 + 4}du$

$= \dfrac{1}{2}\tan^{-1}\dfrac{u}{2} + C$

$= \dfrac{1}{2}\tan^{-1}\dfrac{e^x + 2e^{-x}}{2} + C$

(3 marks)

QUESTION 14

(a) $\dfrac{dy}{dx} = e^{x+y} = e^x e^y$

Separable differential equation:

$e^{-y}dy = e^x dx$

$\int e^{-y}\,dy = \int e^x\,dx$

$-e^{-y} = e^x + C$

Substitute $(0, 0)$:

$-e^0 = e^0 + C$

$-1 = 1 + C$

$C = -2$

$-e^{-y} = e^x - 2$

$e^{-y} = 2 - e^x$

$-y = \ln(2 - e^x)$

$y = -\ln(2 - e^x)$

Domain:

$2 - e^x > 0$

$e^x < 2$

$x < \ln 2$

Range:

As the inner function e^x and outer function $\ln x$ are always increasing, consider the end behaviour.

as $x \to -\infty, e^{-\infty} \to 0$, so $y \to -\ln 2$

as $x \to \ln 2, (2 - e^x) \to 0$, so $y \to \infty$

$\therefore y > -\ln 2$

(4 marks)

(b) $f(x) = \dfrac{kx}{1+x^2} + \arctan x$

Find turning points in terms of k:

$f'(x) = \dfrac{k(1+x^2) - 2x(kx)}{(1+x^2)^2} + \dfrac{1}{1+x^2} = 0$

$\dfrac{k + kx^2 - 2kx^2}{(1+x^2)^2} + \dfrac{1+x^2}{(1+x^2)^2} = 0$

$\dfrac{k - kx^2 + 1 + x^2}{(1+x^2)^2} = 0$

$(1 - k)x^2 + (1 + k) = 0$

$x^2 = -\dfrac{1+k}{1-k} = \dfrac{k+1}{k-1}$

$x = \pm\sqrt{\dfrac{k+1}{k-1}}$

There will be no turning points if $\dfrac{k+1}{k-1} < 0$.

$(k + 1)(k - 1) < 0$

When $k = -2$, $(k + 1)(k - 1) > 0$.

When $k = 0$, $(k + 1)(k - 1) < 0$.

When $k = 2$, $(k + 1)(k - 1) > 0$.

Also there will be no turning points if $f'(x) = 0$:

$(1 - k)x^2 + (1 + k) = 0$

so for $k = \pm 1$

There are no turning points and hence the function is invertible, for values of k in the interval $[-1, 1]$.

(3 marks)

(c) (i) $\tan^{-1}(ax)$ is always increasing, and the sum of two increasing functions is also increasing, hence the function will only have the value of θ once.

(1 mark)

(ii) $\tan^{-1}(3x) + \tan^{-1}(10x) = \dfrac{3\pi}{4}$

$$\tan(\tan^{-1}(3x) + \tan^{-1}(10x)) = \tan\left(\frac{3\pi}{4}\right) = -1$$

Tangent angle sum identity:

$$\frac{\tan(\tan^{-1}(3x)) + \tan(\tan^{-1}(10x))}{1 - \tan(\tan^{-1}(3x))\tan(\tan^{-1}(10x))} = -1$$

$$\frac{3x + 10x}{1 - (3x)(10x)} = -1$$

$$13x = 30x^2 - 1$$

$$30x^2 - 13x - 1 = 0$$

$$(15x + 1)(2x - 1) = 0$$

$$x = -\frac{1}{15} \text{ or } x = \frac{1}{2}$$

As $\dfrac{3\pi}{4} > 0, x = \dfrac{1}{2}$ is the only solution.

(2 marks)

(d) $D(t) = \sqrt{(Vt\cos\theta)^2 + \left(Vt\sin\theta - \dfrac{gt^2}{2}\right)^2}$

$D(t)$ is increasing when $[D(t)]^2$ is increasing.

$$[D(t)]^2 = V^2t^2\cos^2\theta + V^2t^2\sin^2\theta - Vgt^3\sin\theta + \frac{g^2t^4}{4}$$

$$= V^2t^2(\cos^2\theta + \sin^2\theta) - Vgt^3\sin\theta + \frac{g^2t^4}{4}$$

$$= V^2t^2 - Vgt^3\sin\theta + \frac{g^2t^4}{4}$$

$$\frac{d}{dx}[D(t)]^2 = 2V^2t - 3Vgt^2\sin\theta + g^2t^3 = 0$$

$$= t(2V^2 - 3Vgt\sin\theta + g^2t^2) = 0$$

$t = 0$ or

$$t = \frac{3Vg\sin\theta \pm \sqrt{9V^2g^2\sin^2\theta - 4(g^2)(2V^2)}}{2(g^2)} = \frac{3Vg\sin\theta \pm Vg\sqrt{9\sin^2\theta - 8}}{2g^2}$$

As $t > 0, \dfrac{d}{dx}[D(t)]^2 > 0$ if $2V^2 - 3Vgt\sin\theta + g^2t^2 > 0$.

This will occur when there are no roots, i.e.

$$\Delta = 9\sin^2\theta - 8 < 0$$

$$\sin^2\theta < \frac{8}{9}$$

$$\sin\theta < \sqrt{\frac{8}{9}}\left(0 < \theta < \frac{\pi}{2}\right)$$

$$\theta < \sin^{-1}\sqrt{\frac{8}{9}}$$

(4 marks)

NSW Education Standards Authority

2024 HIGHER SCHOOL CERTIFICATE EXAMINATION

Mathematics Advanced
Mathematics Extension 1
Mathematics Extension 2

REFERENCE SHEET

Measurement

Length

$$l = \frac{\theta}{360} \times 2\pi r$$

Area

$$A = \frac{\theta}{360} \times \pi r^2$$

$$A = \frac{h}{2}(a + b)$$

Surface area

$$A = 2\pi r^2 + 2\pi rh$$

$$A = 4\pi r^2$$

Volume

$$V = \frac{1}{3}Ah$$

$$V = \frac{4}{3}\pi r^3$$

Financial Mathematics

$$A = P(1 + r)^n$$

Sequences and series

$$T_n = a + (n - 1)d$$

$$S_n = \frac{n}{2}\left[2a + (n - 1)d\right] = \frac{n}{2}(a + l)$$

$$T_n = ar^{n-1}$$

$$S_n = \frac{a(1 - r^n)}{1 - r} = \frac{a(r^n - 1)}{r - 1}, r \neq 1$$

$$S = \frac{a}{1 - r}, |r| < 1$$

Functions

$$x = \frac{-b \pm \sqrt{b^2 - 4ac}}{2a}$$

For $ax^3 + bx^2 + cx + d = 0$:

$$\alpha + \beta + \gamma = -\frac{b}{a}$$

$$\alpha\beta + \alpha\gamma + \beta\gamma = \frac{c}{a}$$

and $\alpha\beta\gamma = -\frac{d}{a}$

Relations

$$(x - h)^2 + (y - k)^2 = r^2$$

Logarithmic and Exponential Functions

$$\log_a a^x = x = a^{\log_a x}$$

$$\log_a x = \frac{\log_b x}{\log_b a}$$

$$a^x = e^{x \ln a}$$

Trigonometric Functions

$$\sin A = \frac{\text{opp}}{\text{hyp}}, \quad \cos A = \frac{\text{adj}}{\text{hyp}}, \quad \tan A = \frac{\text{opp}}{\text{adj}}$$

$$A = \frac{1}{2}ab\sin C$$

$$\frac{a}{\sin A} = \frac{b}{\sin B} = \frac{c}{\sin C}$$

$$c^2 = a^2 + b^2 - 2ab\cos C$$

$$\cos C = \frac{a^2 + b^2 - c^2}{2ab}$$

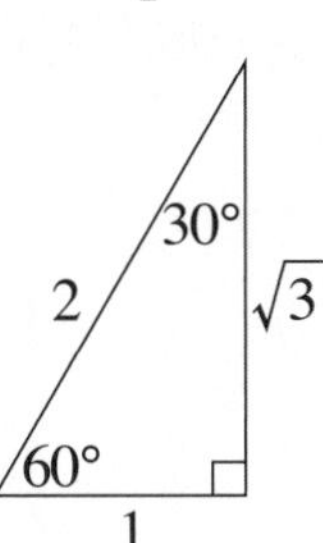

$$l = r\theta$$

$$A = \frac{1}{2}r^2\theta$$

Trigonometric identities

$$\sec A = \frac{1}{\cos A}, \ \cos A \neq 0$$

$$\operatorname{cosec} A = \frac{1}{\sin A}, \ \sin A \neq 0$$

$$\cot A = \frac{\cos A}{\sin A}, \ \sin A \neq 0$$

$$\cos^2 x + \sin^2 x = 1$$

Compound angles

$$\sin(A + B) = \sin A\cos B + \cos A\sin B$$

$$\cos(A + B) = \cos A\cos B - \sin A\sin B$$

$$\tan(A + B) = \frac{\tan A + \tan B}{1 - \tan A\tan B}$$

If $t = \tan\frac{A}{2}$ then $\sin A = \frac{2t}{1 + t^2}$

$$\cos A = \frac{1 - t^2}{1 + t^2}$$

$$\tan A = \frac{2t}{1 - t^2}$$

$$\cos A\cos B = \frac{1}{2}\left[\cos(A - B) + \cos(A + B)\right]$$

$$\sin A\sin B = \frac{1}{2}\left[\cos(A - B) - \cos(A + B)\right]$$

$$\sin A\cos B = \frac{1}{2}\left[\sin(A + B) + \sin(A - B)\right]$$

$$\cos A\sin B = \frac{1}{2}\left[\sin(A + B) - \sin(A - B)\right]$$

$$\sin^2 nx = \frac{1}{2}(1 - \cos 2nx)$$

$$\cos^2 nx = \frac{1}{2}(1 + \cos 2nx)$$

Statistical Analysis

$$z = \frac{x - \mu}{\sigma}$$

An outlier is a score
less than $Q_1 - 1.5 \times IQR$
or
more than $Q_3 + 1.5 \times IQR$

Normal distribution

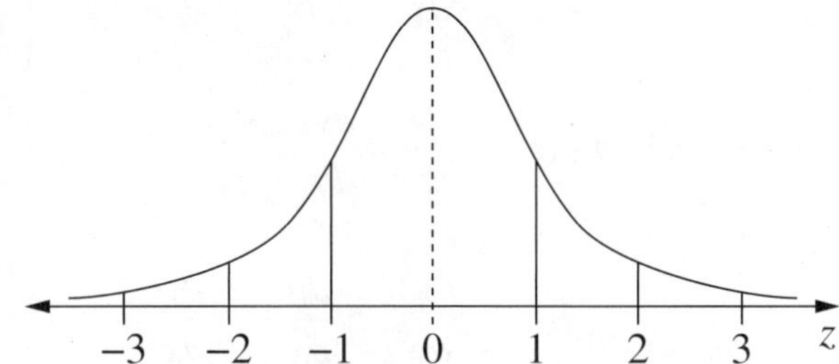

- approximately 68% of scores have z-scores between −1 and 1
- approximately 95% of scores have z-scores between −2 and 2
- approximately 99.7% of scores have z-scores between −3 and 3

$$E(X) = \mu$$

$$\text{Var}(X) = E\left[(X - \mu)^2\right] = E\left(X^2\right) - \mu^2$$

Probability

$$P(A \cap B) = P(A)P(B)$$

$$P(A \cup B) = P(A) + P(B) - P(A \cap B)$$

$$P(A|B) = \frac{P(A \cap B)}{P(B)}, \ P(B) \neq 0$$

Continuous random variables

$$P(X \leq r) = \int_a^r f(x)\,dx$$

$$P(a < X < b) = \int_a^b f(x)\,dx$$

Binomial distribution

$$P(X = r) = {}^nC_r p^r(1 - p)^{n-r}$$

$$X \sim \text{Bin}(n, p)$$

$$\Rightarrow \quad P(X = x) = \binom{n}{x}p^x(1 - p)^{n-x}, \ x = 0, 1, \ldots, n$$

$$E(X) = np$$

$$\text{Var}(X) = np(1 - p)$$

Differential Calculus

Function	Derivative
$y = f(x)^n$	$\dfrac{dy}{dx} = n f'(x)\left[f(x)\right]^{n-1}$
$y = uv$	$\dfrac{dy}{dx} = u\dfrac{dv}{dx} + v\dfrac{du}{dx}$
$y = g(u)$ where $u = f(x)$	$\dfrac{dy}{dx} = \dfrac{dy}{du} \times \dfrac{du}{dx}$
$y = \dfrac{u}{v}$	$\dfrac{dy}{dx} = \dfrac{v\dfrac{du}{dx} - u\dfrac{dv}{dx}}{v^2}$
$y = \sin f(x)$	$\dfrac{dy}{dx} = f'(x)\cos f(x)$
$y = \cos f(x)$	$\dfrac{dy}{dx} = -f'(x)\sin f(x)$
$y = \tan f(x)$	$\dfrac{dy}{dx} = f'(x)\sec^2 f(x)$
$y = e^{f(x)}$	$\dfrac{dy}{dx} = f'(x)e^{f(x)}$
$y = \ln f(x)$	$\dfrac{dy}{dx} = \dfrac{f'(x)}{f(x)}$
$y = a^{f(x)}$	$\dfrac{dy}{dx} = (\ln a)\, f'(x)a^{f(x)}$
$y = \log_a f(x)$	$\dfrac{dy}{dx} = \dfrac{f'(x)}{(\ln a)\, f(x)}$
$y = \sin^{-1} f(x)$	$\dfrac{dy}{dx} = \dfrac{f'(x)}{\sqrt{1-\left[f(x)\right]^2}}$
$y = \cos^{-1} f(x)$	$\dfrac{dy}{dx} = -\dfrac{f'(x)}{\sqrt{1-\left[f(x)\right]^2}}$
$y = \tan^{-1} f(x)$	$\dfrac{dy}{dx} = \dfrac{f'(x)}{1+\left[f(x)\right]^2}$

Integral Calculus

$$\int f'(x)\left[f(x)\right]^n dx = \frac{1}{n+1}\left[f(x)\right]^{n+1} + c$$

where $n \neq -1$

$$\int f'(x)\sin f(x)\,dx = -\cos f(x) + c$$

$$\int f'(x)\cos f(x)\,dx = \sin f(x) + c$$

$$\int f'(x)\sec^2 f(x)\,dx = \tan f(x) + c$$

$$\int f'(x)e^{f(x)}dx = e^{f(x)} + c$$

$$\int \frac{f'(x)}{f(x)}dx = \ln\left|f(x)\right| + c$$

$$\int f'(x)a^{f(x)}dx = \frac{a^{f(x)}}{\ln a} + c$$

$$\int \frac{f'(x)}{\sqrt{a^2 - \left[f(x)\right]^2}}dx = \sin^{-1}\frac{f(x)}{a} + c$$

$$\int \frac{f'(x)}{a^2 + \left[f(x)\right]^2}dx = \frac{1}{a}\tan^{-1}\frac{f(x)}{a} + c$$

$$\int u\frac{dv}{dx}dx = uv - \int v\frac{du}{dx}dx$$

$$\int_a^b f(x)\,dx$$

$$\approx \frac{b-a}{2n}\left\{f(a) + f(b) + 2\left[f\left(x_1\right) + \cdots + f\left(x_{n-1}\right)\right]\right\}$$

where $a = x_0$ and $b = x_n$

Combinatorics

$${}^nP_r = \frac{n!}{(n-r)!}$$

$$\binom{n}{r} = {}^nC_r = \frac{n!}{r!(n-r)!}$$

$$(x+a)^n = x^n + \binom{n}{1}x^{n-1}a + \cdots + \binom{n}{r}x^{n-r}a^r + \cdots + a^n$$

Vectors

$$\left|\underset{\sim}{u}\right| = \left|x\underset{\sim}{i} + y\underset{\sim}{j}\right| = \sqrt{x^2 + y^2}$$

$$\underset{\sim}{u} \cdot \underset{\sim}{v} = \left|\underset{\sim}{u}\right|\left|\underset{\sim}{v}\right|\cos\theta = x_1x_2 + y_1y_2,$$

where $\underset{\sim}{u} = x_1\underset{\sim}{i} + y_1\underset{\sim}{j}$

and $\underset{\sim}{v} = x_2\underset{\sim}{i} + y_2\underset{\sim}{j}$

$$\underset{\sim}{r} = \underset{\sim}{a} + \lambda\underset{\sim}{b}$$

Complex Numbers

$$\begin{aligned} z = a + ib &= r(\cos\theta + i\sin\theta) \\ &= re^{i\theta} \end{aligned}$$

$$\begin{aligned} \left[r(\cos\theta + i\sin\theta)\right]^n &= r^n(\cos n\theta + i\sin n\theta) \\ &= r^n e^{in\theta} \end{aligned}$$

Mechanics

$$\frac{d^2x}{dt^2} = \frac{dv}{dt} = v\frac{dv}{dx} = \frac{d}{dx}\left(\frac{1}{2}v^2\right)$$

$$x = a\cos(nt + \alpha) + c$$

$$x = a\sin(nt + \alpha) + c$$

$$\ddot{x} = -n^2(x - c)$$

Notes

Notes

Notes